AF181524

Restoration of Forests

NATO ASI Series

Advanced Science Institutes Series

A Series presenting the results of activities sponsored by the NATO Science Committee, which aims at the dissemination of advanced scientific and technological knowledge, with a view to strengthening links between scientific communities.

The Series is published by an international board of publishers in conjunction with the NATO Scientific Affairs Division

A Life Sciences	Plenum Publishing Corporation
B Physics	London and New York
C Mathematical and Physical Sciences	Kluwer Academic Publishers
D Behavioural and Social Sciences	Dordrecht, Boston and London
E Applied Sciences	
F Computer and Systems Sciences	Springer-Verlag
G Ecological Sciences	Berlin, Heidelberg, New York, London,
H Cell Biology	Paris and Tokyo
I Global Environmental Change	

PARTNERSHIP SUB-SERIES

1. Disarmament Technologies	Kluwer Academic Publishers
2. Environment	Springer-Verlag / Kluwer Academic Publishers
3. High Technology	Kluwer Academic Publishers
4. Science and Technology Policy	Kluwer Academic Publishers
5. Computer Networking	Kluwer Academic Publishers

The Partnership Sub-Series incorporates activities undertaken in collaboration with NATO's Cooperation Partners, the countries of the CIS and Central and Eastern Europe, in Priority Areas of concern to those countries.

NATO-PCO-DATA BASE

The electronic index to the NATO ASI Series provides full bibliographical references (with keywords and/or abstracts) to more than 50000 contributions from international scientists published in all sections of the NATO ASI Series.
Access to the NATO-PCO-DATA BASE is possible in two ways:

– via online FILE 128 (NATO-PCO-DATA BASE) hosted by ESRIN, Via Galileo Galilei, I-00044 Frascati, Italy.

– via CD-ROM "NATO-PCO-DATA BASE" with user-friendly retrieval software in English, French and German (© WTV GmbH and DATAWARE Technologies Inc. 1989).

The CD-ROM can be ordered through any member of the Board of Publishers or through NATO-PCO, Overijse, Belgium.

Series 2: Environment – Vol. 30

Restoration of Forests

Environmental Challenges in Central and Eastern Europe

edited by

Richard M. Gutkowski
Department of Civil Engineering,
Colorado State University,
Fort Collins, U.S.A.

and

Tomasz Winnicki
Environmental Engineering Department,
Technical University of Wrocław,
Poland

Springer-Science+Business Media, B.V.

Proceedings of the NATO Advanced Research Workshop on
Science and Technology to Save and Better Utilize Central and Eastern Europe's Forests
Saulberg, Lower Bavaria, Germany
July 15–19, 1996

A C.I.P. Catalogue record for this book is available from the Library of Congress

ISBN 978-94-010-6337-1 ISBN 978-94-011-5548-9 (eBook)
DOI 10.1007/978-94-011-5548-9

Printed on acid-free paper

All Rights Reserved
© 1997 Springer Science+Business Media Dordrecht
Originally published by Kluwer Academic Publishers in 1997
Softcover reprint of the hardcover 1st edition 1997

No part of the material protected by this copyright notice may be reproduced or utilized in any form or by any means, electronic or mechanical, including photocopying, recording or by any information storage and retrieval system, without written permission from the copyright owner.

TABLE OF CONTENTS

Position Papers

Working Group 1 - "Environmental Factors in Forest Depletion"

Position Papers

Working Group 2 - "Forest Growth, Expansion and Management"

Position Papers

Working Group 3 - "Forest Utilization for Raw Materials and Value Added Products"

Position Papers

Working Group 4 - "Socio-Economic Considerations for the Forest Resource"

Preface

This publication is the written record of the NATO Advanced Research Workshop (ARW) on "Science and Technology to Save and Better Utilize Central and Eastern Europe's Forests". The objective of the workshop was to examine the environmental factors contributing to the rapid depletion of Central and Eastern Europe's Forests. The goal was to determine a research action plan to help abate the impact, while sustaining or expanding the forests, the economy and energy sufficiency of the region. The workshop convened a multi-disciplinary group of experts to discuss the broad spectrum of environmental issues related to utilizing the forests of Central and Eastern Europe. It was conceived on the basis of the existing forest damage situation and identified potential for these countries to work together to make improvements and mutually benefit from common foci.

The whole of Central and Eastern Europe is undergoing an enormous transition, almost transformation, in its entire social structure. The end of a communist era government in many CP countries revealed a critical need to solve certain infrastructure problems in the region. Modernization was needed in everything that affects everyday living, particularly housing and commercial building construction, transportation, waterways and ports, and energy production and utilities. Equally critical was the evident need to address extensive environmental concerns created by past neglect of needed modernization. Engineering the repair and replacement of the infrastructure of these changing societies within the setting of fostering a more healthy and secure environment is an enormous challenge, both technically and economically. Science and technology have vital roles in the technical progress needed to fully adapt to a modern world. As a consequence of the recent political changes in the region, the relationships between the countries and states have been altered too. For peace and security in the region, it is critical to foster cooperation in positive directions between the countries themselves as well as with Western nations. One of the opportunities to do so is through joint scientific and technological efforts to address regional environmental challenges.

For nearly 50 years, environmental contamination was rampant in former communist parts of Europe. Old inefficient power and industrial plants demanded considerable energy, principally coming from fossil fuels. Industrial air pollution had drawn negative attention to mining and coal, steel, concrete and other materials producing industries. Indeed, the extent was so grave that the area where the borders of Poland, the former East Germany and the former Czechoslovakia come together became known as the "Black Triangle". Published reports in the early 1980s characterized threats to survivability of forests in Central Europe as acute, with forest damage spreading through the area with frightening rapidity. In Central Europe, air pollutants from combustion of fossil fuels and smelting of metallic ores were seriously undermining forests. About 2.4 million acres were believed to be affected in the former Czechoslovakia and Poland. Scientists predicted that 7 million acres in Poland would be lost if increased burning of the nation's high-sulfur brown coals occurred. Austria, Switzerland, the former East Germany, and Romania were described as facing an unprecedented extent of forest devastation.

Energy production and use is a key aspect of needed directions to relieve the aggravated forest depletion situation. Energy issues range from the modes of energy production, the economic factors in meeting present and future demand, the effluent abatement, the role of forests as a bio-mass fuel source and many other implications. The linkages between industry, energy, deforestation, agriculture and value-added forest products, economic development are complex and unavoidable. These linkages must be balanced, if not intertwined, within the public at large perspective of the broader social resource the forest itself represents.

The preceding description provides an overview of the prevailing situation of 10–15 years ago. In the late 1980s, dramatic political events occurred throughout Central and Eastern Europe. To differing degrees, a new focus on the need to modernize the infrastructure and evolve market based economies arose in many countries and states. Since that time, some monetary and technical efforts have been made to improve industry and abate environmental contamination. Thus, a primary motivation for the ARW was to examine the current status of this effort with regard to forest environmental concerns.

The compelling aspect of the above challenges is that the needs are both extensive and common in all the countries. Regionality is introduced by the potential use of regional forest resources for cost effective, sound construction materials and energy practices to solve enormous housing needs and to produce high quality employment. The associated regional nature of the environmental challenges to save and better utilize the forest resources necessitates cooperation among the various nations and states. The consequences of past neglect and fiscal incapacity to address them are widespread in the region. For resolution of the greater environmental problems, unilateral solutions do not exist. The environmental issues related to the forest resource require the consensus of countries to achieve action. Regionality introduces a need to pool capabilities and resources to interact positively to attenuate and reduce the causes and effects of a poor environment. The ARW was conducted to develop such recommendations, actions and human networking needed to begin to implement them.

An international delegation of invited experts was organized within a matrix of four desired Working Groups (WGs) versus four areas of related forest-based work activity (reflecting expertise, experience and/or responsibility). The Working Groups chosen were:

WG 1 – Environmental Factors in Forest Depletion (Leader: D. Ellsworth, USA)
WG 2 – Forest Growth, Expansion and Management (Leader: A. Brady, UK)
WG 3 – Forest Utilization for Raw Materials and Value-Added Products (Leader: M. Gizejowski, Poland)
WG 4 – Socio-Economic Considerations for the Forest Resource (Leader: P. Pellicane, USA)

The four areas of professional activity included were: (1) ecology, environment and science, (2) forest ownership and management, (3) forest utilizers, and (4) government and public agencies. The goal was a diverse delegation whose collective expertise and knowledge spanned the full spectrum of critical influences and issues.

The venue for the ARW was located in Saulburg, Germany in the heart of Lower Bavaria. On route to the venue, delegates from the CP countries in Central and Eastern Europe assembled in Wroclaw, Poland and participated in a pre-ARW study tour of industrial sites and forests within the Black Triangle area, conducted by Co-Director, Tomasz Winnicki.

The four-day long ARW was conducted with a carefully ordered set of events. Each invited expert was assigned to an appropriate WG required to submit, in advance, a position paper on a subject related to his/her WG topic, reflecting his/her expertise and viewpoint. These were provided to all participants at the outset of the ARW.

Day 1 of the ARW included the presentation of the following situation papers:

Central European Environmental Hot Spots
Jozef Pacyna, Norwegian Institute for Air Research, Norway

Extent and Causes of Forest Depletion in Europe
Kazmierz Rykowski, Instytut Badawczy Lestnictwa, Poland

Sustainable Economy of Forestry
Julius Natterer, Ecole Polytechnic Fédérale de Lausanne, Switzerland

and the following state-of-the-art papers:

Options for Power Generation and Emission Control Technologies
Tomasz Winnicki, Poland

Forestry Area Recovery and Restructuring – Experience from Sherwood Forest
Austin Brady, Forestry Commission, United Kingdom

Increasing Public Awareness of the Contribution of Forestry and Wood Utilization to Ecology
Gerd Wegener, Universität München, Germany

Value Added Utilization of Forests
Julius Natterer, Ecole Polytechnique Fédérale de Lausanne, Switzerland

On the second day, a technical excursion was organized to the private forest of Thurn and Taxis within the Bavarian Forest. This was conducted by Dr. Winfried Duffner of the Forest of Thurn and Taxis in Regensburg, Germany. This extensive tour provided the delegation with a first hand observation of the most modern forest growth, management and harvesting practices and of equipment presently possible.

Day 3 involved discussion meetings of the individual WGs who prepared discussion documents which were distributed to all participants at the end of the day.

Day 4 began with the first meeting of the whole delegation to hear presentations of summaries by the WG leaders. This was followed by a round-robin set of joint meetings of paired WGs, whereby each possible pair of WGs had an opportunity to discuss each other's summaries and arrive at joint observations. The WG leaders then met to prepare collective Draft Preliminary Conclusions and Recommendations. These were presented and discussed at a second meeting of the whole delegation. A type of 'open space' discussion format was used in which every participant had the individual (one by one) opportunity to provide comment and input and then react to that of the others, and all delegates did so contribute. This concluded the on-site activities.

In a post ARW process, Mr. Austin Brady prepared a set of Final Conclusions and Recommendations based on the input of the entire delegation, both on-site and via mail (and electronic mail) in the month following the event.

Position Papers and written outcomes of the above events are included in this publication.

The ARW was a cooperative program between Colorado State University (USA) and Wroclaw Technical University (Poland). Funding was provided through the Scientific and Environmental Affairs Division of the North Atlantic Treaty Organization (NATO). That funding and the support of Dr. L. Veiga da Cunha, Director, Priority Area on Environmental Security, are acknowledged and appreciated. The Co-Directors express their appreciation to Professor Albert Yates, President – Colorado State University, Professor Andrzej Mulak, President – Wroclaw Technical University for their assistance in the organization of the ARW. Numerous government officials within Lower Bavaria provided support and assistance with the local arrangements. In particular, we thank Mr. Heinrick Förk, Regional Council Member – Saulburg, Mayor Anton Drexler – Wiesenfelden, Lord Mayor Perlak – Straubing for their commitment of time, effort and resources.

Co-Directors
Professor Richard M. Gutkowski
Professor Tomasz Winnicki

DIRECTORATE AND STAFF FOR THE ARW

Directors

Professor Richard M. Gutkowski
Colorado State University

Professor Tomasz Winnicki
Technical University of Wroclaw

Organizing Committee

Professor Richard M. Gutkowski
Colorado State University-USA

Professor Tomasz Winnicki
Technical University of Wroclaw-Poland

Professor Julius Natterer
Ecole Polytechnique Federale de Lausanne-Switzerland

Professor Wolfgang Winter
Technical University Wien-Austria

Secretariat

Milena Valencia

Planning Committee

Sybille Hilgeland
Pierre-Aime Favre
Professor Reinhard Hoepfl
Christa Müller
Johannes Natterer
Dr. Winfried Duffner
Christa Haitzer
Christian Hilgeland
Werner Müller
James Anhalt-Birch

Local Officials

Mr. Heinrick Förk
Lord Mayor Perlak
Mayor Anton Drexler

We would like to extend our sincere gratitude to Mr. Christian Bachner for hosting the post workshop tour through the Bavarian Motor Works plant in Regensburg.

Invited Experts and Organizers

Opening Presentation

OPENING PRESENTATION AND CHARGE TO THE WHOLE DELEGATION

Professor Richard M. Gutkowski, Ph.D., P.E.
Director-NATO ARW
Science and Technology to Save and Better Utilize Central and Eastern Europe's Forests
Saulburg, Lower Bavaria, Germany
July 15-19, 1996

Ladies and gentlemen, after a warm welcome by Professor Julius Natterer and the officials of Saulburg and other communities in Lower Bavaria and other organizational matters, let me begin the official technical proceedings.

On behalf of the North Atlantic Treaty Organization, it is my distinct privilege to convene this NATO Advanced Research Workshop on the important topic of "Science and Technology to Save and Better Utilize the Forests of Central and Eastern Europe." My Co-Director, Professor Tomasz Winnicki, and I and the Whole Delegation express our warm gratitude to the local and regional officials of Lower Bavaria for hosting this important event. We are particularly appreciative of the opportunity to be in Saulburg an be among the kind people of this community. It is exciting to be your guests, indeed residing in your homes, and we look forward to our time with you. As our workshop progresses we will also be visiting and hosted by the nearby communities of Straubing, Wiesenfelden and Regensburg, and we express our deep gratitude for their gracious welcome, too.

This ARW is sponsored by the North Atlantic Treaty Organization (NATO) and its Scientific and Environmental Affairs Division. Dr. L. Veiga da Cunha, Director of the NATO Scientific Committee on the Environmental Security, is participating in a planning meeting of that committee and deeply regrets he is unable to be present today. He offers his enthusiastic support and encouragement as the delegation begins its deliberations on this significant and complex topic. He also expresses his sincere gratitude to our hosts in Lower Bavaria.

We are assembled in a very special venue, the Altes Forsthaus, about which I will comment later. We come here as a multi-disciplinary group of knowledgeable experts to discuss the broad spectrum of issues related to saving and utilizing the forests of central and eastern Europe. At this time, I officially convene this NATO Advanced Research Workshop and bring the Invited Experts to order as a Whole Delegation. As Invited Experts, each of you is distinguished and accomplished in your technical field and have important responsibilities, and it is a great honor to begin working with you.

The Director's statement in your folders describes the justification for the ARW and is taken from the proposal to NATO. As some salesmanship was involved, it includes some extreme scenarios described in other writings. This ARW is an opportunity for a reality check about them. In proposing this ARW topic the Directors sought to focus

R. M. Gutkowski and T. Winnicki (eds.), Restoration of Forests, 3–8.
© 1997 *Kluwer Academic Publishers.*

on the interrelationship of industrial pollution, energy demands and the depletion of the forest.

Let me read part of that statement into the record. For nearly 50 years, environmental contamination was rampant in some parts of central and eastern Europe. Industrial air pollution and costs of industries such as coal mines, steel mills and concrete plants. Air pollutants from the combustion of fossil fuels and the smelting of metallic ores were and still are undermining forests and soils. Acid-forming sulfates under extreme stress due to acid rain. Some forests receive thirty fold more acidity than from untainted snow and rain. Ozone levels are regularly in the range that damages trees. Power plants and industrial production plants demand considerable energy, principally fossil fuels. CO_2 is an almost inevitable waste product. Concentrations in the air of CO_2, methane, nitrous oxide and CFC's have made the "greenhouse effect" a focus of political and scientific debate. Emissions from inefficient or high speed automobiles, trucks and other transport aggravate the problem.

Recently the United Nations Food and Agriculture Organization estimated that half the world's forests have disappeared since 1950. According to the Pan American Development Foundation, trees now cover only 8% of the Earth's land. This depletion of the forest natural resource is the outcome of significant man-made environmental factors and has compelling implications to the human condition. Deforestation sets off a chain reaction on incalculable loss and harm. Forests retain moisture during periods of drought and absorb water during floods. When land is deforested, the loss of evaporation to runoff approaches 75%. Rangeland, irrigation and rainfed agriculture suffer extensively. Sediment can fill dams and rivers, causing them to silt up, overflow and flood the adjacent countryside.

In the late 1980's and early 1990's the following was reported. Threats to survivability of forests are acute in central Europe with frightening rapidity. Half the forests in Germany are injured, with no sign of abatement. Hardest hit are the Bavarian and Black Forests, which some say could be extinguished in the next decade. About 2.4 million acres are affected in Czech Republic, Slovakia and Poland. Some scientists predict that 7 million acres in Poland will be lost if increased burning of the nation's high-sulfur brown coals occurs. Austria, Switzerland,Eastern Germany, and Romania are facing unprecedented forest devastation.

Forest damage has led to a unified determination to reduce emissions of sulfur dioxide. Broader type solutions (energy efficiency, recycling forest products, generating more power from alternative energy sources, etc.) are needed to reduce other pollutant levels. National and regional energy policies could determine the extent and pace of forest survival in Europe. The European Community (EC) consumes one-quarter of the total world energy and is the largest energy importer. Russia and the Ukraine consume another one-sixth. Europe is a major consumer of coal. Germany and eastern Europe burn large quantities of lignite in electric power plants. Oil accounts for 40% of Europe's energy consumption, but some reports say reserves will last only fifteen years. Gas and oil will probably last but 50-100 years. Understandably, substitution of nuclear power plants for fossil fuel energy plays a major role in Europe, but is a controversial reality.

Deforestation causes ecological deterioration, agricultural decline due to loss of the water restoration ability of forests, climatic changes, accelerated CO_2 concentrations due to loss of trees which absorb CO_2 (photosynthesis), and unsustainable supply of raw materials for valuable forest products. Regeneration and expansion of the forest and increased, but wise and sustainable, utilization of wood must be accomplished, no matter how challenging.

The issues are known and have scientific, social, political and moral dimensions. Environment and energy are interrelated. Without energy there is no survival to lead a dignified life. Humans need energy. Adaptation to a dying forest environment is untenable to modern life. Broad restriction of tree harvesting is contentious due to economic consequences to jobs and forest products industry and development. Prevention of forest depletion and technological compensation for its effects are the workable approaches. Replantation, development of productive agricultural land after reforestation, and development of agro-forestry industries are realizable and sustainable if energy pollution issues can be abated. Sustainable forests provide a valuable bio-mass source of heating. However, the security of forests is perhaps better served by value-added use of forest products.

For example, the low energy of production and low cost as a construction material has made wood products the predominant material for housing and low-rise construction in the United States. Over half our population (125 million) owns its home, and we rarely see them burn down. Indeed, after long service we often sell them for great profit or readily borrow capital against them to send our children to college. Wood framed houses are our most valued capital asset-banks prove it continuously. Perhaps this ability to capitalize on property values, which allows us to "pay our own way" is why our unemployment is only 3-5% and our social programs have been manageable for so long.

This workshop involves experts from technical and non-technical disciplines that bear on addressing the impacts of the depletion of European forests. Restoration and improvement of the forests is critically needed and fundamental to success in the above endeavors. Wood is the most important renewable resource of the region. The workshop objectives are to develop action plans needed to research and put into place restorative and preventative processes including wiser use of materials; upgrading manufacturing to be clean and modern; future planning, management and operations via high technology such as expert systems and geographical information systems; and many other considerations.

Wood is the most important renewable resource of the region. The common regional goal is to determine how to better care for and use the forest natural resource for the benefit of the societies of Central and Eastern Europe and their neighbors. The Whole Delegation will develop conclusions and recommendations and set priorities for future research. A regional action plan is an essential imperative and this workshop will try to develop one for scientific and engineering research and development to expand forests and expand their multiplicity of benefits to mankind and nature. A key result will be the improvement of technology transfer and sharing via future networking and joint research.

Let me move aside from my formal and technical comments, to the personal and altruistic (human spirit) aspects I see as motivation for us to be here. The primary mission of NATO is the maintenance of security. Within the public eye, NATO is predominantly, if not solely, visible in its role in maintaining security vis-a-vis ongoing or perceived threats of armed conflict. However, from my NATO sponsored activities, I see the organization's extensive involvement in and its biggest challenge and endeavor is the less visible application of the peaceful setting of its role. NATO views peace as a cornerstone of security.

To me the ARW is an excellent example of what is called the Partnership of Peace initiative. The Partnership for Peace initiative followed the step to independence of the Central and Eastern European societies. The initiative facilitates the networking and exchange of creative people and ideas across the cultural boundaries of the United States and these changing societies. This ARW is an example of how it is being extended to all the NATO nations. Typically, the ARW is concluded on narrowly focused scientific topics. I have fostered the use of the ARW to address inter-cooperation in science and technological needs of mutual interest in the region.

With recent changes in the political geography of Central and Eastern Europe, I hope and believe the positive future of central and eastern Europe is the arena of economic competition, as competition for markets can and should be friendly and fair. Certainly better than conflict. The map of Central Europe and Eastern Europe has been redrawn by the socio-political events of the past decade. The entire spectrum of the societies within and between merging and independent new states has been dramatically altered. All are restructuring in the face of some common challenges: political stability; a stronger economy; enhanced commerce; modernization of manufacturing industries, abating serious environmental conditions. Science and engineering; differing political structures, laws and regulations and economic states; financial considerations; and historical factors and relationships all come in to play. Hence, the collective issues are multidisciplinary and regional in scope. The common forestry issues are a clear example! The linkage between industry, energy, deforestation, agriculture and economic development is complex and unavoidable. Regionality and globality of the problems require consensus of countries to achieve action.

The Altes Forsthaus (Altes forest house) is important as a venue for it is in fact a home, the summer home of Professor Julius Natterer and his family. Part of the selection of this special forest home is its proximity to the Bavarian forest, it being just outside its doors. And its existence and that of its residents (the family) are the products of the forest. As I understood, his father and grandfather were foresters. He carries their love and value system for the vital resource it is; they in the peaceful and plentiful past, he in the present a time of threat, constraint and contention. What better medium for our workshop then to take steps to restore peace to the forest? As you will learn, he believes the forest is the only way to save the world. and, I am certain that, to different degrees, he will convince you of it later today. And he will befriend you, and you him, as he does it. I know it because he and I have bonded in that way.

During 16 years of developing professional collaboration, Professor Julius Natterer and I have become, above all, friends. Our friendship also extends to a common care for the forest and its plight and the role of technology in making invaluable, sustainable utilization of its forest products possible for bettering the human condition. Is not the drive for a better life an underlying present theme of the inhabitants of the newly developing nations and states of Central and Eastern Europe?

Lastly, let me comment on how it is that each of you, specifically, is here.

First, the NATO has its rules about the number of delegates and representation of NATO, Cooperation Partner and other nations. I will not describe that, but we tried to oblige them.

Second, we have a mix of scientists, technologists, industry and government. We need all of you. Our issues involve the quality of life and these types of individuals all influence our quality of life.

Quality of life has many facets. One of them is the overriding goal of NATO, namely security. A critical aspect of individual security is employment. It is well known that unemployment in EU countries is 10-11%, and Germany is an exact example of that situation. In CP nations the job situation has even more extreme needs. So a compelling issue is "jobs" and can forests provide them? The answer is - YES, absolutely and emphatically YES.

The ready answer is at the labor force level, carpenters, fabricators, loggers, transporters, and so on. This is the obvious and visible level since, in some countries, there is contention over saving these jobs versus objectives of forest protection proponents. But let us consider the next level, that of professional careers. It seems to be overlooked greatly. The disciplines needed to address today's energy environment and forest utilization challenges are extensive. They include botany, biology, chemistry, forestry, engineering, economics, finance, genetics, manufacturing, physics, agriculture, atmospheric science, computer science, business, political science, education, and even being an astronaut helps. In civil engineering alone we need environmental, geotechnical, hydraulic, structural, and construction engineers and others. Indeed, we even have wood engineers! I am one. Some of you are, too. Just take a look at our Whole Delegation, you the Invited Experts, prove the point. Forests lead to high level, rewarding careers!

One question several of you have asked the Directors is, how did you locate me or select me?

Well, given NATO rules and a fixed budget, we could not approach it like Noah and his ark, picking two from every relevant discipline. So, as in events of this nature, we had a pool of specialists either known to us or recommended to us by others. We did an invitation and had alternates. We had acceptances, declinations, other commitments interfered, last minute cancellations, too. We used alternates for replacements. This established our final participant list- the Invited Experts. So given the breadth and complexities of the subject, one might wonder, have we got the right people? I am certain we do. Let me explain.

The forest-energy environment interface is an intricate puzzle, so during our stay we must try to put a few pieces in place. But we must also document what we do, write a story and we will. Our story is a mystery novel, so it must contain clues and evidence for the readers. And of course, we must create a path to the solution to at least of a part of the puzzle. Can we do it? In my view, there is no doubt we can make such a contribution. For I believe that for such workshops the following points are indisputable:

It is only at this time that this particular group is to be assembled as it is and it won't happen again. So we must and will capture this single opportunity to accomplish our tasks and make our unique contribution to the vital issues we are deliberating.

Now, as to why we are the ones to do it, I say

Whoever has come here are the right people, for only we can know our particular story.

Whatever happens is the only thing that can happen, for we are meant to be here for that purpose.

Whatever happens is the right thing to happen, for only we can write our particular story.

One striking characteristic of some of the current debate about forests is the prevalence of pessimism, if not panic. Even my own proposal makes it sound that way. We need much more optimism. We are not amoeba, we are humans, the highest order of life. We have intellect, soul and innate ability to meet our needs - and in time, we will do it with care and cooperation.

With the preceding positive encouragement, I thank you for listening to my perspective. You are outstanding and experts in your field and Professor Winnicki and I look forward to working with you and learning much from each of you. NATO very much appreciates your commitment to be here and looks forward to having your guidance in the matters we are addressing.

Please enjoy the next five days and the challenge it brings. Your collective spirit will make it very exciting and rewarding. Lastly, at the end, I am very confident we will all regret leaving Altes Forsthaus, Saulburg and each other very much.

Report
of the
Joint Study Tour

REPORT OF THE JOINT STUDY TOUR - The Black Triangle Area

Working Group Topic:

Visit to the Sudety Mountains forest area under anthropogenic pressure (from power generation area).

Participants:

Dr. Marian A. Gizejowski (Poland)
Professor Audronis Kvedaras (Lithuania)
Professor Kazimierz Rykowski (Poland)
Dr. Svetlana Sergeichik (Belarus)
Ing. Otakar Schwarz (Czech Republic)
Professor Tomasz Winnicki (Poland)

Excursion Route:

On the Polish side of the border, two groups visited the most devastated upper parts of the mountains in the vicinity of Szklarsksa Poreba. The group at the border pass of Jakuszyce/Harrachov was met by Ing. Schwarz and guided through the forest damage on the Czech-side above the town of Harrachov. The area of damage on both sides of the border line belongs to a large extent to the Czech and Polish Karkonose National Parks.

Observations:

The forest area examined is almost homogeneously a coniferous type mainly composed of spruce. Numerous hills are either deforested or covered by very low density forest consisting of trees completely dead, broken at various heights. Some specimens had fallen down (exposing roots) as a result of wind. Some other parts of the mountains present earlier stages of devastation and are covered by trees at various stages of damage or exhibit a more homogeneous situation of forest degradation. We observed some evidence of silvicultural activity to reforest the entirely bald areas. Various technologies have been used for planting there. The main principle applied is implementation of new species, more tolerant to environmental pollutants of acidic character.

Comments:

Some Czech and Polish members of the study tour have been permanently engaged in examining sources of the described forest degradation. Existing technical and observational systems of forest monitoring were pointed out in the power generation zone of the so-called "Black Triangle," a name which is based on the local brown coal used by all three neighboring countries (Czech Republic, Germany and Poland) which

R. M. Gutkowski and T. Winnicki (eds.), Restoration of Forests, 11–12.
© 1997 *Kluwer Academic Publishers.*

is the dominating source of acidic components damaging the mountain forestry.

Recommendations:

The German side has begun the process of closing some of the oldest plants and modernizing the other power plants. The only Polish plant Torow has also been in the process of extensive modernization in both combustion and abatement. The same approach should be widely implemented on the Czech side and carried out to reach the level of acceptable pollutant deposition on specially protected zones of national parks (which have the most demanding environmental standards). The reforestation process should be intensified.

Notice:

Two videotapes presenting the situation on the Polish and Czech side of the damaged zone (totaling 30 minute presentation time) were shown to all participants during the workshop.

Situation Papers

CENTRAL EUROPEAN HOT SPOTS OF AIR POLLUTION

J.M. PACYNA and K. TØRSETH
Norwegian Institute for Air Research (NILU)
P.O. Box 100
N-2007 Kjeller
Norway

1. Introduction

At the beginning of the 1980's major concern of the public was drawn on unusually high degradation of forests in Europe, particularly in most Central and Eastern European countries. Major symptoms observed were reduced foliar densities and foliar discoloration of tree crowns. For the first time forest damage was observed on a large, regional scale and not only on a local scale, caused mainly by extreme weather conditions, insect attack, or local pollution. This regional phenomenon of forest degradation coincided with the acidification of precipitation over Europe. With limited knowledge on dose-effect relationships at that time it was concluded that acid rain was the major factor contributing to the forest degradation. This stand point on the causes of forest degradation was refined starting in the 1980's when extended research on the subject was launched in Europe, particularly in Germany and the Nordic countries (EEA, 1995). It became clear that air pollution is only one of the factors contributing to the forest degradation. This impact can be either direct through the gaseous pollutants and solutes on needled and leaves, or indirect through changes in the soil system.

The results of recent progress made in understanding the complicated processes leading to the forest degradation have been presented during the 5th International Conference on Acid Deposition - Science and Policy, held in Gøteborg in June 1995 (e.g. Rodhe *et al.*, 1995). It was concluded that in high emission regions, forest damage may occur from direct impact of gaseous pollutants. Apart from these regions, acidification may cause adverse effects on forest through induction of base cation deficiency. Increased nitrogen input can, on one side, enhance the acidification when there is saturation in soil. On the other side, this increased input may result in the increase of biomass growth, directly related to nutrient deficiency, and finally to susceptibility to drought, wind, frost.

In addition, high concentrations of ozone in surface air, caused by anthropogenic emissions of nitrogen oxides, non-methane volatile organic compounds (NMVOCs), and carbon monoxide occur regularly during the summer season over large areas of Europe. The uptake of ozone in trees and plants may damage plant tissue by affecting various anatomical, biochemical, and/or physiological functions. Reduced photosynthesis and growth, aging, and increased susceptibility of trees to climatic stress have been documented.

R. M. Gutkowski and T. Winnicki (eds.), Restoration of Forests, 15–35.
© 1997 *Kluwer Academic Publishers.*

Increased acidification of precipitation would also result in enhanced mobilization of metals, including cadmium and lead, in soil solution and their higher uptake by root system, thus adding an additional strain on forests.

One of the most important actions in Europe with respect to the problem of acidification of the environment, defined at the above mentioned Conference, is to reduce emissions of nitrogen and sulfur oxides, and ammonia to a level which brings their deposition below critical levels. This action requires international cooperation and coordination. The UN Economic Commission for Europe (ECE) has adopted the Convention on Longe-range Transboundary Air Pollution (LRTAP) to identify the effects of transboundary air pollution and to undertake the necessary abatement actions. In 1985 the International Co-operative Programme on the Assessment and Monitoring of Air Pollution Effects on Forest (ICP-Forest) was established within the UN ECE LRTAP (e.g. Lorenz, 1995). The major objectives of the ICP-Forest are to (1) improve our knowledge of the spatial and temporal variations in forest conditions in the ECE region, and (2) study cause-effect relationships in forest ecosystems.

The main goal of this work is to review our knowledge on current levels, as well as historical and future trends of emissions of acidic compounds, NMVOCs, and selected metals in Europe, and to define hot spot regions of these emissions. Then, the spatial distribution of emissions is compared with the maps of atmospheric deposition of the studied compounds and exceedance of their critical loads. The latest information on the subject, obtained through various UN ECE programs and other international and national programs in Europe is used here.

2. Atmospheric Emissions in Europe

Information on atmospheric emissions of various pollutants in Europe has been collected by a number of international organizations and programs. Emission inventories for sulfur dioxide, nitrogen oxides, NMVOCs, and ammonia are among those prepared by national authorities for the UN ECE LRTAP. They are reported within the same source category split recommended by the European Commission (EC) program CORINAIR and the ECE European Monitoring and Evaluation Programme (EMEP). The European countries are requested to report their total emissions, as well as emissions from major source categories. They are also obliged to present the spatial distribution of emissions within the EMEP grid system.

2.1. CURRENT EMISSIONS

Almost all solid and liquid fuels contain sulfur. The combustion of sulfur containing fuels and wastes is by far the major source of sulfur oxide emissions to the air. Other sources include processing of sulfur-containing materials, (e.g. roasting of ores or sintering plants), production of sulfur compounds, (e.g. sulfuric acid manufacturing), using sulfur compounds to produce other industrial goods, (e.g. cellulose production with the use of the sulfide process), and processes of sulfur removal, (e.g. from liquid and gaseous fuels). The sulfur content of fuels and sulfur retention in ashes, along with the type and efficiency of flue gas desulfurization (FGD) installations are the major parameters affecting the amount of emissions from combustion processes.

As much as 34 Mtonnes of sulfur dioxide were emitted in Europe in 1993. The 1993 emissions of sulfur dioxide in Europe within the EMEP grid system of 150 km by 150 km are presented in Fig. 1 for point sources above 100 m high (Fig. 1a) and area sources (Fig. 1b), separately (EMEP, 1995). The largest amounts of sulfur dioxide in 1993 were emitted in Central Europe, particularly in the Black Triangle area.

Combustion of fuels in stationary and mobile sources is also a predominant category of nitrogen oxide emissions (excluding nitrous oxide). Three major mechanisms are responsible for the formation of nitrogen oxides during combustion: fixation of atmospheric nitrogen in the combustion air (thermal nitrogen oxides), conversion of chemically bound nitrogen in the fuel (fuel nitrogen oxides), and interaction of reactive radicals in the primary combustion zone (prompt nitrogen oxides). Similar processes result in the formation of nitrogen emissions during the waste combustion. The amount of emissions is highly dependent on the parameters related to fuel or waste characteristics (e.g. nitrogen content), the apparatus (e.g. design), the operation conditions, and the type and efficiency of control equipment. Other sources of nitrogen oxide emissions include: production of nitrogen-containing chemicals (e.g. nitric acid), open thermal processes involving oxygen (e.g. production of steel in open-hearth and electric arc furnaces), and use of nitrogen compounds (e.g. for surface treatment).

In 1993 the total emissions of nitrogen oxides were estimated to be more than 21 Mtonnes. The spatial distribution of these emissions within the EMEP 150 km by 150 km grid is presented in Fig. 2 (EMEP, 1995). The largest emissions of nitrogen oxides were generated in Western Europe, mainly due to emissions from traffic.

Ammonia is not always recognized as an acidifying compound. Although gaseous ammonia is alkaline and, thus, neutralizes airborne acids, when deposited to terrestrial ecosystems, ammonia may be converted to nitrate and become acidifying in its effect on soil and ground waters. Biological degradation processes within animal husbandry wastes, commercial fertilizers, sewage treatment, and landfills are the major anthropogenic sources of ammonia emissions to the atmosphere. Of particular importance is decomposition of nitrogen compounds in domestic animal wastes during housing of the animals, application of manure and its storage, and the period when animals are out to pasture. Other sources of ammonia include the manufacture of certain nitrogen compounds, such as ammonia and artificial fertilizers, handling of ammonia in cooling installations, and combustion of solid and liquid fuels and wastes.

In 1993 the ammonia emissions in Europe were estimated to be over 6.5 Mtonnes (EMEP, 1995). Their spatial distribution, presented in Fig. 3 is rather even with higher values in the Benelux countries and Denmark, and lower values in northern Scandinavia.

Non-methane volatile organic compounds are often inventoried as a group, though scientific teams generally require some degree of speciation breakdown. Emission sources of NMVOCs are numerous. They include incomplete combustion (particularly in engines); incineration and open burning of wastes and biomass; industrial production processes, mostly in the chemical and petrochemical industries; end-use of products and solvents; processing of various organic products using solvents;

evaporation during storage, transfer and handling operations of volatile organics (such as vehicle fuels); evaporation from cars; and fermentation in the production of food and alcoholic beverages. Obviously, many parameters, such as temperature, fuel composition, and combustor type, affect the amount of NMVOC emissions from these sources.

High temperature processes, such as coal and oil combustion in electric power stations and heat and industrial plants; gasoline combustion; roasting and smelting of ores in non-ferrous metal smelters; melting operations in ferrous foundries; refuse incineration, and kiln operations in cement plants emit various trace metals into the atmosphere. The amounts of trace metal emissions depend on: the contamination of fossil fuels and other raw materials by these compounds, trace metal physico-chemical properties affecting their behavior during industrial processes, the technology of industrial processes, and the type and efficiency of control equipment.

Two trace metals were selected for this study: cadmium, emitted mostly from industrial sources, and lead, emitted mainly during the combustion of leaded gasoline. The total emissions of Cd and Pb in Europe at the beginning of the 1990's were calculated to be almost 0.9 for Cd and over 58 ktonnes for Pb (Pacyna, 1996). Spatial distribution of these emissions within the EMEP grid is presented in Fig. 4a for Cd and in Fig. 4b for Pb. While the Pb emissions are spread around the continent, the Cd emissions are more centered towards regions in Central and Eastern Europe, and Spain.

2.2. HOT SPOTS OF ATMOSPHERIC EMISSIONS

A number of studies have been carried out in order to present a list of major point sources of emissions of acid pollutants (e.g. Barrett and Protheroe, 1994), and toxic trace metals (e.g. Pacyna, 1996) in Europe. The largest 100 point sources make up some 42% of total sulfur emissions in the ECE region. Their location and sulfur emission are presented in Fig. 5a. Again, the major emission regions include Central and Eastern Europe, and some areas in Spain. The Black Triangle area is of particular interest in this respect. Location of major sources in this region is presented in Fig. 5b (Barrett and Protheroe, 1994).

Non-ferrous metal smelters and refineries in Eastern Europe are the major point sources of Cd and other trace metal emissions in the ECE region (Pacyna, 1996). They are located in northern part of Russia, in the Upper Silesia region in Poland, and in Bulgaria.

2.3. HISTORICAL TRENDS OF EMISSIONS

The historical trends of total emissions of sulfur (decreasing), nitrogen dioxide (approximately constant), and ammonia (approximately constant) in Europe over the last 15 years are presented in Fig. 6a (EMEP, 1995). The decrease of sulfur emissions is related mostly to the installation of FGD equipment in major power plants and smelters over Europe. Washing or switching of fuels in Europe is much less used than in the United States.

A historical review of emissions of Cd, Pb, and Zn in Europe during the period from 1955 through 1987 has been approached at the International Institute for Applied System Analysis (IIASA) (Olendrzynski *et al.*, 1995). A summary of these data is also shown in Fig. 6b together with the emissions at the beginning of the 1990s. The largest European emissions of Cd and Zn, emitted mostly from industrial sources, are estimated for the late 1960's. At the beginning of the 1970's the first efficient electrostatic precipitators (ESPs) were employed at major electric power plants, smelters, and cement plants. As a result, emissions of particles and thus heavy metals decreased significantly from these sources. Concerning lead emissions, obviously, the part of lead emissions from industrial sources followed the decreasing trend described for Cd and Zn. However, the major source of Pb, combustion of leaded gasoline, was still increasing its emissions at the beginning of the 1970's, as the low-leaded gasoline was introduced in Europe only in the end of the 1970's (mostly in Germany). Unleaded gasoline appeared in the European market in small amounts only in the first half of the 1980's.

3. Atmospheric Deposition in Europe

While emitted into the atmosphere, air pollutants undergo various physical and chemical transformations, participate in chemical reactions, and finally undergo various removal processes. Wet and dry deposition are the main removal processes for acid and toxic pollutants from the atmosphere.

Major analysis of the European transboundary acidifying air pollution has been carried out at the EMEP Meteorological Synthesizing Centre-West (EMEP, 1995). The model estimates of the total deposition of oxidized sulfur and total deposition of nitrogen in Europe in 1993 are presented in Figs. 7a and 7b, respectively. Oxidized sulfur deposition mirrors the spatial distribution of sulfur emissions with localized peaks in the Black Triangle area in Central Europe and in Spain. Rapid dry deposition of sulfur dioxide and its solubility, together with the solubility of sulfates, which are formed through conversion from sulfur dioxide explain the similarities of spatial distributions of emissions and deposition of sulfur compounds.

Total nitrogen deposition has much less variable pattern than the sulfur deposition. Most of the products of gas-to-particle conversions of nitrogen species are more readily deposited and soluble than their gaseous precursors, however, this conversion takes some time, and thus some transport from the source. Of the two nitrogen components individually, reduced nitrogen (ammonia and ammonium) is of equal or greater contribution across much of Europe than the oxidized nitrogen (nitrates) (EMEP, 1995).

The EMEP studies concluded that in 1993 sulfur deposition represented the largest single component in deposition of compounds with an acidity creating potential. However, the combined oxidized and reduced nitrogen deposition may represent greater contribution than the sulfur deposition (EMEP, 1995). The EMEP model estimates were then compared with measurements carried out at more than 100 stations in Europe. A good comparison has been obtained in most of the regions of the ECE area.

The results of modeling the total deposition of Cd and Pb in Europe in 1985 are presented in Fig. 8a nad 8b, respectively (Olendrzynski *et al.*,1995). The largest Cd deposition was estimated for the Black Triangle area and the Upper Silesia region, and northern Spain, thus, resembling the Cd emission pattern. The Pb deposition pattern also resembles the metal emission pattern, however, is different than the Cd patterns. The Pb deposition is more evenly spread over Europe with highest values in the Benelux countries and the U.K.

4. Critical Loads of Air Pollutants

The critical load has been defined as "the highest deposition of a compound that will not cause chemical changes leading to long-term harmful effects on ecosystem structure and function" (Nilsson, 1986). Thus, the critical load appears to be an indicator for sustainability of an ecosystem. Very often the deposition data for various pollutants are compared with critical loads values to determine which areas currently receive deposition levels which exceed the area's critical load. These areas of "exceedance" indicate where present levels of pollutant deposition increase the risk of damage to the ecosystem. Mapping of critical loads for various pollutants in Europe has been done within the UN ECE Task Force on Mapping. The Coordination Center for Effects (CCE) was established by the Netherlands Ministry of Housing, Physical Planning and the Environment with major goal to assist in the development and production of methods and maps of critical loads on a European scale, in cooperation with the Task Force on Mapping (e.g. CCE, 1993).

Sulfur was given priority in mapping the exceedance of critical load of air pollutant deposition in Europe. The 5 percentile critical deposition in Europe is presented in Fig. 9a (CCE, 1993). The effects of base cation uptake and base cation deposition on critical load of sulfur were taken into account when preparing the map presented in Fig. 9a. northern part of Europe is much more sensitive to sulfur deposition than the rest of the continent. A comparison of the 1990 deposition of sulfur with the critical sulfur deposition is presented in Fig. 9b. The highest exceedances were calculated for the Black Triangle region. This high exceedance is also presented in the map of exceedance of critical loads of sulfur in Poland, presented in Fig. 10a. For comparison, the exceedance of critical loads of nitrogen, presented in Fig. 10b is much lower with more even pattern (CCE, 1993).

Ozone seems to be an important component contributing to the forest damage. There are theories indicating that ozone, through its interaction with acid mist, can induce or speed up this deficiency, particularly in mountainous areas in summer. In general, measured concentrations seem to indicate a linear increase of ozone by 1 to 3 % per year over the last decades (e.g. Hov *et al.*, 1986).

Various critical levels of ozone have been defined. With respect to forests and other plants, it should be indicated that ozone is taken up through stomata and then reacts with a number of enzymes and antioxidants. When ozone exceeds certain threshhold, plants respond by reduced photosynthesis and carbon dioxide uptake, the factors affecting the tree growth.

A concept of accumulated ozone exposure above threshhold of 40 ppb (AOT40) was introduced in order to compare the ozone effects. For forests, the AOT40 of 10 000 ppbh was introduced, calculated for the growing season (1st of April through 31st of September) during night and day (24 hrs). The measurements carried out in 1992 povided information needed to present the AOT40 values at least for a part of Europe (Hjellbrekke, 1995). This information is presented in Fig. 11. The highest values were obtained for sites in Central Europe. It was also pointed out that the 1992 values were generally higher than the 1990 and 1991 values.

Results of modeling the deposition of Cd and Pb accumulated in Europe over the period from 1955 through 1987 are presented in Figs. 12a and 12b, respectively. The largest Cd deposition was estimated for the Silesia region, the Black Triangle area, and the Ruhr region. The largest Pb deposition values were those calculated for the Ruhr area.

Aluminium is another important element which may contribute to changes in the health of forests. Aluminium that is liberated from the soil as the pH decreases is toxic to spruce trees. However, the quantitative assessment of this process is lacking.

5. Future Projections

About 170 treaties, most of them drafted in the past 20 years, are designed to safeguard our environment. Many of these treaties deal with emission reductions in the near future. In Europe, the ECE Convention on LRTAP is the major body dealing with the international agreements on emission reductions.

In July 1985, the Protocol on the Reduction of Sulphur Emissions or their Transboundary Fluxes by at least 30% was signed in Helsinki (the so-called "Helsinki Protocol"). This reduction of the 1980 levels of emissions as a basis was expected to be achieved by 1993 at the latest. However, it is still too early to discuss the effectiveness of this agreement because not all countries have yet reported their 1993 and 1994 emission data to the Convention. A new Protocol on Further Reduction of Sulphur Emissions after 1993 was adopted in June 1994. This new agreement was negotiated to obtain emissions reductions of the 1980 levels, ranging from 30 to 87% within 3 schedules: years 2000, 2005, and 2010. In addition, requirements have been set for certain stationary combustion sources and for the sulfur content of petroleum products.

To moderate acid rain and ozone formation, the ECE Convention provided a platform for emission reduction protocols on nitrogen oxides and volatile organic compounds (VOCs). In October 1988, the Protocol concerning the Control of Emissions of Nitrogen Oxides or their Transboundary Fluxes was signed in Sofia, Bulgaria (so-called "Sofia Protocol"). The countries agreed to reduce or to control emissions "so that these, at the latest by 31 December 1994, do not exceed their national annual emissions of nitrogen oxides or transboundary fluxes of such emissions in the calendar year 1987." Alternatively, a country could choose upon signature any year previous to 1987 as a base year provided that in addition the average annual national level of emissions between 1987 and 1996 does not exceed the 1987 level.

In November 1991, in Geneva, the Protocol concerning the Control of Emissions of Volatile Organic Compounds or their Transboundary Fluxes was adopted (so-called

"Geneva Protocol"). The emission target set out in the basic obligations of the Geneva Protocol is to reduce national annual VOC emissions by at least 30% by 1999, using either 1988 levels as a basis or any other annual level between 1984 and 1990. There are, however, alternatives to this obligations related to either specific regions or situations where the national annual emissions of VOCs in 1988 were below 500,000 tonnes, 20 kg per inhabitant, and 5 tonnes per square kilometre.

At present, a multi component emission reduction protocol is being prepared under the ECE Convention to further reduce emissions of nitrogen compounds and VOCs. Separate legal measures are also discussed to reduce emissions of heavy metals, particularly lead, cadmium, and mercury. The 2010 emission estimates by Pacyna (1996) indicate a decrease of 40 % of the 1990 Cd emissions. The 1990 Pb emissions should be lower by a factor of 4 in the year 2010.

Information on current emission reduction plans is used to estimate critical loads values. For example, the critical load of nitrogen at 2010 sulfur deposition (current reduction plan), yielding protection of 95 % of the ecosystems against acidification has been presented by Hettelingh et al. (1996).

6. Concluding Remarks (Pertinent observations from the ARW)

Most of the hot spots of atmospheric pollution in Europe are located in Central Europe, particularly in the Upper Silesia and the Black Triangle area. Major point sources of emissions are located in this region, including power plants, smelters, and iron foundries. Although the atmospheric emissions of acid compounds, photo-oxidant precursors, and toxic metals have decreased in the region during the last 5 years due to general decline in the production of industrial goods and the consumption of raw materials, these emissions are still higher than the emissions in other parts of Europe.

Hot spots of atmospheric deposition, as well as regions with high exceedances of critical loads of acid compounds and ozone are also located in Central Europe. The maps of emissions and critical load exceedances in Europe coincide very well with the maps informing on forest damage (e.g. the maps presented in EEA, 1995). Therefore, one can expect adverse effects of air pollutants on the degradation of forest in the region. However, it is still premature to conclude to what extent different air pollutants contribute directly to forest damage. Further research is needed on the combined impact of air pollutants and natural stress factors for forests, including insects, fungal diseases, drought, and frost.

Almost all issues presented in this paper need further examination. A few examples can be given. More information is needed on the role of both oxidized and reduced nitrogen compounds in overall acidification. Balances and even sources of nutrients in the forest ecosystems are rather poorly evaluated. Mobilization of metals in forest soils and their enhanced uptake by root system need further investigations. In recent years a number of major point sources in the Upper Silesia region have been equipped with highly efficient electrostatic precipitators which remove most of the particles, also those containing major cations. This process has caused additional acidification of precipitation in the region, because the installation of FGD equipment is still sporadic in this part of Europe. Enhanced acidification of precipitation has caused solution of toxic metals from soil to soil solution and further migration of metals with surface waters, as well as higher uptake by plants.

More research is needed in order to explain limitations and uncertainties, as well as to improve the concepts of critical loads and critical levels of air pollutants. This improvement is needed specially for better understanding the impact of air pollution on the forest ecosystem in polluted regions, such as the Upper Silesia and the Black Triangle area.

7. References

1. Barrett, M. & Protheroe, R. (1994) Sulphur emission from large point sources in Europe, Swedish NGO Secretariat on Acid Rain, 2nd Ed., Gøteborg, Sweden.
2. CCE (1993) Calculation and mapping of critical loads in Europe: Status Report 1993. Convention on Long-range Transboundary Air Pollution, Coordination Center for Effects, National Institute of Public Health and Environmental Protection, Bilthoven, The Netherlands.
3. EEA (1995) *Europe's Environment. The Dobris Assessment.* European Environment Agency, Copenhagen, Denmark.
4. EMEP (1995) European Transboundary Acidifying Air Pollution. The Co-operative Programme for Monitoring and Evaluation of the Long range Transmission of Air Pollutants in Europe. The EMEP/MSC-W Rept. 1/95. The EMEP Meteorological Synthesizing Centre-West, the Norwegian Meteorological Institute, Oslo, Norway.
5. Hettelingh, J.-P., Posch, M. & de Smet, P. (1996) Mapping critical thresholds and stock at risk. Background paper. Convention on Long-range Transboundary Air Pollution, Coordination Center for Effects, National Institute of Public Health and Environmental Protection, Bilthoven, The Netherlands.
6. Hjellbrekke, A.-G. (1995) Ozone measurements 1990-1992. EMEP/CCC Rept. 4/95. Norwegian Institute for Air Research (NILU), Kjeller, Norway.
7. Hov, Ø., Becker, K.H., Builtjes, P., Cox, R.A. & Kley D. (1986) Evaluation of the photooxidants-precursor relationship in Europe 1986. The CEC Air Pollution Research Rept. 1, The Commission of the European Commmunities, Brussels, Belgium.
8. Lorenz, M. (1995) International Co-operative Proagramme on Assessment nd Monitoring of Air Pollution Effects on Forests - ICP Forests. *Water, Air, and Soil Poll.* **85**, 1221–1226.
9. Nilsson, J. (1986) Critical Loads for Nitrogen and Sulphur. Miljørapport 1986:1. Nordic Council of Ministers, Copenhagen, Denmark.
10. Olendrzynski, K., Anderberg, S., Bartnicki, J., Pacyna, J.M. & Stigliani, W. (1995) Atmospheric emissions and depositions of cadmium, lead and zinc in Europe during the period 1955-1987. IIASA WP-95-35 Rept. International Institute for Applied Systems Analysis (IIASA), Laxenburg, Austria.
11. Pacyna, J.M. (1996) Atmospheric emissions of heavy metals in Europe. Improvements, updates, historical data and projections. A report for the International Institute for Applied Systems Analysis (IIASA), Hagan, Norway.
12. Rodhe, H., Grennfelt, P., Wisniewski, J., Ågren, C., Bengtsson, G., Johansson, K., Kauppi, P., Kucera, V., Rasmussen, L., Rosseland, B., Schotte, L. & Sellden, G. (1995) Acid Reign' 95? - Conference summary statement. *Water, Air, and Soil Poll.* **85**, 1–14.

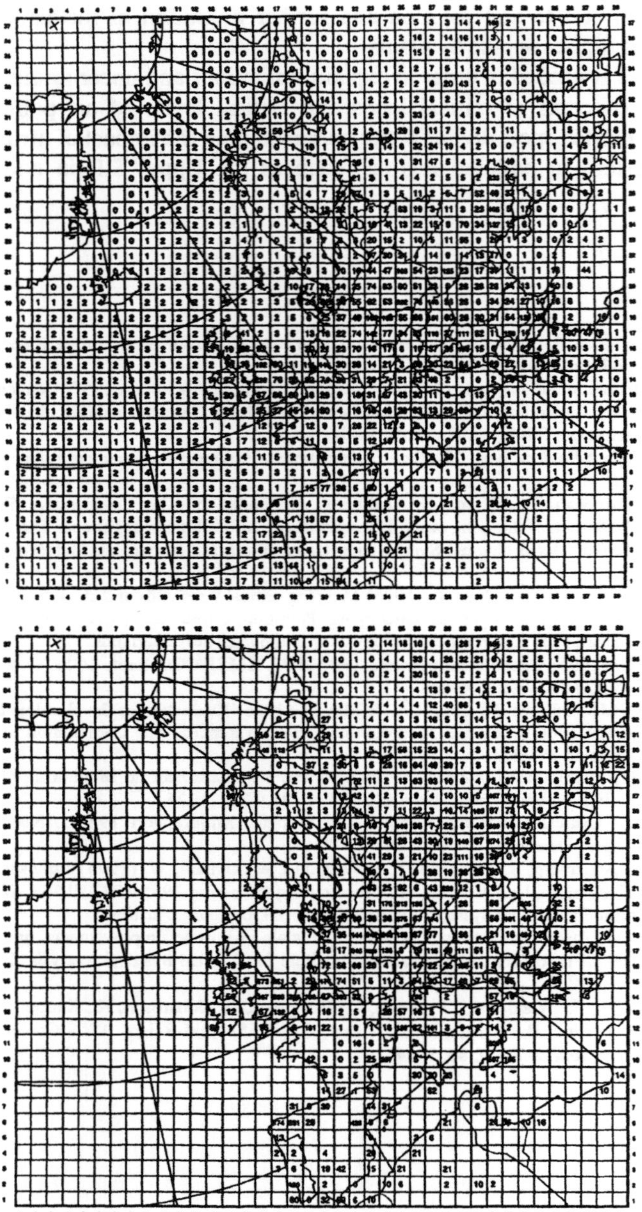

a)

b)

Figure 1. Spatial distribution of the European emissions of sulfur dioxide from sources above 100 m high (a), and below 100 m high (b) in 1993 within the EMEP grid system of 150 km by 150 km (EMEP, 1995), (1000 tonnes of sulfur dioxide).

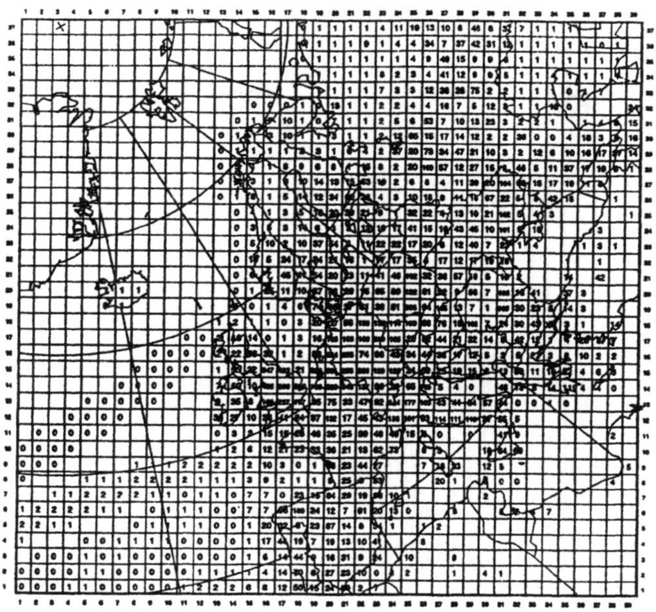

Figure 2. Spatial distribution of the European emissions of nitrogen oxides in 1993 within the EMEP grid system of 150 km by 150 km (EMEP, 1995), (1000 tonnes of nitrogen dioxide).

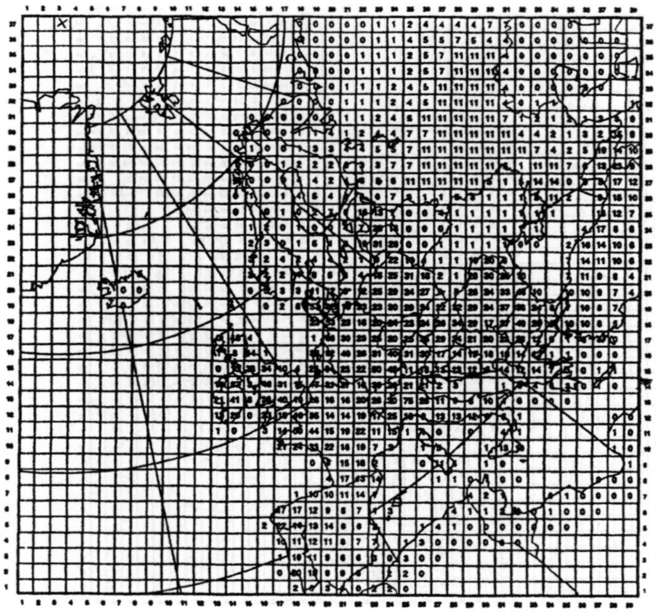

Figure 3. Spatial distribution of the European emissions of ammonia in 1993 within the EMEP grid system of 150 km by 150 km (EMEP, 1995), (1000 tonnes of ammonia).

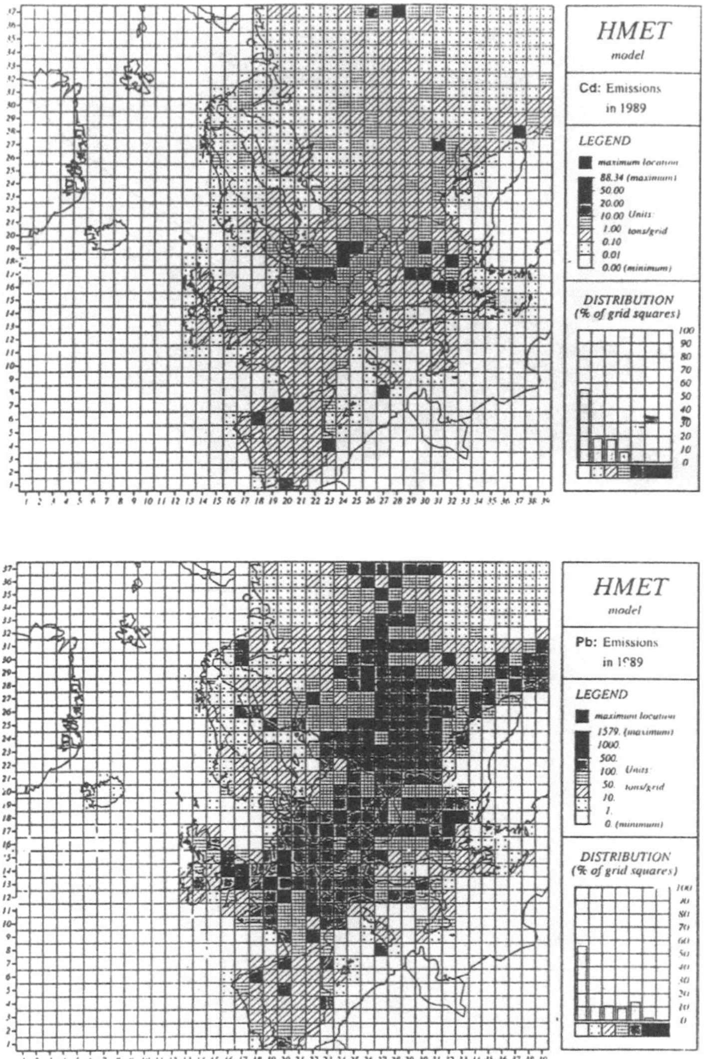

a)

b)

Figure 4. Spatial distribution of the European emissions of cadmium (a), and lead (b) in 1989 within the EMEP grid system of 150 km by 150 km (Pacyna, 1996), (tonnes).

Figure 5. Hot spots of sulfur emissions in Europe (a), and the Black Triangle (Barrett and Protheroe, 1994), (1000 tonnes of sulfur).

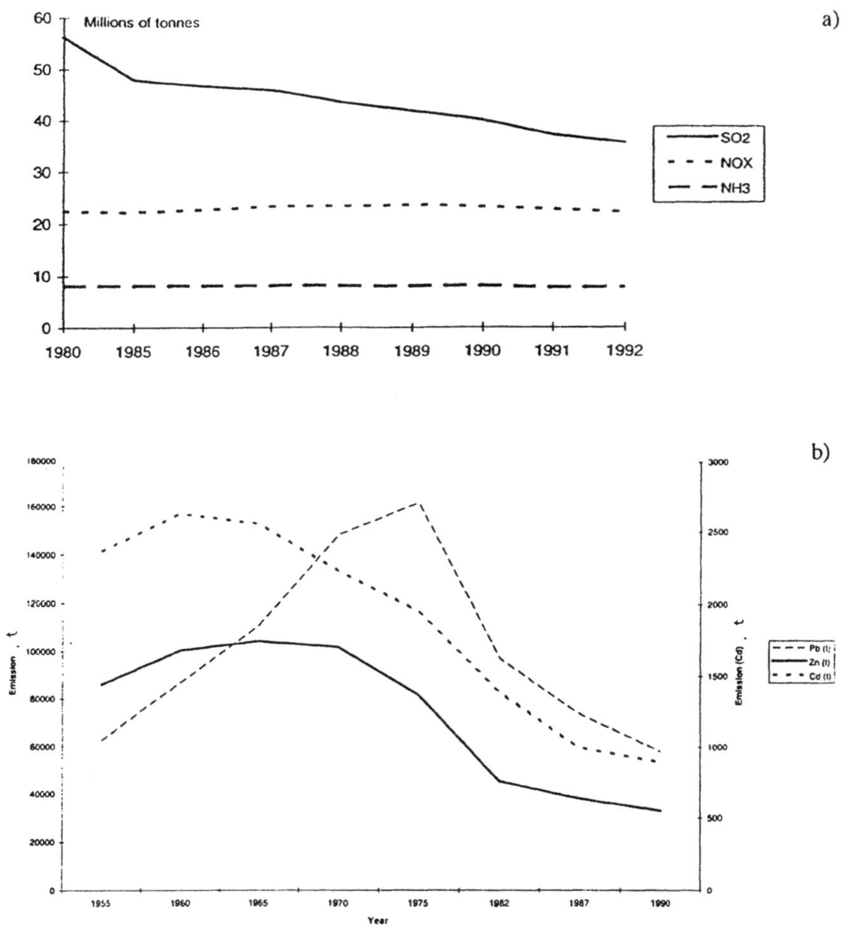

Figure 6. Historical emissions of sulfur dioxide, nitrogen oxides, and ammonia (a), and lead, cadmium, and zinc (b) in Europe.

a)

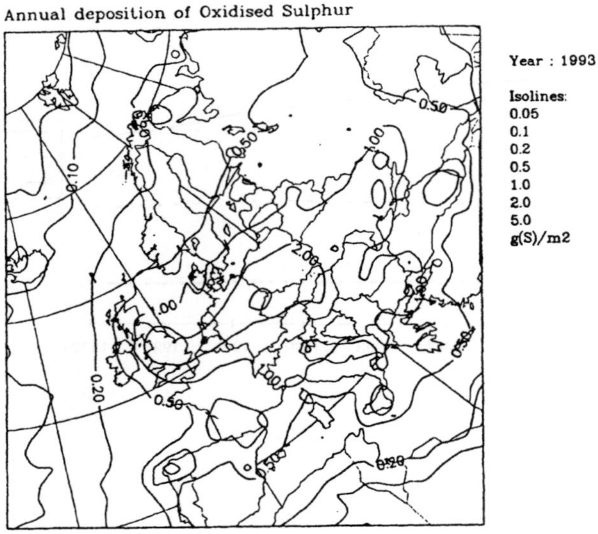

b)

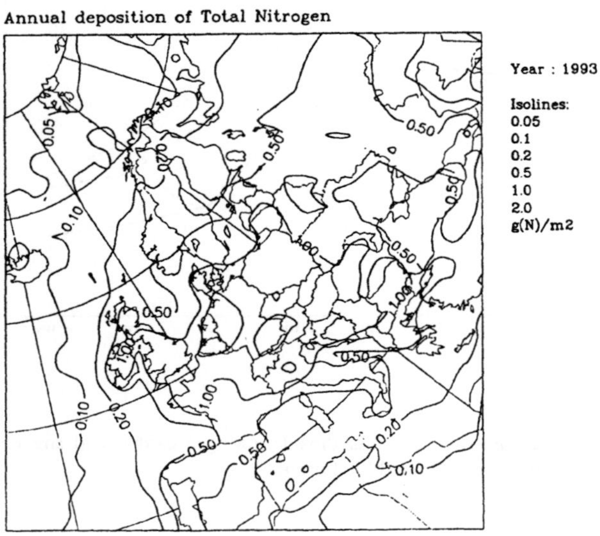

Figure 7. Deposition of oxidized sulfur (a), and total nitrogen (b) in Europe in 1993 (EMEP, 1995).

a)

b)

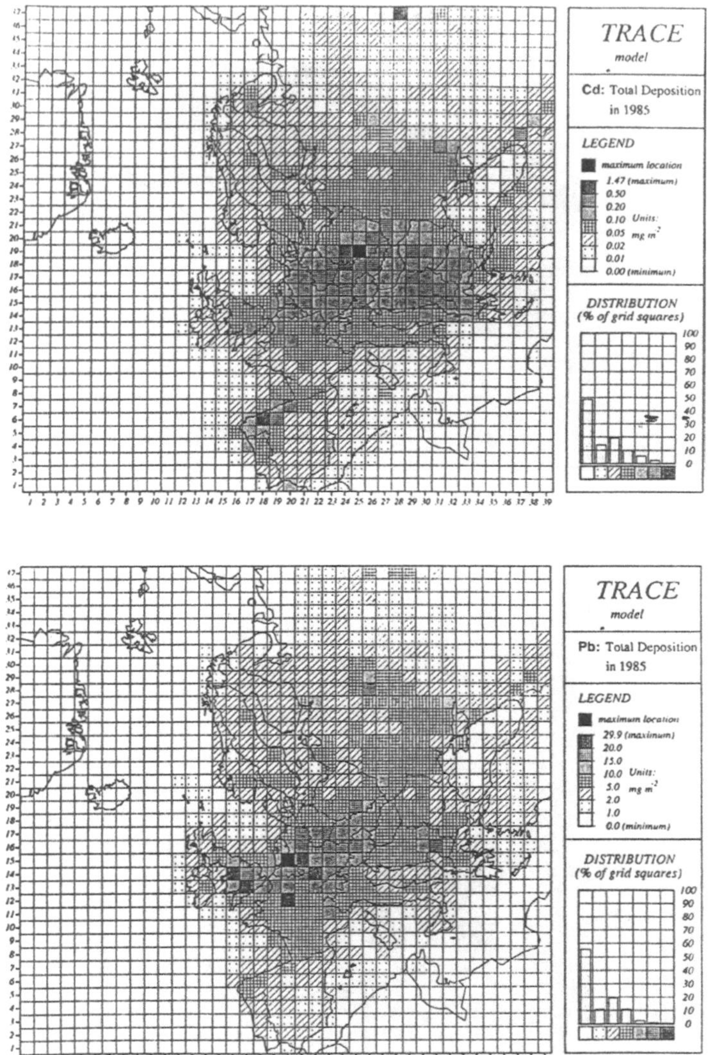

Figure 8. Deposition of cadmium (a), and lead (b) in Europe in 1985 (Olendrzynski et al., 1995).

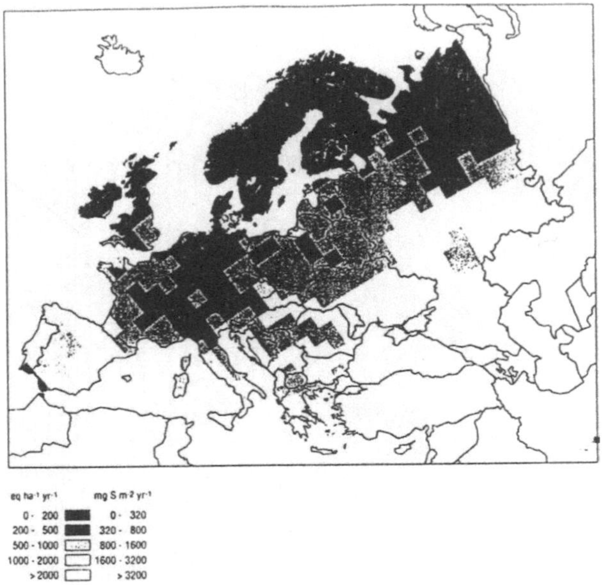

a)

b)

Figure 9. Critical sulfur deposition (5 percentile) (a), and exceedance of critical sulfur deposition (5 percentile) in Europe in 1990 (CCE, 1993).

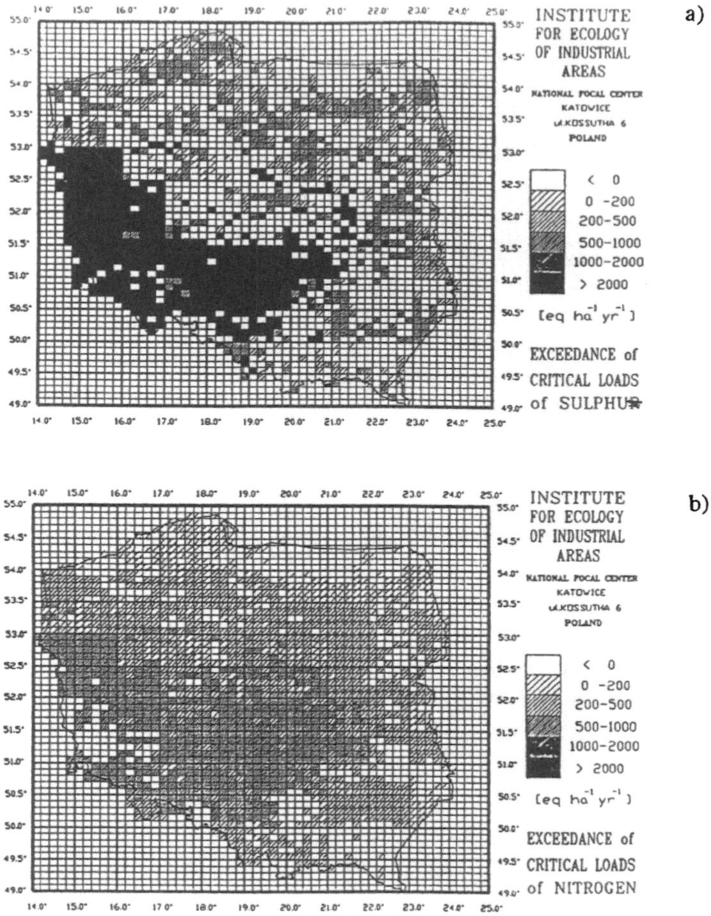

Figure 10. Exceedance of critical loads of sulfur (a), and nitrogen (b) in Poland in 1990 (CCE, 1993).

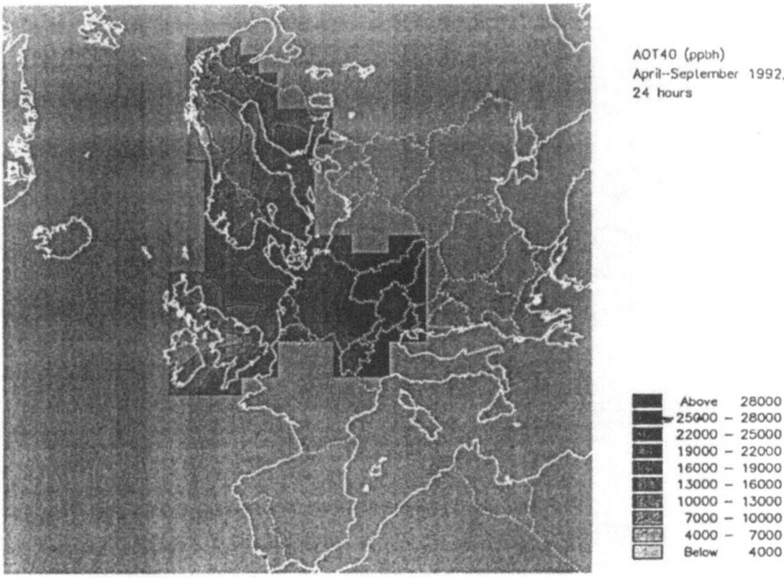

Figure 11. Accumulated ozone exposure above treshold of 40 ppb (AOT40) values in Europe in 1992 (Hjellbrakke, 1995).

a)

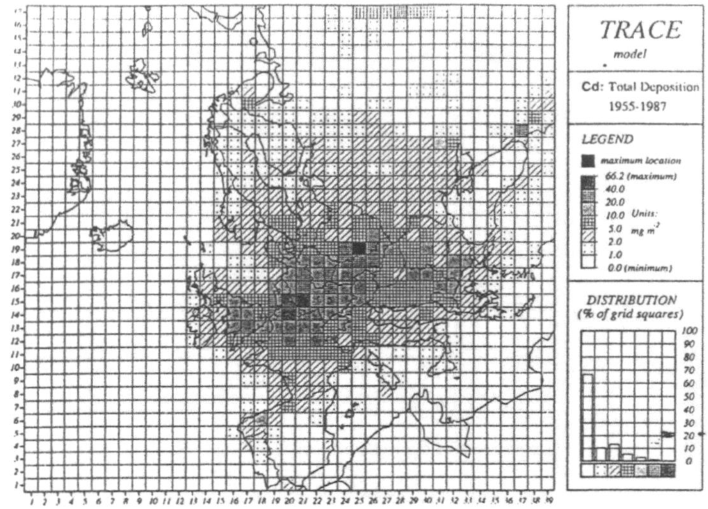

b)

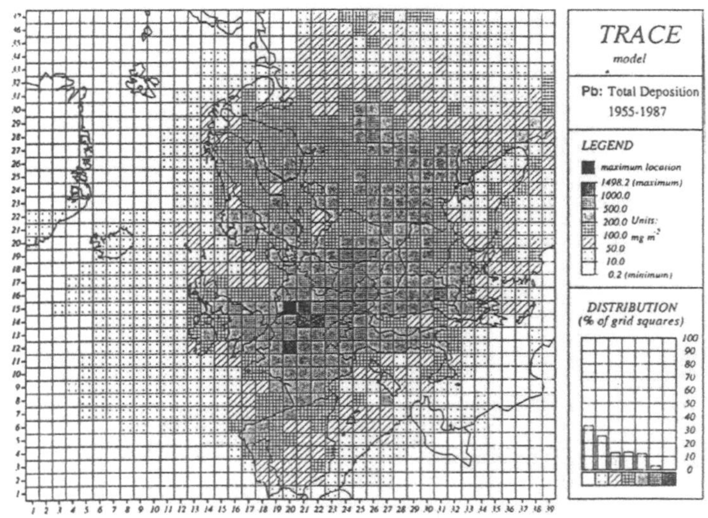

Figure 12. Accumulated deposition of cadmium (a), and lead (b) in Europe for the period from 1955 through 1987 (Olendrzynski et al., 1995).

FOREST DEPLETION IN EUROPE: EXTENT, CAUSES AND INTERPRETATIONS

K. RYKOWSKI
Forest Research Institute
Warsaw, Poland

1. Introduction

In the 1970s the first reports of "Waldsterben" hit the headlines in Germany. This apparently new and widespread type of forest diseases was appearing in many places in Europe and was called "Forest Decline" or "Deperissement des Foret." After Germany, the countries in Central-Eastern Europe reported about "Forest Decline"especially in mountainous regions. Some researchers, public, policy makers, NGOs and others in all Europe have started to accuse industrial civilization of ecological disaster and announced the end of european forests. At the same time passionate debate over the causes and possible disease factors and interrelations between them have started producing many confusions and controversy.

The rapid decline witnessed in Germany led to the setting up of a national survey in 1982. This demonstrated that, by 1985, only about half of the country's forests were considered to be "healthy." Concerned by these findings, other European countries set up their own regional or national surveys.

Pan-European assessment of forest health using standard procedures was initiated by the United Nations Economic Commission for Europe (UNECE) and the European Commission (EC). The survey covers 34 countries and involves analysis of 33,000 sample plots and 620,000 trees.

Some increase in recorded damage is due to more careful surveying techniques. It is clear that about one-fifth of Europe's trees are suffering more or less serious decline, and that in the last few years there has been an average increase in the severity of effects. Tree health is especially poor in Central and Eastern Europe. In spite of some decrease of emissions in Poland (fig. 1), tree damages are still high with a tendency to be stabilized during the last 2 - 3 years (fig. 2).

R. M. Gutkowski and T. Winnicki (eds.), Restoration of Forests, 37–53.
© 1997 *Kluwer Academic Publishers.*

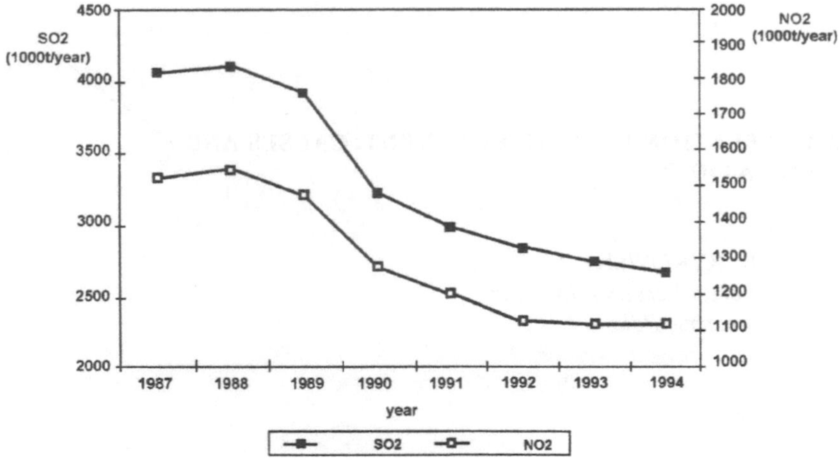

Figure 1. Quantity of emissions for years 1987-1994

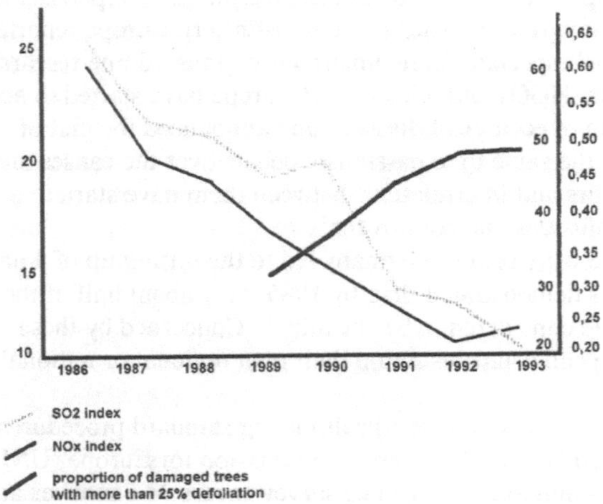

Figure 2. SO_2 and No_x indices ($mg/m^2/day$) for years 1986-1993 and the proportion of damaged trees with more than 25% of defoliation (%) for years 1989-1993

2. Healthy State of European Forests

Some pathological processes in the European forests occurred all over Europe before the "Forest decline" occurred. Let me analyze some of them.

One of the most typical problems in this part of Europe is the so-called "forest on post-agricultural land." It concerns mostly coniferous species and at first - Scots pine. Ecological unadaptability is the source of permanent stress to this forest - of artificial forest association to ecological site or more general - unadaptability of biocenosis to biotope. Many pathological phenomena caused by biotic and abiotic factors occur in these conditions.

In the case of Pinus we can make a list (it concerns not only pine stands established on post-farming lands): 32-50 species of fungi occur in the crown on needles and branches, 25-38 of fungi species were found on the stem and inside wood, 75-120 in the root system.

In the case of Norway spruce (Picea abies) the literature says about 15 - 34 species of fungi on needles and branches; 27 - 34 on stem and 40 - 120 in the root system. Some of them are harmful, while others - beneficial. Root rot Fomes annosum (Heterobasidion annosum) is undoubtedly the most harmful fungi pathogen on post-farming lands. The classical methodology to analyze this pathological process was as follows (fig. 3): after thick planting (mostly one species) on a large area, stress of density occurs, caused by very high competition in relation to light, nutrients, life space, and so on.

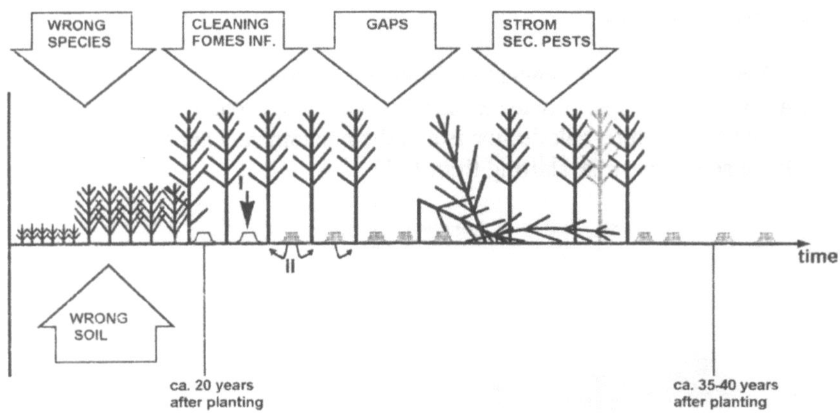

Figure 3. Pine stand on the post-farming lands

After the first cleaning, the entrance for the primary F. annosum infection is open. Secondary infection ways enlarge the gaps. The gaps open the stand to wind and storm, and to secondary pests as well. In the age of 35 - 40-years-old the stand is made away.

Many other pathological processes could be observed in the same situation of stands established on post-farming lands. In 1982 - 87 the “pine shoot decline” passed through Europe, especially in central and eastern parts of Europe. The diagnosis made in that time indicated the role of Gremmeniella abietina as a biotic agent and the climatic - weather condition as an abiotic one (fig. 4).

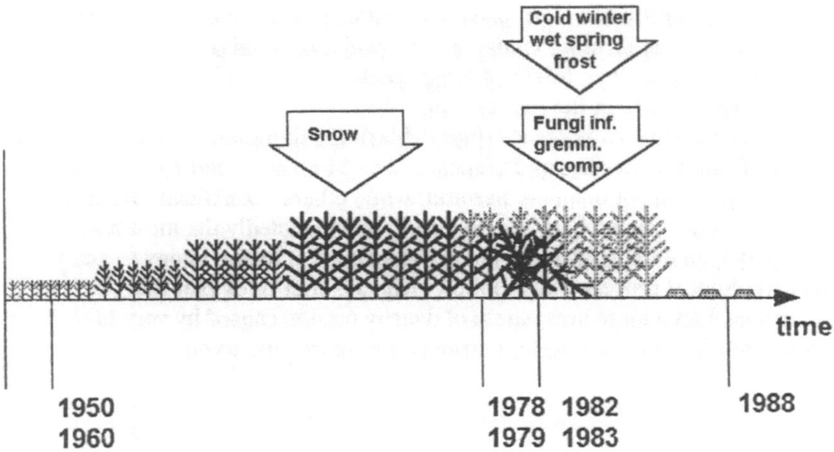

Figure 4. Pine shoots decline

Let me focus for a moment on another pathological phenomenon, typical for Europe, especially for eastern Europe, called "oak decline" (fig. 5). According to literature, oak decline is a cyclic phenomenon appearing in Europe in some intervals, starting from 1877.

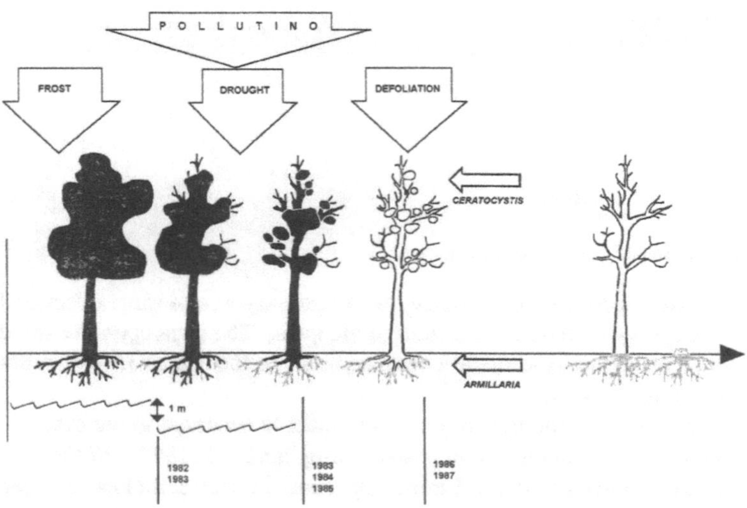

Figure 5. Oak decline

It was usually analyzed in the context of the occurrence of unfavorable climatic factors - drought, low temperatures, changes in the level of soil water.

The infection hypothesis developed during the last "oak decline" in Europe focused the attention on the so-called "vascular fungi" of the Ceratocystis genus. Phytopathological research showed that 207 fungi species belonging to 131 genera occur on six declining oak species. They represent different ecological functions: parasites, saprophytes with behavior looking like pathogen, endophytes. Many other pathogenic organisms were isolated - bacteria, viruses, nematodes, mycoplasms and so on.

More than 90 species of folio-cambio and xylofagous insects were found on dying oaks. Entomofauna on oaks is extremely abundant and it contains over 800 species.

Other factors were registered either as the cause of oak decline or as coacting agents: improper soil site condition, unduly silvicultural treatments, inappropriate cutting and regeneration methods, mechanical injuries and so on.

Of course, the relationship between "oak decline" and the general state of environment, especially industrial pollution of air (acid rain) was established.

3. The Causes

Not surprisingly, given the potential ecological, social and economic impacts of forest decline, there has been considerable speculation over possible causes. Clearly, no single factor is responsible and a picture is emerging of interactions between a number of different stress factors. Some of them act over long periods to weaken trees, thus increasing susceptibility to periodic stress which might otherwise have had limited impact.

Although there is a multitude of factors contributing to forest decline, there are four areas of particular concern: climate, air pollution, forest management practice and biotic factors (insects and fungi).

3.1. CLIMATE

More than half of national rerports in the European survey consider changes in forest condition to be due to weather conditions, particularly droughts and low or high temperatures. Water stress is particularly important - it is suggested that the exceptionally dry summer of 1976 triggered the decline in Europe.

More controversial is the potential impact of climate change. Scenarios which predict warmer and wetter winters and warmer and drier summers would exacerbate tree health problems associated with water stress but it seems to be more evident the the impact of climate change on the occurrence of harmful insects and pathogenic fungi was significant.

3.2. AIR POLLUTION

The main pollutants considered are sulphur dioxide (SO_2), nitrogen oxides (NO_x), ammonia, ozone and other products of photo-oxidation process, carbohydrates, acid deposition, heavy metals as well as deficiencies in certain nutrients and magnesium. The pollutants alter directly or indirectly main tree physiological activity like photosynthesis, respiration, transpiration, tree physiological transport, reproductive process and others.

3.3. FOREST MANAGEMENT

Intensive, industrial forestry, which has spread across of Europe, has had direct ecological implication. Choice of species has been determined by the productivity rather than the best ecological match for the site. Forestry practices like use of chemicals, fertilizers as well as the omission of silvicultural treatments have had impacts on forest health.

3.4. BIOTIC FACTORS

The simplification of forest practice and changes of environment favors the occurrence of the pests and fungi diseases which, until recently, were known to have only secondary or third class of importance or even organisms that were regarded as saprophytes. There occur qualitative and quantitative changes in the populations of microorganisms dwelling on the assimilatory organs and on shoots, and the activity is growing of more than 50 species of insects and 25 - 30 species of pathogenic fungi.

Other organisms like nemathods, bacteria, virus, mycoplasma-like organisms are becoming more and more important.

4. Hypotheses (more important still)

To try to explain the phenomenon of "forest decline," more than 35 hypotheses have been developed. Each of them tends to explain only one possible way of declining and all of them are probably true. Some of the hypotheses exclude industrial pollution as a stress factor involved in this process. It is worth mentioning some of them.

4.1. HYPOTHESIS 1: ACID RAIN

The theory attributes forest decline to the strong acids generated when sulfur dioxide and nitrogen oxides react with sunlight, oxygen and water vapor to form acid rain (or dry acidified particles). The acids alter the chemical structure of soils and plants drastically. Magnesium, potassium and manganese are easily leached from the upper root horizons of the soil as well as from the damaged leaves. At the same time, increased amounts of soluble aluminum, manganese and other metals suddenly reach toxic concentrations causing damage to root systems - particularly the mycorrhiza - thus decreasing the plants ability to take up nutrients and water. The effects appear most noticeably during dry periods when there is an overall increase in concentrations of acids and aluminum.

4.2. HYPOTHESIS 2: OZONE

This theory attributes most of the observed leaf damage to ozone. Ozone is formed when nitrogen oxides and hydrocarbons react with oxygen in the air during periods of bright sunshine, when the UV radiation is high. The products of these process damage the assimilation apparatus directly, causing the cell membranes to weaken and nutrients to be lost from the damaged cells. The leaves become more sensitive to attacks from fungi and insects. Of course the synergistic effects of SO_2 and NO_x might also account for some of the leaf damage.

4.3. HYPOTHESIS 3: GENERAL STRESS

This hypothesis was developed by a group of botanists, plant pathologists and physiologists and it is based on field and laboratory studies of damage on various tree species especially spruce and beech. According to this hypothesis deposition of atmospheric pollutants including sulfur dioxide with nitrogen oxides, hydrocarbons and heavy metals and other toxic substances, reduces photosynthetic activity and the formation of important secondary metabolites. Taken separately, these pollutants are usually not found in sufficient concentrations to cause the damage. The "cooperation" between them and synergistic effect results in poor development of the fine roots and decreased vitality of micorrhizae and finally reduced assimilate supply to the root system and nutrients to the leaves. Water and nutrient uptake is diminished and the crown no longer receives sufficient supplies. The result is premature senescence of needles and leaves and other tree organs which become more vulnerable to the biotic disease factors.

4.4. HYPOTHESIS 4: AMMONIUM HYPOTHESIS

This hypothesis is based an the observation concerning the greater than normal atmospheric deposition of biological available nitrogen compounds, principally ammonium and nitrate ions. The trees have become "over-saturated" with nitrogen; they have grown fast and formed large cells with a high volume and they are easily attacked by wind, drought, insects and parasites. The inner hormone balance is easily disturbed by this type of nitrogen stress.

4.5. HYPOTHESIS 5: COMPLEX HIGH-ELEVATION DISEASE

This hypothesis combines the theories of ozone damage and occurrence of magnesium deficiencies in high elevation of spruce stand. The combined impact consists of the partial breakdown of chlorophyll, causing decreased assimilation and reduced root growth.

4.6. HYPOTHESIS 6: RED-NEEDLE DISEASE OF SPRUCE

This hypothesis explains damage symptoms in spruce on sites with differing bedrock and different soil types at elevations up to 700 m. It was a result of extensive field observations, needle analyses, site classifications and of the study of endophytic fungi, (found mainly on the older needles of mature trees).

4.7. HYPOTHESIS 7: ORGANIC AIR POLLUTANTS

From this hypothesis it is possible that some among the many thousands of organic compounds present in the atmosphere can have toxic or growth altering effects on forest trees. Organic compounds are emitted in large quantities (about 20 million metric tons per year in the US) because of their extensive use as solvents as well as in numerous chemical processes. Nature contributes large amounts of organic compounds to the environment. Three organic compounds that have biological effect, in the context of forest damage, have been seriously discussed - ethylene, aniline and dinitrophenol.

5. Theoretical Generalizations

All hypotheses mentioned above attempt to describe the dysfunction of a living system on different organization level - cell, tissue, needle or leaf, tree or forest ecosystem. What is meant by "dysfunction?"

Plant pathologists do not agree on a precise definition of what is a plant disease. Many of them consider disease as any deviation from the normal functioning of a plant, caused by some types of persistent agents (MANION 1981[1]). However the question arises how can the "normal functioning" of a plant be defined? What is the standard of health ?

The "health standard" is commonly considered as a statistical mean or as a offshoot of "stability." But the term of "stability" is even more unclear than that of "disease." The problem of "stability" has a very rich literature base, not only in ecology. It became a standard problem in the system theory, as well as in the dissipative structure theory. According to the previous theories, "stability" or "balance" are something unusual for a living system. If the system is thermodynamically stable, it means that it is dead. In the light of that statement, an unstable forest ecosystem could be considered as a normal state of a living system in a changing environment.

[1] MANION, P.D., 1982: Tree disease concepts, Prentice-Hall, Inc., Englewood, New Jersey 07632

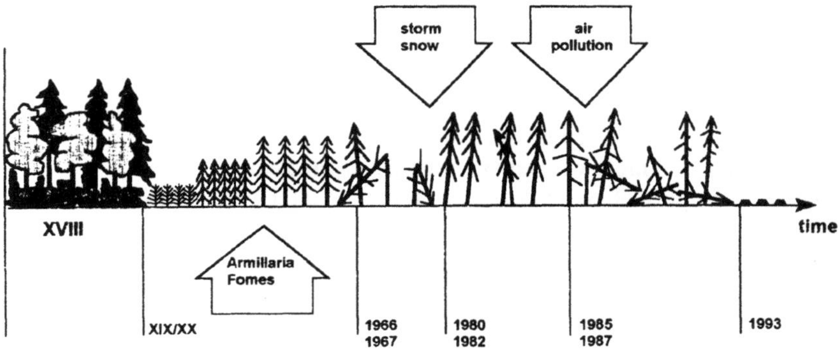

Figure 6. Forest decline in Sudety Mts

"Forest decline" has become a symbol of the global threat to the European forests. We can observe the typical process in the Sudety Mountains in Poland (fig. 6). It is widely known that the natural state of Sudety forest ecosystems had not been disturbed until the 15th century. In the course of 15th and 16th centuries, human activities (mines, glass works, etc) caused a rise of consumer demand for wood and a large area of forest stands was cut down. Fir and beech regeneration on large clear-cut areas was reduced, spruce regeneration was successful and share of this tree species grew up. Controlled forestry management was introduced in those regions at the second half of the 18th century or at the beginning of the 19 century. Intensive, artificial forestry, which started at that time, brought planting stock with inconvenient genetic properties to many locations. At the second half of the 19th century and at the beginning of 20th century, forest tree seeds were imported exclusively from Austria and Germany. Tree species composition changes, including genetic quality changes, considerably disturbed the ecological stability of forest ecosystems. Spruce share in species composition increased 2-7 times more, fir and beech share decreased. The native broadleaved or mixed stands were replaced to conifer monocultures for purely economic purposes. Natural forest ecosystems have been changed into managed ones and naturally heterogeneous forest communities gave place to man-made, homogeneous, ecologically simplified associations. The natural source of stability such as species diversity, age and space structure diversity, as well as genetic diversity have become reduced.

This historical change is to be regarded as the crucial factor responsible for a high vulnerability of conifers in Europe. These are primary sources of eco-physiological stress, which then cooperates with different other stress factors of biotic and abiotic nature, occurring in different parts of Europe. The four pathological processes mentioned above, i.e.;

- root rot forest on post-farming land;
- pine shoot dying;
- oak decline;
- forest decline,

characterized quite well unstable forest ecosystems and interactions between forest pest factors and hosts-trees under stress conditions. But this kind of analysis gives us rather only a descriptive image of forest disease, without any sufficient explanation. Even on the contrary - the term of "disease" is becoming more and more washed out, ambiguous, unclear.

Forest pathologists have to deal more and more often with complex disease brought in by the combination of biotic and abiotic agents. It is not always possible to distinguish between sources of stress and disease symptoms, and separate them for the diagnostic purpose. In individual cases it becomes increasingly difficult to distinguish between a parasite and a pathogen or among a saprophyte, epiphyte and endophyte, parasite of weakness and parasite of injury, primary or secondary disease agent or the obligatory or facultative one.

It is almost impossible to separate, in the course of pathogenesis, the influences of silvicultural mistakes, climatic factors or stress induced by industrial pollution. Many phenomena and disease categorization systems already described and known in forest pathology have become unclear.

We are in the situation that the quantity of data and figures increases (many forest health monitoring systems are working), but information, explanation and forecasting are, lacking.

We confront, probably, the problem of methodological nature, which is a symptom of a crisis of the Cartesian paradigm: searching for immediate cause - effect relationships, and considering that we deal with a linear process, mono-dimensional problems, that are continuos and deterministic.

5.1. "CHAIN DISEASE"

The concept of "chain disease" is the classical model based on this deterministic background (fig. 7). The model which describes individual events in the form of a sequence of causes and effects that can be specified and analyzed separately.

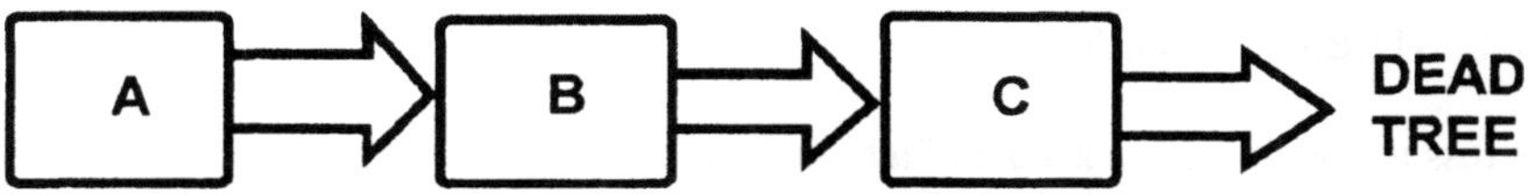

Figure 7. "Chain disease" model

5.2. "SPIRAL DISEASE"

Using the model of "chain disease"it is hard to explain multifactoral disease. It seems to be better to analyze the model of "spiral disease" (fig. 8).

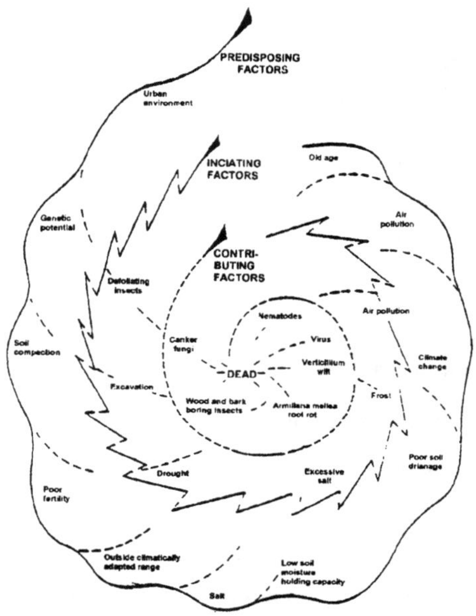

Figure 8. Model of "Spiral disease" (after MANION 1981)

This concept separates causal agents into 3 categories: (A) predisposing factors, (B) inciting factors, (C) contributing factors. We can easily identify those individual factors for each of our pathological processes described previously. Some of them are as follows:

Predisposing factors: inappropriate site, climatic conditions, industrial pollution, improper management treatment.
Inciting factors: drought, defoliation by insects,frost, snow.
Contributing factors: fungi, secondary pests, other biotic factors.

The concept of "spiral disease" is related to the works on the stress theory and system analyses.

5.3. STRESS THEORY

Stress is understood as a state of a biological system provoked by a continuos strain, then the system is ready to refrain or counteract aggressive, unfavorable external factors. The stimuli received by the system from outside are "interpreted" as a demand to an organism to be adapted to disturbances. It is felt as the request to remove the strain.

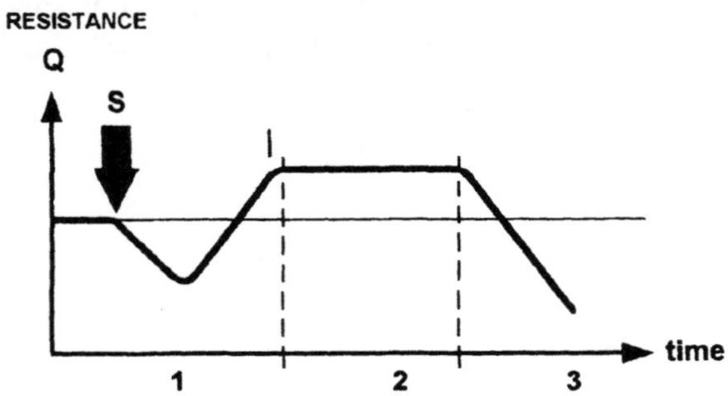

1 - alarm reaction
2 - resistance reaction
3 - exhaustion reaction

Figure 9. Model of General Adaptation Syndrom (GAS) (after SELYE 1978)

In this approach the stress is an imminent element of adaptation and survival - the basic strategy of all living beings.

When seeking a definition of the multi-caused disease, SELYE (1978[2]) described "syndrome caused by various harmful agents" as a GAS - General Adaptation Syndrome (GAS). GAS has three phasal characters (fig. 9):

1 - alarm reaction;
2 - resistance reaction;
3 - exhaustion reaction.

[2] SELYE, H. 1978: Stress, Montreal Acta, Inc.

Under the impact of a strong stressor the three phases occur in rapid succession. In other cases they may be "prolonged" in various ways and the proportion between them can be different.

The stress theory predicts the state when the system does not reach the exhaustion state. Two states can then be differentiated (fig. 10):

(1) reversible (elastic stress, resilience);

(2) irreversible (plastic).

"Reversible stress" (fig. 10a) means such a state, when the system reacts with elasticity and it comes back to the initial state after a deviation from the standard (N) ((t could be its measure).

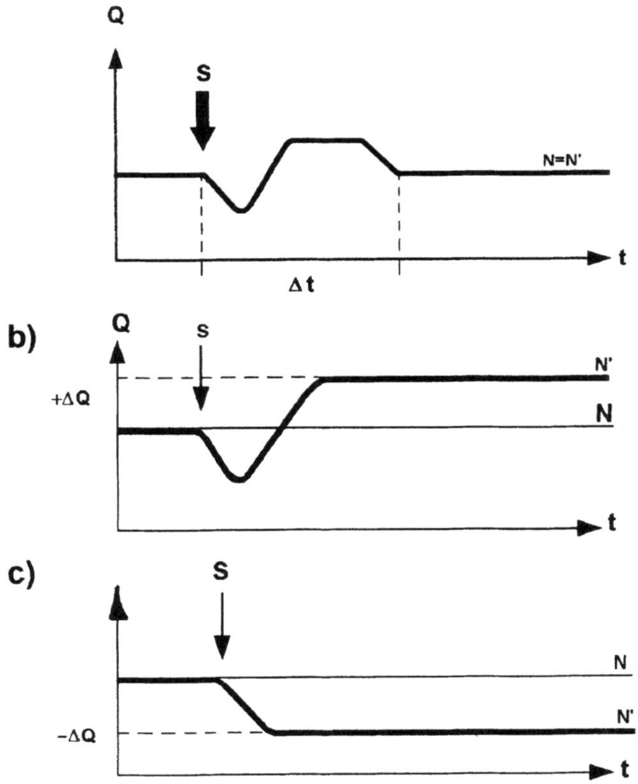

Figure 10 (a,b). Model of reversible (a) and irreversible (b,c) stress

Irreversible stress means plastic deformation and signifies that system properties change in a durable way. Two situations are possible at least (fig. 10b):

(1) the system gets a new level of resistance, lower than previous one (-Q);

(2) new level is above of the previous (+Q).

Adaptation processes represent plastic deformations. From the viewpoint of evolutionary flows, adaptation processes are of basic importance for sustainability of nature systems. Elastic systems, crossing some level of stress, collapse frequently, suddenly and they deteriorate.

5.4. THEORY OF DISSIPATIVE STRUCTURE

A new light on the forest disease and stress phenomenon may be projected from the thermodynamic interpretation of living systems - the theory of dissipative structures. Contrary to the reduction paradigm, as mentioned earlier, the theory of dissipative structure (structure arising in the effect of energy dissipation - entropy) states, that living systems do not develop and sustain in linear, continuous, and definite processes, but exactly reversely - just non-continuous, non-linear, multi-meaning and multi-directional processes determine the development of living systems. The state of each living system is a result of external forces that tend to disintegrate this system and of internal forces, that organize a given system. The state in which those forces are at equilibrium is the desirable state and it is called the Optimum Operating Point (OOP).

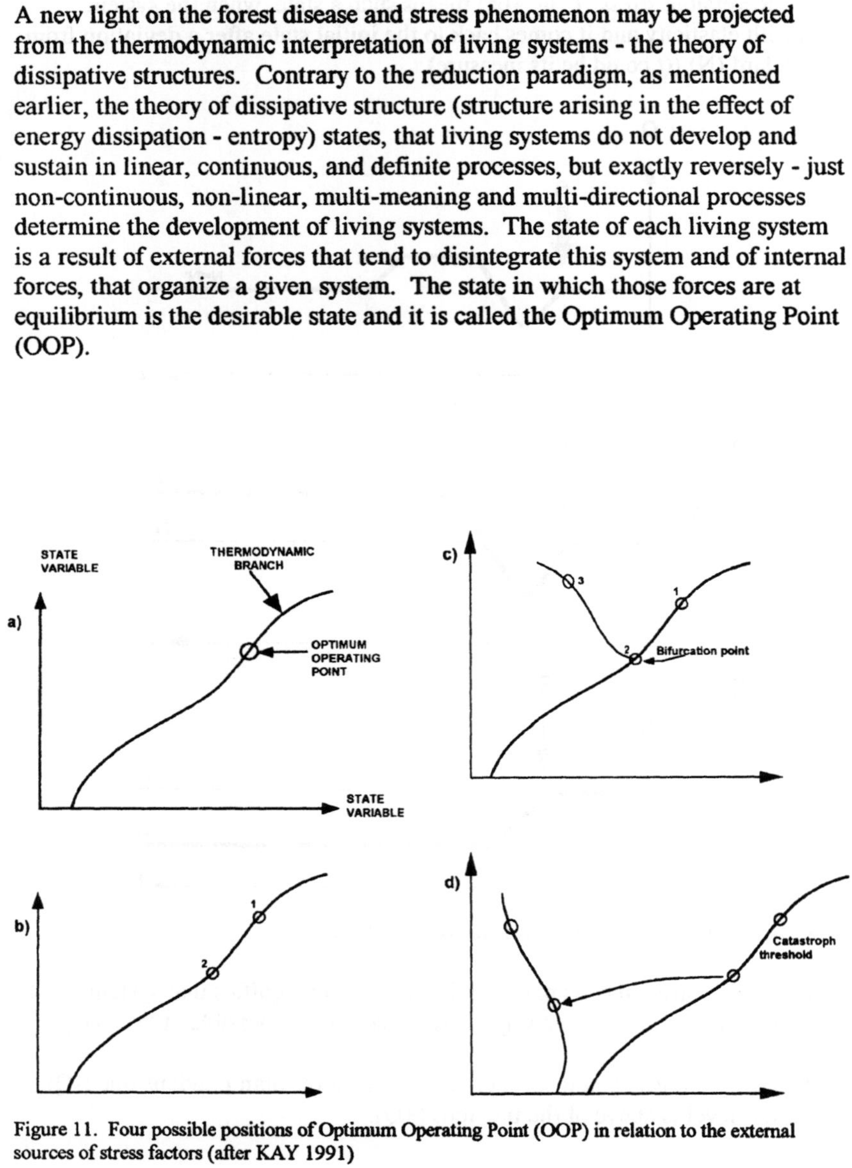

Figure 11. Four possible positions of Optimum Operating Point (OOP) in relation to the external sources of stress factors (after KAY 1991)

Environmental changes could create four situations - four positions of the system according to the position of OOP (KAY 1991[3]) (fig. 11a, b, c, d):

1) OOP remains on the same place of the thermodynamic trajectory in the space of state;
2) OOP goes back from its initial point but it stays on the same trajectory;
3) the system, in the result of bifurcation, creates a new trajectory, along which it moves;
4) the system "jumps" on another thermodynamic way and it develops as a quite another system.

After the change of trajectory the system moves to another type of stability.

5.5. CATASTROPHE THEORY

The sudden collapse of natural system structures is described by the catastrophe theory (fig. 12). This theory can be useful in interpretation of sudden and non-linear changes resulting from accumulation of slow and small changes of one or more parameters. Catastrophes are the ways of achieving a new balance by stressed live systems. These changes are called respectively - a disease, disturbance, calamity, disaster, and so on.

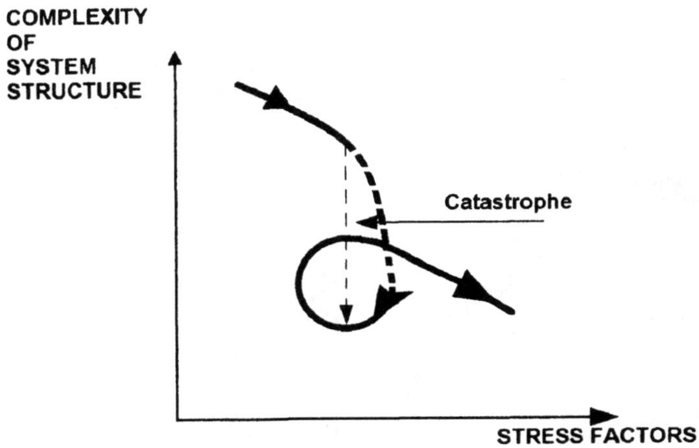

Figure 12. Model of "catastrophe"

5.6. CHAOS THEORY

A new interpretation of phenomena that proceed in living systems could be given by one of the most general concepts of nature understanding, i.e. theory of chaos.

[3] KAY, J.J., 1991: A nonequilibrium Thermodynamic Framework for Discussing Ecosystem Integrity. Environmental Management, vol. 15, No. 4 pp. 483-495

Medical sciences started to say in the frame of that theory, about "dynamic diseases," considering disordering of the system and breakout of coordination or control over the course of processes. The basic concept of chaos theory is that very small change causes very big effects. Chaos occurs when the important forces act against each other and keep the system in the dynamic balance. The "move" of this balanced system depends on the unpredictable, difficult to record and often neglected slight factor. This situation is illustrated by ... the motion of a butterfly's wings which cause the storm.

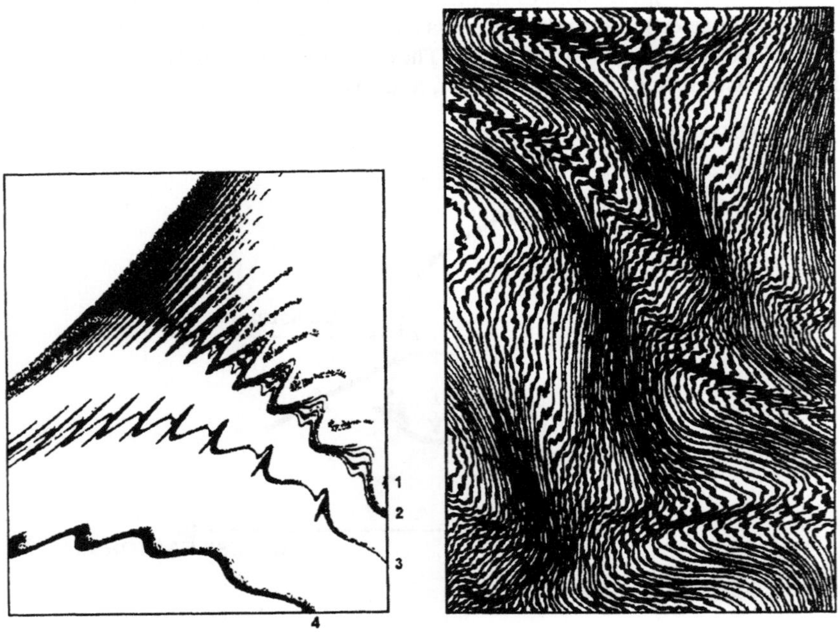

Figure 13. The four factors coacting (a) and the result(b) - image generated by computer (after GLEICK, 1993)

One can see the degree of complication of co-acting phenomena on the image (generated by computer) presenting the cooperation or coexistence of 4 variables involved in the pathological process (GLEICK, 1993[4]) (fig. 13a).

[4] GLEICK, J., 1993: Chaos. Making a new science. Abacus, UK

In the case of forest diseases we can imagine 4 factors co-acting in the same time and in the same place:

(1) acidification;
(2) eutrofication of the soil;
(3) defoliation by insects;
(4) climatic change.

Is that the image of "unstable forest ecosystems" (fig. 13b)? How to diagnose the forest disease? How to predict it?

6. Conclusions

The author started to analyze some typical and the most important pathological phenomena in European forests and finished by the concept of disorganization of all living systems independently on what level of organization it occurs. Some conclusions and questions seem to be important:

1. Stress is a normal state of nature or living system. It is necessary to their evolution, persistence, and survival.
2. Nature does not know the state of disease. Disease is an anthropogenic category. Forest disease is the category of forest economy. If we want to keep our forest as close as possible to the nature, or natural state, we have to forget our anthropogenic position.
3. Disease of a part does not mean the disease of the whole, e.g. the deterioration of cells does not mean the disease of tissues; the death of tissues does not mean the death of organism. The presence of sick individuals does not signify that the whole population or species is ill.

If so, the question arises:

- Is such forest healthy where diseases are absent?
- How many diseased trees are needed lest forest ecosystem should be healthy?

OPTIONS FOR POWER GENERATION AND EMISSION CONTROL TECHNOLOGIES

T. WINNICKI
Environmental Engineering Department
Technical University of Wroclaw
Wyb.Wyspianskiego 27, 50-370 Wroclaw, Poland
phone & fax: +4871 229892
E-Mail: WINNICKI@pwr.wroc.pl

Abstract

Based on the evidence, that major damages to forestry are due to the air pollution caused by energy generation, main options for combustion control and emission abatement have been presented. The current air pollution situation in Poland, as an example of a country in an economic transition, as well as predictable global energy generation scenarios have been discussed to forecast the future situation of forests.

1. Introduction

There is no doubt power generation is directly and indirectly responsible for extensive damages in forestry. The combustion of fossil fuels containing up to two per cent sulfur, sometimes even exceeding that amount, is directly producing sulfur dioxide. The parallel use of air supplying oxygen for burning of a fuel is, simultaneously introducing the 80% of a ballast-nitrogen present in the air. At high temperatures of fuel combustion, nitrogen is converted into nitrogen oxides. Both acidic gases, while passed to emission, produce with atmospheric humidity or rain (so called "acid rain"), which are very often transferred over large distances to precipitate on agricultural and forested land.

Another indirect detrimental effect linked to power generation is attributed to open pit mining. Water drainage, sometimes covering large areas, dehydrates the environs and leads to a - slower or faster - death of forests adjacent to mining excavations. Power generation authorities, aware of the described situation and severely punished by local environmental inspections, are searching for various solutions able to meet even stricter environmental standards.

To be realistic a new technology should fulfill several criteria: (1) to solve problems by "preventing" rather than "repairing" environmental damage; (2) to implement a modern "at the end of the pipe" technology to solve problems still remaining due to the imperfection of presently applied prevention means; (3) to meet international and national environmental standards in both short- and long-term perspectives, (4) to

R. M. Gutkowski and T. Winnicki (eds.), Restoration of Forests, 55–66.
© 1997 *Kluwer Academic Publishers.*

resist expected changes in fuel and energy prices, in other words, to be economical in a predictable period.

Probably the greatest progress could be made in completing recommendations of criterion (1), consisting of: (a) changes in furnace construction; (b) changes in combustion technology; (c) substitution of higher sulfur content fuels with lower- or no-sulfur combustion substrates; (d) substituting fossil carbon-related fuels with "environment-friendly" fuels, such as hydrogen, or fuels that are at least less impacting the environment, such as well controlled nuclear power generation technology.

A lot could be done on criterion (2), especially in developing countries, where highly complete particulate abatement technology is not accompanied by acid gases neutralizing and managing solid wastes. High investment costs of such technology transfer should be facilitated by developed countries owning the required technology and by international capital ventures supporting those undertakings of transboundary environmental effects.

In regard to criterion (3), the unlimited pathway of air pollutants across political and economic borders should bring local national problems to the consciousness of the international community. The fatal case of Chernobyl emphasized the idea of "Only One Earth" expressed in slogan at the Stockholm '72 congress.

2. Poland - The Case of a Country in Economic Transition

The political turnover of 1989 created in Central and Eastern European countries, among them Poland, a completely new environmental situation and a pro-ecology approach. As a result of a deep crisis in all sectors of the national economy there evolved a much improved situation in the environment. Correspondingly, the already quite adequate environmental laws in Poland were modernized and standards brought to the level typical for Europe.

Just a few data related to the main issue of this ARW, show that forests cover about 28% of Poland's area, with the domination of coniferous species (close to 80%) and pine prevailing in this category (70%). The annual mining of fossil fuels and main minerals consists of: about 180 million tons of bituminous coal, over 70 million tons of lignite and over 5 million cubic meters of natural gas as well as roughly 30 millions tons of copper ore and over 5 million tones of zinc & lead ores. Other major productions affecting the quality of the environment include power generation of over 150 TWh, pig iron over 15 million tons, cement over 17 million tons and sulfur 5 million tons.

The south-western corner of Poland, neighboring with Germany and Bohemia and very rich in lignite resources, was for the last 30 years highly saturated with power plants at all three countrysides. Due to the emission of flue gases, this area was known under derogatory name of "The Sulfur Triangle" or "The Black Triangle". Fortunately, the reunification of Germany caused the closing of some the most polluting power plants and equipped the others with adequate modern combustion and abatement technology.

Unfortunately, the situation in the Czech Republic is developing much slower. The only large Polish plant in Turow (of a capacity of 800 MW) has entered upon a thorough restructuring of its technology, but it can't solve local problems, since monitoring stations installed in the last decade along the borderline showed that close to 80% of air pollution originating from power plants in the described region is imported to Poland.

All those managerial means and technological innovations could stop but cannot convert the process of massive damages to local woods, resulting - at least at some areas - in a total deforestation at the higher elevations of the Sudety Mountains.

Apart from exhaust of acidic gases the combustion of fossil fuels generates carbon dioxide recognized as a victim for the so called "greenhouse effect", which is not harmful to forests, but seriously taken into account while considering a substitution of carbon-related fuels by any other source of energy.

The still worsening situation in the global atmosphere could be halted and perhaps converted by the following scenario consisting of three subsequent phases, which are very difficult to be dated in a prcise manner:

I. Current - including: (a) changes in the existing technology of power generation; (b) limiting of energy consumption by changes in other production technologies and the introduction of a new generation of products consuming less energy; (c) partial substitution of fossil fuels by other sources of energy, mainly by conventional (diffusion) nuclear power plants, but also by solar, geothermal, biomass and others.

II. Transition - consisting of: (a) slow exhaustion of crude and gas resources and their substitution by coals; (b) large scale implementation of diffusion-type nuclear power plants; (c) greater introduction of solar energy technology.

III. Future - expecting: (a) complete exhaustion of fossil fuels and their substitution by already mentioned and other sources of energy, with domination of so called "solfus" - parallel large scale use of solar and nuclear fusion energies.

The above scenario, which is just one out of several possible, introduces many question marks related to R & D progress and is, no doubt, controversial taking into account a great resistance exerted by local, national and partly global communities against any use of nuclear energy.

Since phase II will, sooner or later, come into effect, an increased combustion of coals could be expected in the future. It is most likely the remaining coals will generally have a higher content of sulfur than coals which developed societies use presently. Also this important element is justifying a dramatic need for changes in existing technology for power generation, including both combustion of fuels and abatement of flue gas.

3. Possible Changes in Power Generation Essential for the Reduction of Air Pollution

3.1. DESULPHURIZATION OF FOSSIL FUELS

Desulphurization of crude oil has not been a technical and economic problem for a long time, but the same process applied for hard-coal or brown-coal is much less effective due to the solid state of the substrate. The total volume of coals used for power generation creates, of course, a separate technical and managerial problems.

Progress in the construction of furnaces (described below) and in the abatement technology of sulfur dioxide from flue gas, as well as the above mentioned problems in desulphurization of coals, are dalaying a wide implementation of that process, dispute several technological options. The variety of typically chemical technologies used for that purposeis to extensive to present them and discuss in the framework of this abbreviated review.

3.2. CHANGES IN THE CONSTRUCTION AND OPERATION OF FURNACES

Limiting nitrogen oxides emission should soon be based mainly on controlling of combustion process. A long evolution of the construction of burners led to several concepts of controlled fuel burning, attempting to keep the furnace temperature in a range providing low oxidation of nitrogen originating from three sources - air-nitrogen, intermediate air-nitrogen compounds and nitrogen chemically bound in fuel.

In the past 15 years a new combustion technology, based on circulating fluidized bed boiler concept (CFBB), was introduced to the power generation. Over 200 boilers of a new type have been built since 1979 all over the world. The new technology (Fig. 1) was transferred in the beginning of 90s to Poland by RAFAKO company in cooperation with the German EVT company in Stuttgart. Without going into any technical details, it is worth mentioning that the system consists of the furnace enabling a mutual combustion of coal and injection of a pulverized limestone neutralizing sulfur compounds "in statu nascendi". Next, the flue gas is passed through a cooling tower to an electrostatic precipitator and exhausted through the chimney.

Commercial folders will enable reduction of sulfur dioxides from 2000-6000 in traditional boilers to 200-400 mg/l in CFBB system (permissible standard - about 500).The same time the content of nitrogen oxides could drop from 600-1400 to 150-240 mg/l (permissible standard - about 450). 90 to 95 % reduction of sulfur dioxide in CFBB compared to conventional boilers (at the average sulfur content in coal of 1 - 1,5%) could be achieved. A mutual reduction of nitrogen oxides by 65 to 85%, creates a new strategy of environmental protection by power generation industry, moving from the development of flue gas abatement to combustion control. The CFBB process is meeting the basic environmental control recommendation - a prevention prior to damage repair.

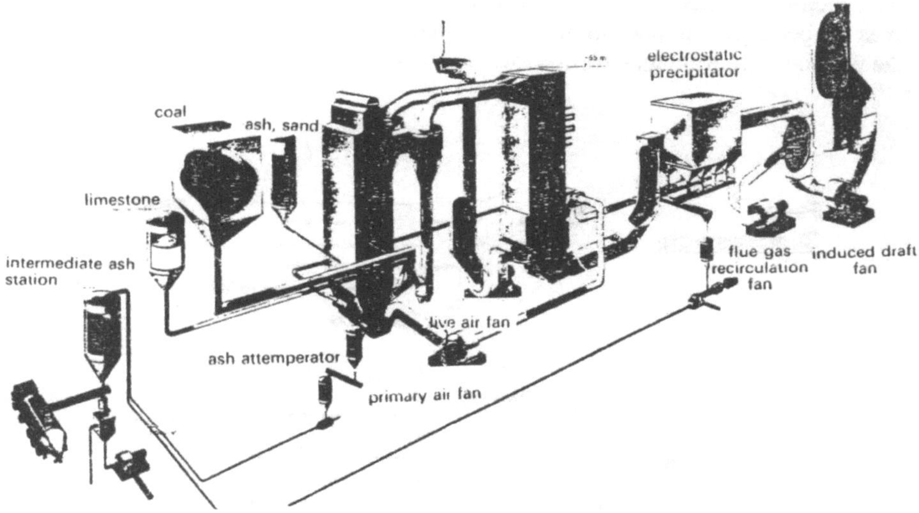

Figure 1. Circulating Fluidized Bed Boiler

Apart from the above mentioned advantages, the CFBB technology also enables the utilization of low caloric and high ash waste fuels, which were previously stored in piles. On the other hand a reduction of nitrogen oxides is possible due to the lowering of combustion temperature to the range of 850°C, which, of course, decreases the combustion efficiency.

3.3. MODIFICATION OF COMPOSITION OF AIR FOR FUEL COMBUSTION

The high investment cost of any replacement of old equipment by a new equipment will substantially limit a transition from conventional boilers to CFBB systems in a full required range. Therefore, the problem involving nitrogen oxides will exist for a long period of time. Also the extent of NO_x reduction is not sufficient even in the CFBB technology.

The only universal solution for a deep reduction of nitrogen oxides is to minimize the amount of nitrogen in air "running" the combustion process. The substitution of air ballast with 4/5 of nitrogen by pure oxygen won't be economically feasible for a long time. The only satisfactory option, therefore, would be an enrichment of combustion air in oxygen. Out of commercially available methods only the membrane separation could meet large volume and low cost requirements of fuel combustion processes.

The current status of the membrane separation technology allows one-stage enrichment of air up to 45% of oxygen. Two modes of operation have been investigated: One is “feed compression” - in which inlet air is passed through a membrane at a high pressure and the enriched oxygen air is collected at a pressure slightly above the atmospheric pressure Fig.2). Advantages of this option are a smaller required membrane area and, consequently, smaller membrane modules as well as the possibility to send slightly compressed air - on the permeate side - directly to the furnace. Disadvantages are due to a necessity of pumping through membranes air

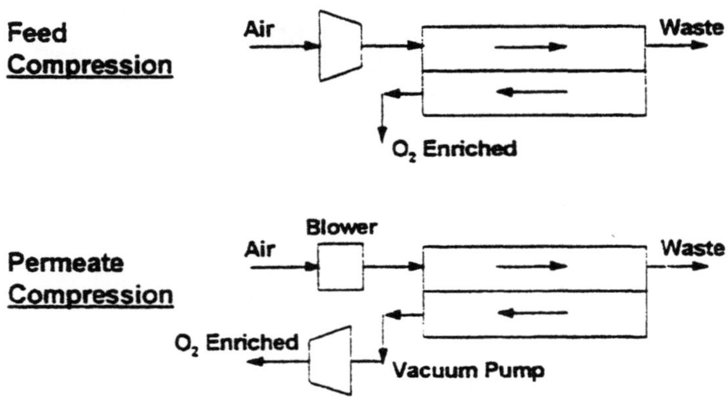

Figure 2. Membrane Air-Enrichment Process

with a nitrogen ballast of 80%. This mode is of a lower investment cost, but a higher operation cost option. The second mode is "permeate vacuum receive" - in which the oxygen enriched air is sucked on the permeate side of membranes. One advantage is the relatively low operation cost (despite of the necessity to pomp the permeate to the furnace), which must be equilibrated by a higher membrane area and, consequently, increased investment costs of the installation.

Incidentally, membrane separation technology could be applied to the abatement of carbon oxides from flue gas. Presently, this is more a theoretical idea than a practical alternative.

4. Flue Gas Abatement Technologies

4.1. PARTICULATE ABATEMENT

Dust capturing technology has been well advanced at least for three decades and, due to its simplicity, quite well transferred to even less developed countries. Availability of over 90% efficiency is easy to obtain and became a minimum level for properly functioning equipment.

Out of various systems and designs, the electrostatic precipitation has been widely implemented for power generation flue ash abatement allowing to reach at least 95% isolation of particulate.

A complementary problem has been created by spent flue ash storing and recycling to any useful and harmless product. A wide application was found in bricks and other construction elements made of flue ash in various compositions with cement, limestone and sand. A limiting criterion for the use of such material for building construction was the radioactivity of flue ash.

4.2. DESULPHURIZATION OF FLUE GAS

Processes of flue gas desulphurization (FGD) - the removal of sulfur dioxide-have been commercially available over the past two decades. The first attempts originated in England prior to WWI and, made use of the alkalinity of the River Thames. Its water neutralized flue gas and was discharged back to the river. Modern FGD technology has been developed mainly in the USA, Japan and Germany. Out of over a hundred of already invented FGD processes only a small fraction has been brought to a full scale and offered commercially. All those methods could be categorized in four already commercially available groups of processes, but the fifth group is in progress: (i) Wet absorption processes - using various kind of alkalinity liquids as well as dilute sulfuric acid as the scrubbing agents. (a) Caustic soda, producing sodium sulfate & sulfite and gypsum. Several processes are applied in parallel to recover sulfur dioxide for the production of sulfuric acid or elemental sulfur, among them a very efficient absorption of SO_2 in concentrated sodium sulfite solution and regeneration of resultant bisulfite which could be converted to required end-products. (b) Lime or limestone, the cheapest out of all desulphurization reagents, producing gypsum (Fig.3)

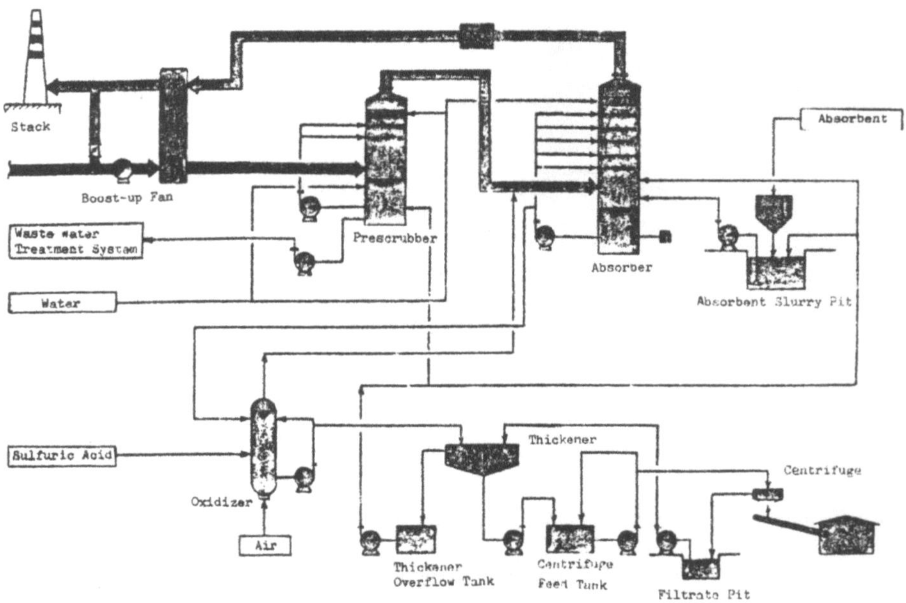

Figure 3. Limestone - Gypsum Process (separate SO_2 and dust removal method)

for direct construction use or as a retardant in cement composition. (c) Ammonia, in contrary a relatively expensive agent, produces ammonium sulfate, a valuable fertilizer (Fig.4). (d) Aqueous slurry of magnesium hydroxide, which leads: to the recovery of sulfur dioxide through calcination of intermediate magnesium sulfite and reuse of magnesium hydroxide or to a simple disposal of spent magnesium sulfate. (e) Dilute sulfuric acid, absorbing SO_2 to be treated with lime to produce gypsum - the process not used very often due to the corrosion problems in the first stage. (f) In ISPRA PROCESS a dilute sulfuric acid with hydrobromic acid and small addition of bromine as an oxidant produces a more concentrated sulfuric acid and hydrogen (from electrolysis of HBr to recover Br2). No wastewater is produced in this process.
(ii) Dry absorption processes - more recently implemented in FGD technology - using the majority of already mentioned alkaline agents, but in a dry, pulverized form. Options are: (a) Lime or limestone injection into hot boiler gases. The process has never exceeded the efficiency of 40% and requires a relatively high excess of lime, therefore it was never used in a large scale. (b) Activated magnesium injection to produce magnesium sulfate & sulfite to be precipitated. A possible transformation of magnesium sulfate to ammonium sulfate has never been explored on a commercial scale. (c) Copper oxide as well as other metal oxides could also be used as dry absorbents, but this process has never reached a full scale.

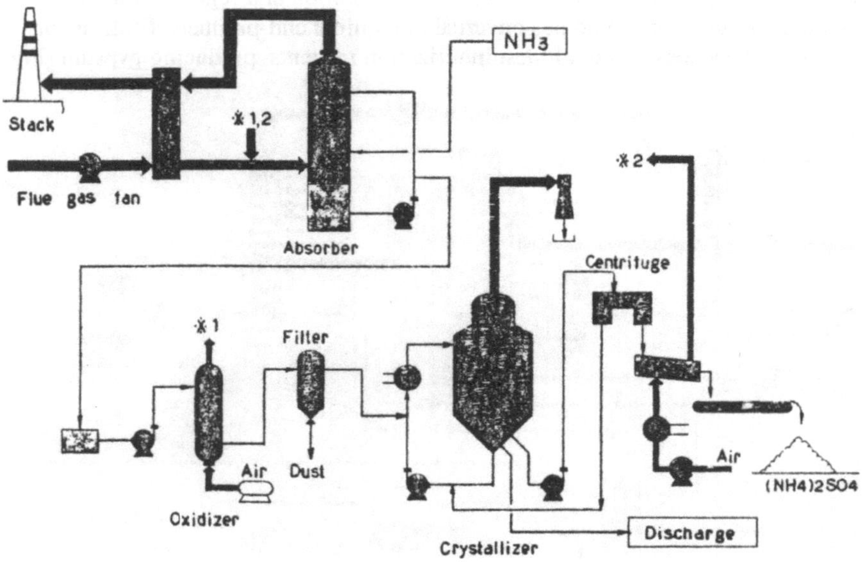

Figure 4. Ammonium Process (leeding to fertilizers production)

(iii) Semi-dry absorption processes - which chemically are similar to already described wet and dry reactions, but using the spray drier atomizer producing first droplets and afterwards particulate of spent products collected together with the excess of reactant

in electrostatic precipitators or bag filters. The process is highly advantageous compared to previous ones. It requires much less reheating, showed minimum scaling and plugging and yields a totally dry product at lower investment cost.
(iv) Dry adsorption processes - using mainly activated carbon as an adsorbent producing sulfuric acid in the presence of water and oxygen. Carbon could be then regenerated by flushing with water or more often by thermal desorption, which is advantageously oxidizing carbon surface increasing its sorptive capacity. The process is still to meet economic requirements and appears to be "too chemical" for a wider use in power generation industry.
(v) Membrane separation processes - being in an intensive R & D progress, could lead to the optimum solution for the recycling of sulfur dioxide for sulfuric acid or elemental sulfur production. No chemical reagents and no spent products are expected, but the investment cost and energy to drive a huge volume of gases through the membrane system, could be for a long period a barrier for an implementation of those processes.

4.3. SIMULTANEOUS REMOVAL OF SULFUR DIOXIDE AND NITRIC OXIDES FROM FLUE GAS

It has been found in prolonged laboratory and industrial practice, that the separate sorption of nitrogen oxides is less effective than its mutual abatement with sulfur oxides. Therefore, various processes have been tested and some were brought to a commercial scale.

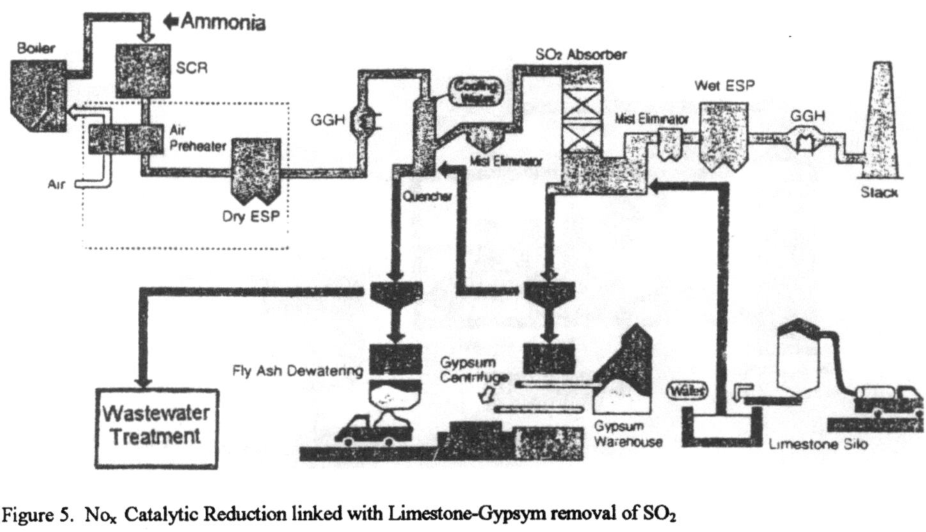

Figure 5. No_x Catalytic Reduction linked with Limestone-Gypsym removal of SO_2

The Limestone-Gypsum Process for SO_2 abatement following the Ammonia Selective Catalytic Reduction of nitrogen oxides (Fig.5) has already become a classic solution.

One of the most dynamically approaching technologies, invented by German and Japanese companies, was named BF/UHDE/MITSUI-PROCESS and is using a specially prepared active coke adsorbing and oxidizing SO_2 and reducing NO, in the presence of ammonia and oxygen, to gaseous nitrogen. Resulting sulfuric acid is then reacting with ammonia to produce ammonium sulfate. In addition, the same process could be applied to clean-up a very troublesome off-gases from waste incineration.

Among the already mentioned desulphurization processes those based on ammonia as an absorbent where also mentioned. One of them is so called WALTHER FGD PROCESS, which could be linked with the well known DENOX method. After the desulphurization stage the NO could be oxidized to NO_2 and N_2O_5 by injecting of ozone into the gas phase which is then absorbed in the ammonia solution. Since oxidation with ozone is still not economic the other tested option is passing desulphurized gas through a catalyst to produce nitrogen oxides as a reagent for ammonia.

A new interesting process called EBDS is based on the application of artificial radicals - OH and O_2H which are generated due to the interaction of accelerated electrons with the main components of the flue gas. Sulfuric acid is forming an aerosol mutually neutralized with nitric acid by injection of ammonia to produce a mixture of ammonium sulfate and nitrate, collected in fabric filters or electrostatic precipitators and used for the manufactue of fertilizers (Fig.6).

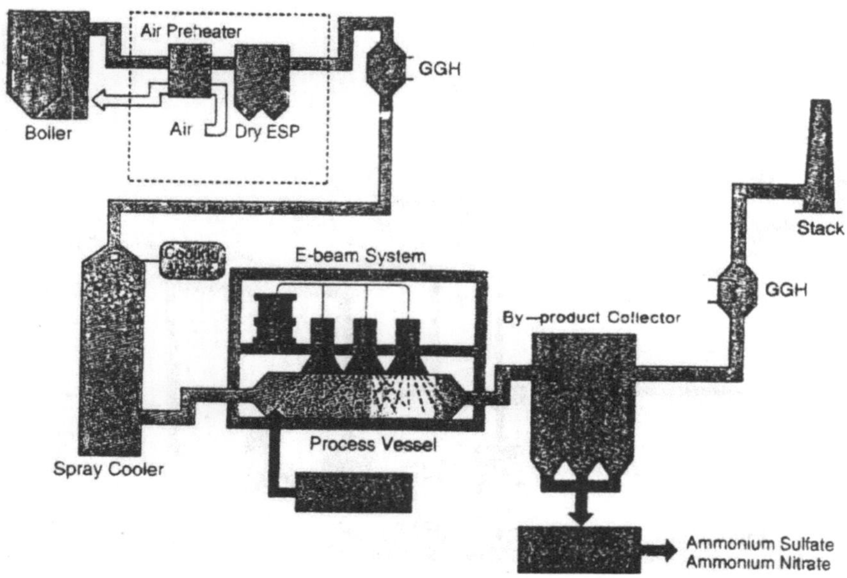

Figure 6. Electron-Beam Flue Gas Treatment Process

5. Conclusions and Recommendations

Power generation is responsible for major damage to forests in Central and Easter Europe. Three parallel reasons have been found to be decisive for that disaster - the quality of fuel, the combustion condition and the technology of the abatement of flue gases.

The substitution of coals by crude oil or gas, of low- or no-content of sulfur, is not always possible in developing countries having large resources of hard and/or brown coal. Desulfurization of fuel should be recommended in such case, regardless of the cost of this operation.

The combustion of fossil fuels, should be controlled either by keeping the low range of temperature, or by supplying oxygen-enriched air to the furnace to prevent the generation of nitrogen oxides.

For the transition period flue gas abatement technology should be developed and implemented. Especially recommended are methods for mutual neutralization of nitrogen and sulfur oxides and converting them into useful and definitely not harmful by-products.

An intensive program to introduce ecology-friendly sources of energy (solar, hydrogen, nuclear fusion, others), should be the main goal of science in modern societies.

6. Pertinent Observations from the ARW

As the Co-Director of the ARW, the author hesitates to present too enthusiastic judgments, but it has been pleasant to hear widely expressed opinion concerning an equilibrated set of experts gathered from both Western and Eastern Europe and competently presenting four areas of the ARW coverage.

As an environmentalist, the author found a splendid opportunity to get familiar with specific problems of forest management as well as utilization of timber. Detailed information about wood's growth and exploitation reinforced his conviction about necessity to relax anthropogenic pressure on nature.

As the citizen of one of those European countries which is undergoing transition from planned to free-market economy, the author had been convinced, that sufficient environmental control to enable protection of nature including forests could be realistic only by mutual action of developing and developed countries, financially supported by the second group.

As a technologist, the author is expecting, that among above mentioned measures, a free of charge, or at least facilitated, technology transfer will play a dominating role.

7. References

1. van Velzen D. (1991) Sulphur Dioxide and Nitrogen Oxides in Industrial Waste Gases: Emission, Legislation and Abatement, Kluwer Academic Publishers

2. Puri S. (1996) Membranes for Gas Separation: Current Status, Proc.UN Seminar on the Ecological Applications of Innovative Membramne Technology in the Chemical Industry, May 1-4, 1996, Cetraro, Italy

3. Kumazawa H. (1996) Air Pollution Control Technologies and Experiences in Japan Proc. Intnl. Conf. Theory and Practice of Atmospheric Air Protection, June 11-13, 1996, Zabrze-Ustron, Poland

4. Gostomczyk M.A. (1996) The Techncial and Economic Possibilities of Pollutant Emission Reduction in Poland, ibid.

5. Jaroszynska J. and Pollo I. (1996) Oxidation of Nitrogen Oxides by Means of Ozone as a Method of Lowering Nitrogen Oxides Content in Flue Gases, ibid.

6. Chmielewski A.G., et al. (1996) Integrated Systems for Simultaneous Multi Pollutants Removal from Combustion Flue Gas, ibid.

7. Reference Book I, (1992) Introduction of Circulating Fluidized bed Boilers into Fabriaction, *Commercial Folder*, RAFAKO, Poland

8. Basic Study for Environmental Cooperation with Eastern European Countries - Poland (1993), Oveseas Environmental Cooperation Center, Japan & Ministry of Environment, Poland

9. Fluidized Bed Boiler and Flue Gas Desulfurization System, (1992) Commercial Material by Ishikawajima-Harima Heavy Industries Co., Ltd, Tokyo, Japan

State-of-the-Art Papers

FORESTRY AREA RECOVERY AND RESTRUCTURING - EXPERIENCE FROM SHERWOOD FORESTS

Reversing the Decline of the World's Most Famous Forest

A. BRADY
Forestry Commission of Great Britain
Sherwood Initiative, Cuckney Road, Carburton,
Worksop, Notts S80 3BP, United Kingdom.
Internet web pages http://www.sherwoodforest.org.uk
Internet e-mail sherwood@dial.pipex.com

Abstract

Sherwood Forest, because of its association with the Robin Hood legend, is probably the best known forest in the world. The real forest has been in decline for many years. The causes of its decline are outlined. The benefits of re-creating parts of the forest are explored. The development of a Trust to implement the restortion of the forest is described.

1. Historical Background

The existence of Sherwood owes much to its underlying geology. Running through the heart of the county is a large sandstone feature known as the Sherwood or Bunter Sandstone. This area is characterised by poor and infertile sandy soil which holds little moisture. As a result, historically, large areas of Sherwood could not support permanent farming, and so a form of shifting cultivation resulted. Parts of the forest cleared for farming were later abandoned when their fertility dropped, or during periods of prolonged drought. Abandoned farmland soon reverted to heathland and scrub and often back to woodland again. By the Norman era (mid eleventh century), Sherwood was a complex mosaic of woodland and open heathland with small areas of farming and settlement. Throughout its recorded history Sherwood has never been wall to wall trees.

The Normans were very fond of hunting, principally deer and wild boar. Large tracts of the countryside were identified as being suitable for the hunt and this often resulted in them being declared as Royal Forests. Such a designation meant that permission was needed to carry out a whole range of farming and woodland management activities, but these areas were still actively managed. However, over time, the rule of forest law in Sherwood became a means of generating income. Fines which had once been aimed at stopping the forest from being damaged became a convenient way of charging for its

R. M. Gutkowski and T. Winnicki (eds.), Restoration of Forests, 69–76.
© 1997 *Kluwer Academic Publishers.*

destruction, and those with power and privilege exploited its resources.

Many of the oak beams in Lincoln Cathedral roof (dating from approximately 1250) came from Sherwood Forest, and the monks of Rufford Abbey were amongst those buying land from the King in the 14th century which they cleared of oaks and turned over to farming. Large areas of the forest remained as open heathland or 'forest waste' as it was called. This land did not always revert back to woodland as it became managed as extensive grazing. Areas of bracken were cut for animal bedding and heather was cut for use as thatch for buildings. Many landowners developed their own deer parks as the Crown's interest in Sherwood declined.

For centuries, the extensive heathlands of the area were the key component of the Sherwood countryside, linking together the woodlands, river valley pastures, farmsteads and settlements. As Sherwood emerged from the medieval period, the influence of crown and church were reducing. Land changed hands after the dissolution of the monasteries and a series of large new Ducal estates were developed. Large areas of Sherwood's wastes were transformed during the eighteenth and nineteenth centuries by new woodland planting, and the development of ornamental lakes and grand houses. The Dukeries estates retained elements of the semi-natural landscapes within their parklands, but also became the driving force for agricultural improvement and new commercial activities further afield.

Suddenly, by the end of the eighteenth century, the pace of change in Sherwood had accelerated like never before. The old economics of heathland management were gone, agriculture had to feed a growing population and the rapid expansion of coal mining was being driven by industrialisation. By the early part of the twentieth century most surviving heathland was turned over to agriculture and also to the new economics of plantation forestry. Large areas of Sherwood's surviving ancient heaths were converted to conifer plantation forest in the early years of the newly formed Forestry Commission. The surviving core of ancient Sherwood Forest at Birklands had became a popular stop on the newly developing tourist trails. The forest was now truly fragmented like never before.

2. Sherwood Forest Today

It is only within the last few decades that the post-war pressure for self-sufficiency in terms of food production has eased off. This has allowed us the luxury of taking a fresh look at the wider countryside and has given us the opportunity to develop new priorities. Since the second World War greater efforts have been made to identify and protect surviving examples of dwindling habitats or disappearing historic features. This resulted in a process where different statutory bodies were working to protect or influence different parts of the countryside for different reasons. This narrowly focused approach continued apace for several decades. Great attention was paid to a relatively small number of key sites, but little was done to influence what happened in the wider countryside. Today, the contemporary, and more enlightened approach to conservation recognises that labelling and protecting a handful of sites is perceived as restrictive and backward looking. A more strategic approach linking conservation, and protection with positive enhancement and management of the wider countryside is emerging. Consequently, the boundaries between the areas of interest of a wide range of statutory bodies are beginning to blur. The Forestry Commission, for example, when first set up in 1919 was principally interested in the production of timber as a raw material. Today, our remit stretches to all the multiple-benefits of forestry and includes a wide range of social and environmental elements.

Consider for a moment the modern landscape of Sherwood Forest. Fragments of the ancient forest and heath survive, more recent woodland has been planted on some of the old heaths, and large areas of land have been tipped on as a result of decades of deep coal mining. In addition, all of this is dominated by extensive arable farmland which accentuates the fragmentation of the landscape. The Sherwood countryside continues to be shaped by a whole series of economic and environmental pressures. This, of course, is not a recent phenomenon, economics have been at the heart of landscape change within the Forest for centuries. Indeed, for the future management of the Sherwood area to be truly sustainable we must look carefully at the costs and benefits of our interventions and the nature of the resources and inputs consumed in undertaking its management. There is little to be gained from trying to shelter Sherwood from economic reality, but a great deal to be gained by properly accounting for the benefits which the Forest delivers. It is most useful to look at the current and future economic and environmental factors at play in the Forest area and to focus our efforts in two key areas; firstly to maximise the generation of real income which can be redeployed within the Forest, and secondly to pursue opportunities to expand the Forest area which achieve additional benefits that are valued and recognised by others.

This can be examined in the context of benefits which are internal and external to the Forest, and benefits which are realisable now or at some future point. This results in a different set of factors to consider in terms of the current management and future expansion of the forest area, although there are also some linkages between the two.

2.1 CURRENT MANAGEMENT

2.1.1. *Plantations*

There is no doubt that plantation forests in the Sherwood area are productive. Plantations of Corsican pine (*Pinus nigra* var. *maritima*) are most common, and achieve yield class of 12 or more, producing good quality softwood for local and regional sawmills. However, the plantation forest area is also expected to provide other benefits for the public in terms of an enhanced landscape, as a wildlife habitat and as a recreational resource. This pursuit of multiple objectives is most apparent within the public sector woodlands, such as those managed by the Forestry Commission, where significant reductions in plantation area and adjustments of felling age are designed to improve the appearance and habitats of the forest.

Recent silvicultural practices following clear felling have allowed the dormant seed bank from Sherwood's ancient heathland to regenerate over parts of the forest. A network of such sites where this recovery has been strongest have been identified for retention as permanent open heath to ensure the survival of this dwindling and locally important vegetation. This forms an important part of a much wider strategy, in conjunction with other agencies, to manage and re-create heathlands right across their former extent over much of Sherwood. Within the plantations a well planned network of glades and permanent open heathland areas, linked by corridors of managed rideside vegetation is not only ideal for habitat continuity, but also provides a mature and stable infrastructure for recreational use which enhances the forest's aesthetic value.

As large areas of plantation neared the end of their rotation at the same time, felling has been staggered so that the forest is restructured across a wider range of age classes. This helps to achieve greater ecological stability, as every stage of plantation development is always present somewhere in the forest. This has resulted in a large and stable population of Nightjar (*Caprimulgus europaeus*) building up in the restructured pinewoods. This bird, usually associated with heathland, nests in very young plantations readily moving to other sites within the forest as any one site closes canopy. Such restructuring also brings major benefits to the internal landscape of the forest, improving its aesthetic qualities.

Protecting and enhancing valued wildlife habitats and providing recreational opportunities (walking, cycling, horse riding etc.) are sufficient justification for the resulting reduction in potential timber income. However, the efficient management of the timber resource still provides the income which is the driving force for the whole forest management cycle. Management of privately owned woodlands is also expected to go some way towards achieving such a balance, a process which is enabled by the use of the Forestry Commission's regulatory powers and grant aid incentives.

2.1.2. *Semi-natural and ancient woodlands*

Broadleaved woodland in Sherwood is dominated by areas of semi-natural and ancient woodland which have long been managed for their aesthetic and heritage value. The poor sandy soils do not produce good quality hardwood timber, but are readily recolonised by the native oak and birch. The surviving ancient woodland of Birklands at the heart of Sherwood still contains several thousand ancient oaks, many of which are in excess of 500 years old. This area is a Site of Special Scientific Interest (SSSI) and has been proposed as a Special Area for Conservation (SAC) under EU legislation. Its association with the Robin Hood legend has long been a fascination with the public, and the site has been a popular place to visit for many years. About 1 million people visit the site each year, and this has dominated the management of the site while also posing a serious risk to its fragile ecosystem.

2.2 FUTURE EXPANSION

Future expansion of the forest area is considered desirable for a range of different reasons, some of which attribute an additional value to the existing forest area which is not accounted for by current valuation criteria. It is convenient to consider these expansion opportunities individually, although they are often related.

2.2.1. *Tourism*

The Sherwood Forest area has long been a popular place to visit, but today's pressures from both the local population and tourists means that some parts of Sherwood are subject to extremely heavy use, with more than three million visits per year to the Sherwood area as a whole. Recent decline in traditional industries, such as coal mining, has meant more emphasis on further developing other economic activities such as tourism. Clearly, the pressure from visitors must be properly managed and spread over a number of sites. This presents opportunities to extend the use of some existing and robust sites as a means of sharing the load, and also justifies the expansion of woodland to both extend the area for recreational use and to improve the aesthetics of those parts of the Sherwood landscape badly damaged by coal mining, intensive agriculture and development.

2.2.2. *Mineral extraction*

At the peak of their production, the coal mines of Nottinghamshire were tipping 16 million tonnes of unwanted spoil on the surface each year. This has dramatically altered the appearance of parts of Sherwood and left a legacy of sites awaiting restoration. Within the last five years, many of the deep coal mines in and around Sherwood have been closed down for economic reasons, but the large scale environmental impact of their spoil heaps is still to be tackled. In the past, many spoil heaps were restored to agricultural use, but today the emphasis is on woodland restoration as this will more convincingly blend the tips into the Sherwood landscape. A programme of work involving the restoration to woodland of seven sites extending in total area to 1500 acres is in preparation. Tackling restoration work on this scale has only been possible because of a close partnership between the County Council (local authority) and the Forestry Commission.

2.2.3. *Water*

The sandstone geology underlying the whole Sherwood region supports a large underground aquifer which is the principal source of drinking water for the city of Nottingham and most of the surrounding county. Nitrates from farm fertilisers have been seeping into the groundwater for some years, resulting in some of the water supply boreholes exceeding the EU nitrates safety limit of 50mg per litre. The local water company has responded by drilling several new boreholes within the surviving woodland and heathland areas as these have never been subject to modern farming and have shielded the groundwater from contamination. This is another benefit which the forest offers to society but for which it gets no economic recognition. Several areas of surrounding farmland have been designated as Nitrate Sensitive Areas, and farmers are offered incentive payments to reduce their nitrogen usage or to alter their farming practices. An expansion of woodland and heathland over the aquifer would reduce nitrate contamination and help to safeguard future water supplies. This is likely to have an additional economic benefit as the land returned to forest and heath would also be removed from agricultural support payments.

2.2.4. *Conservation*

The ancient oak woodlands and heaths of Sherwood, and their associated flora and fauna, are recognised as being of national and international importance for conservation, several areas are designated as Sites of Special Scientific Interest, and the Birklands is listed for Special Area for Conservation status. The large population of ancient oaks represents a unique gene pool and as such is a key biodiversity asset.

2.2.5. *Farming*
The free draining sandy soils of Sherwood are ideally suited to modern agriculture and highly productive when supplemented with fertiliser and irrigation systems. Potatoes, sugar beet, carrots and cereals are all efficiently grown. There are, however, questions of sustainability here because of the high inputs involved. Fertiliser input may be restricted because of nitrate contamination, and continued use of groundwater for irrigation may have to be reduced because of over abstraction. In combination with pressures to reform the Common Agricultural Policy of the EU, and other impact on world markets such as GATT, land which is currently economic to farm may become increasingly marginal and consequently available for forestry use.

2.2.6. *Forestry*
Sherwood's plantations produce good quality timber close to ready markets, but the income this generates cannot compete with agriculture. In the Forestry Commission we continue to offer advice and cash grants to farmers but with little take up. There is no sound economic basis for tree planting as long as land prices are held up by the availability of farm price supports through the CAP. Most recent plantings on farmland have been small scale either for aesthetic reasons alone, or perhaps to improve the shooting potential. There is no longer a tradition of woodland management, with many farmers and landowners neglecting the small woodlands on their land. There is, however, a need to diversify agricultural businesses and provide rural employment. The redevelopment of small scale productive management of farm woodlands could help to fill that need.

2.2.7. *Economic development*
Large areas of Nottinghamshire have suffered from economic decline due to the shrinkage of the coal industry, and textile manufacturing. The quality of the local environment in terms of its aesthetic appeal can be a major factor in attracting inward investment to an area. This relates not only to the immediate setting of business premises but also to the quality of life which senior executives and other key employees can expect if they take up residence in the area.

3. A Vision For The Future Of Sherwood Forest

A whole series of statutory and non-statutory bodies have long held an interest in Sherwood, but those interests have tended to run in parallel with each other, often failing to come together where their combined efforts could achieve much greater impact. The Sherwood Initiative was set up by the Forestry Commission in an attempt to bring about some convergence between those interests. This grew from the recognition that modern multi-purpose forestry could itself accommodate many of these pressures and that consequently an expansion of woodland in Sherwood would be desirable. Through the work of its Steering Group, the Sherwood Initiative has helped to set the context for a more holistic view of Sherwood Forest, and served as a focus for a range of public and private sector interests to work together.

The culmination of this process has been the formation of the Sherwood Forest

Trust, a non-profit making body which will help to identify opportunities to regenerate Sherwood Forest and, more importantly, turn those opportunities into action on the ground. The Trust has grown out of the Sherwood Initiative's partnership approach, and includes Trustees nominated by; The Forestry Commission, English Nature, The Nottinghamshire Wildlife Trust, Nottinghamshire County Council, Newark and Sherwood District Council, Bassetlaw District Council, The Country Landowners Association, The National Farmers Union, with Sir Andrew Buchanan (the Lord Lieutenant of the County) as Chairman of Trustees.

There is no doubt that a more strategic view of Sherwood is now being taken, but there are still many programmes which are strong on advice and weak on implementation. The Trust will be able to act as an implementation vehicle for a number of other programmes and facilitate the uptake of a wide range of existing resources, as well as tapping into new sources of support. It is the Trust's ability to target particular parts of the Sherwood area that is its real strength. The Trust will shape Sherwood's future by working with those who can effect real changes. With the ability to enter into management agreements, to lease land and to buy land, the Trust will put the emphasis on practical action, backed up by awareness raising and education. Using Geographic Information Systems (GIS), such as those developed by Nottinghamshire County Council and Nottingham University, we can identify the key parts of Sherwood where our efforts are likely to get best results, an approach which is particularly important when constrained by limited resources.

We can maximise our impact by using our understanding of the complex ecological and economic relationships. Creating a new hectare of woodland or heathland on arable farmland is likely to be a fairly standard cost across most of the Sherwood area. However, the benefits which that new feature achieves could vary enormously:

- Does it take land out of farming in a Nitrate sensitive Area, and hence protect the aquifer?
- Does it act as a linkage between existing habitat patches?
- Is it improving a degraded landscape zone?
- Is it improving the setting of an established tourist attraction, or improving the landscape of a key road corridor?
- Is it drawing visitors away from a more sensitive site?

This list could go on and on. The more 'points' a site scores, the more impact we are having with our limited resources. Previously, these issues were being pursued separately by different organisations, now the Trust is championing a shared vision. With the help of GIS we can undertake a series of analyses; sieve mapping, determining least cost paths for linking sites, using fragmentation and connectivity analyses to prioritise activity. We can look also to the principles of landscape ecology to help us understand the continuing dynamics of the countryside. The Trust, as a neutral and non-partisan body, driven by consensus, has finally provided the solution to the problem of Sherwood Forest's long decline.

4. Pertinent Observations from the ARW

Many of the views and ideas expressed in the above paper were echoed by the comments and experience of others at the ARW. Of particular interest was the widely shared concern that the benefits and positive impacts of forestry were poorly understood by those outside the discipline. Also, the lack of a recognised means of attributing value to the range of non-monetary and non-timber outputs achieved by forestry tends to reinforce traditional economic decision making. The ARW also highlighted the importance of public involvement and understanding of forestry. The importance of economic utilisation of timber was perhaps underplayed in the paper. The ARW clearly confirmed this as a key element of sustainability for forest management.

INCREASING PUBLIC AWARENESS OF THE CONTRIBUTION OF FORESTRY AND WOOD UTILIZATION TO ECOLOGY

G. WEGENER
Institute for Wood Research,
University of Munich
Winzererstrasse 45, 80797
Munich/Germany

1. Global Developments and Problems

Worldwide, forestry and wood utilization are facing great challenges and are interlinked in very special and complex ways with the pressing global problems of mankind.

Progress in technics all over the world is frequently concomitant with severe adverse effects on the environment. Catchwords such as Tshernobyl, ozone hole, greenhouse climate, forest decline, global special waste "tourism" as well as depleting fossil resources are indicative of man's failure to achieve a successful synthesis between nature, human actions and economic systems. It is becoming increasingly evident that we have gone beyond the tolerance limits of many ecosystems (Fig. 1).

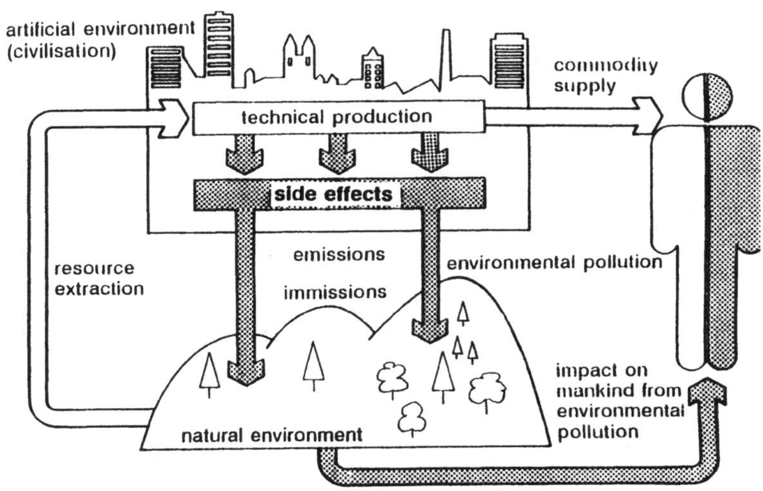

Figure 1. Interactions of technical production, natural environment and man

The impact of technical production processes and products on the natural environment and mankind cannot be allowed, in the long run, to grow to such a great and damaging extent that the price to be paid for any further rise in standard of

R. M. Gutkowski and T. Winnicki (eds.), Restoration of Forests, 77–96.
© 1997 *Kluwer Academic Publishers.*

living will turn out to be irreparable consequences for man, society and its natural foundations. Figuratively speaking, mankind is living on the "Capital Earth" and not, as it should, on the "interests" i.e. on the yields from human economic operations.

In recent years this situation has escalated, as developments are under way which no longer take a merely linear but, in many instances, an exponential course. In other words they are becoming too fast for us to fully comprehend their ultimate implications and we are therefore no longer capable of keeping them under control.

The driving force behind many developments is the global population explosion which is following a classical exponential function and which will lead to an expected world population of over 8 billion people by the year 2020.

Next to the immense demands on land and energy, associated inter alia with great losses in forest area caused by the pressures of exploding population growth in many third world countries, there is a dynamic increase in global urbanization, particularly in the less industrialized parts of the world. Today about 50 % of the world's population lives in cities, with an expected rise to as much as 75 % by the end of the 21st century. This, in turn, will lead to an increase in global industrial production, which has already seen a threefold increase in the past 30 years. Naturally, in this chain of events the emissions of nitrous oxides, methane and chlorofluorocarbons (CFCs) have developed out of all proportion, while primary energy consumption rose by about the factor 3 in the same period, entailing a rapid depletion in fossil resources, e.g. in natural gas. Wood and other plant and animal products as renewable energy resources are able to meet circa 10 % of the energy demand, hydroelectric power about 6 %, while about 2 % are provided by so-called gentle energies (sun, wind, geothermal and tidal energies) (1).

Since mainly fossil energy resources and fuels are being used, atmospheric CO_2 has increased to such an extent that all climate specialists agree in blaming the phenomenon of longterm global warming on the rise in CO_2. At present, highly industrialized countries with per capita emissions of 10 to 20 t/a contribute the lion share towards global CO_2 emissions. On the other hand, by the year 2020 emissions from third world countries will have more than doubled, in direct relation to population growth and the developments outlined above.

On account of the CO_2 problem, in particular, there is a pressing need for action at all political, socio-economic and scientific levels.

At the 1992 Rio Conference over 160 states addressed these global problems for the first time and declared their willingness to press for political action. The Rio Conference's main aim is sustainable development. This implies a development in extension of the forestry principle of sustainable development which commits us to take care of today's needs without jeopardizing the ability of future generations to meet theirs.

One of the essential strategies used in this context is the transition from "High-Tech" to "Hyper-Tech," i.e. from what is "technically possible" to what is "ecologically sensible." Thus, we must be made to realize that the manufacture e.g. of a single automobile catalyst - the requisite instrument for reducing hazardous emissions - is burdened with an "ecological backpack" due to the production or extraction of platinum, ceramics, steel and other materials which cause the "disturbance" of 1 ton of the environment.

Reduction in world population growth
Guarantee of the world food supply
Reduction in greenhouse effect
Sustainable energy supply
Afforestation and sustainable wood utilization

Figure 2. Rio Earth Summit 1992: Future Global Strategies

In formulating the most essential global tasks to be undertaken in the immediate future (Fig. 2), as well as in the "Agenda 21," a whole range of actions with a direct bearing on forests and wood was outlined at the Rio Conference:

- maintenance of the manifold tasks and functions of all kinds of forests, forested areas and groves;
- improvement of the protection, sustainable management and conservation of all forests and restoration of degraded areas through re-forestation and other stabilizing measures;
- support of efficient utilization and assessment of the state of forests to restore the full benefits through goods and services rendered by forests, forest areas and groves;
- creation and/or reinforcement of existing capacities in the planning, systematic observation and assessment of the state of forests as well as of related programs, projects and activities including commerce and processing.

2. Ecological Aspects of Forestry and Wood Utilization

In the above context, the system "Forests and Wood" has acquired a unique role. The world's forests are not only irreplaceable on account of their well-known beneficial ecological and socio-cultural functions and as raw material resource, but also as carbon stores and CO_2 sinks which effectively counteract the greenhouse effect.

Very often the importance and the need for the production of wood as a raw, structural and working material as well as an energy resource is being considerably underrated.

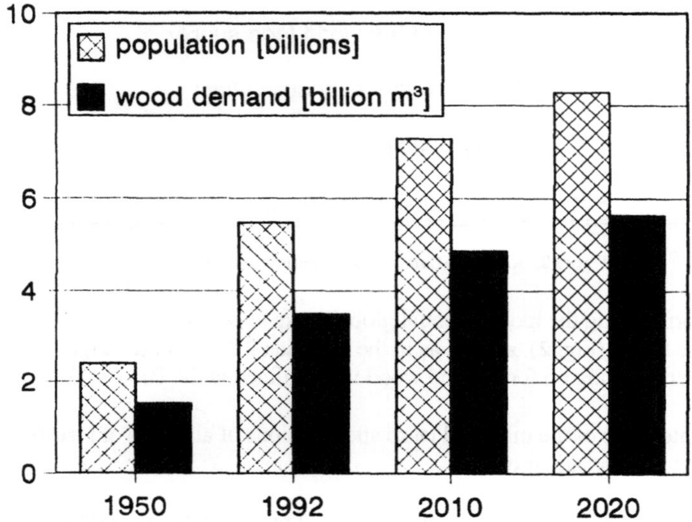

Figure 3. World population development and wood demand

Wood consumption is bound to experience a dynamic increase, in direct relation to the population explosion outlined above (Fig. 3) (2). This is true for industrialized countries and, in particular, for developing countries where fuelwood will remain, for the time being, the most important source of energy (Fig. 4) (3).

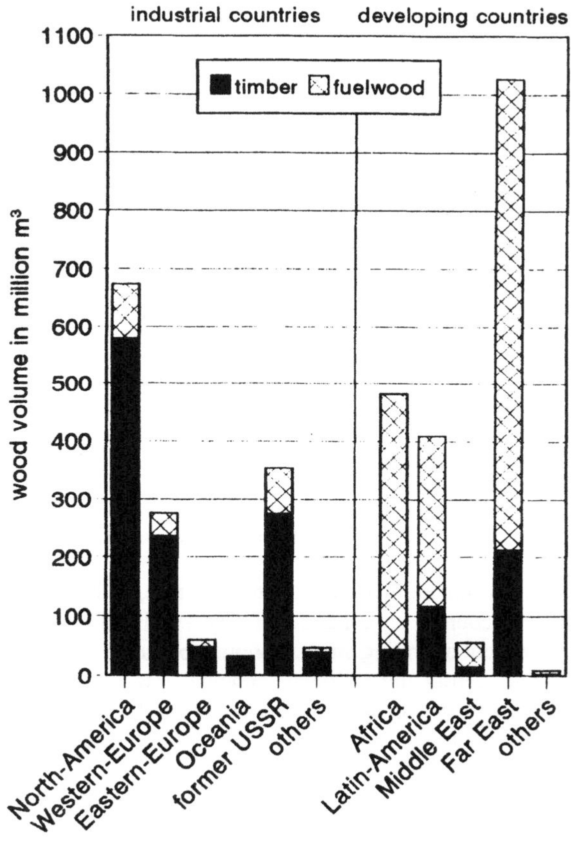

Figure 4. Fuelwood portion in relation to total wood volume

Even in highly industrialized countries such as Germany wood, compared with other materials, constitutes an irreplaceable resource (Fig. 5).

Despite this fact, however, German forestry, as an essential land use system other than agriculture and areas set aside for traffic or industry, has become caught up in a difficult socio-political conflict between ecology and economy.

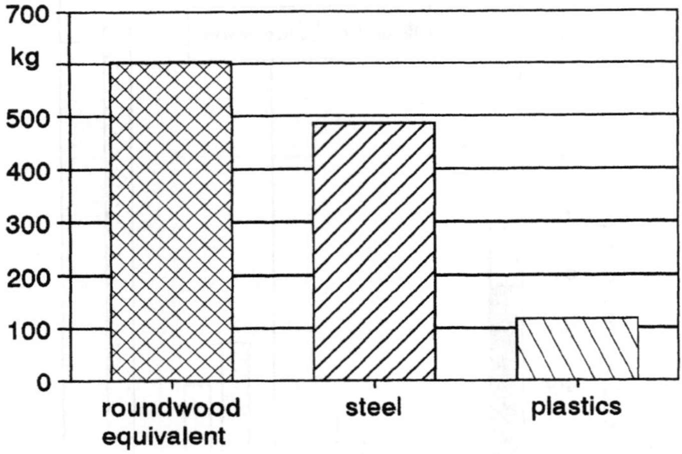

Figure 5. Consumption per head in Germany - 1992

Economic management in agriculture has become, in the course of time, an irrefutable matter of course. By contrast, forestry is obliged to procure, through its forest areas but without operating on an economic basis, i.e. without wood production and wood utilization, the ecological benefits already produced at a very high rate by its other sectors, despite severe handicaps from traffic, industrial emissions and agricultural nitrogen inputs (Fig. 6).

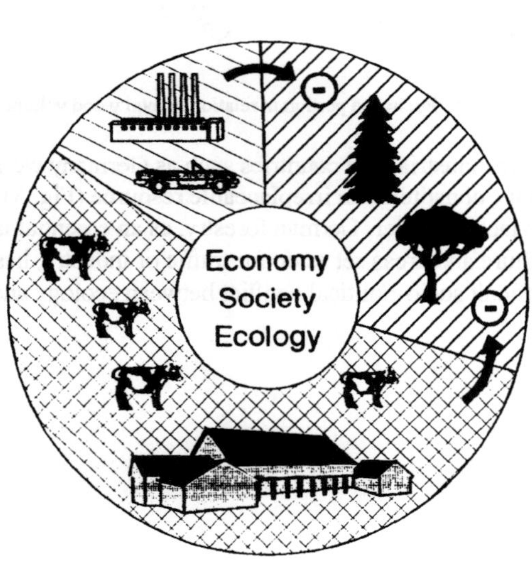

Figure 6. Society's conflict regarding economic and ecological land use

Moreover, forestry in Germany is exemplary and works on a sustainable basis, since only 2/3 of the annual increment is being used, i.e. less than the interests on the capital (2.8 billion m^3 growing stock) are being consumed (Fig. 7 and Table 1), while the carbon storage system forest continues to grow.

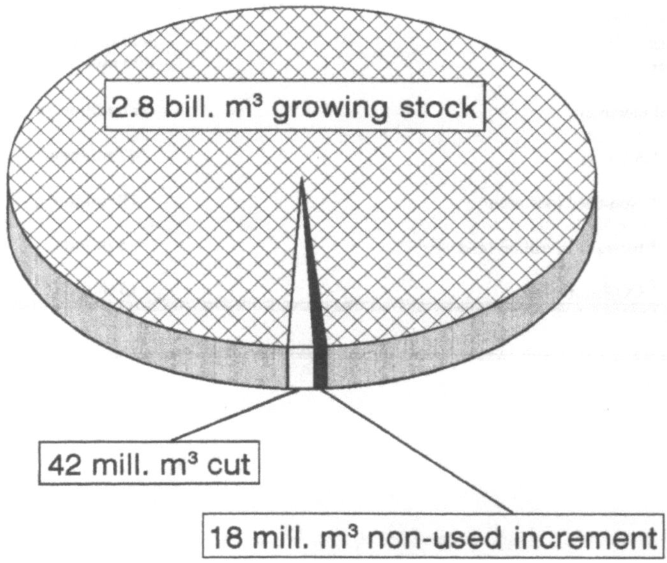

Figure 7. Relationship between growing stock, annual cut and non-used increment

It is imperative that greater efforts be made to make society fully aware of the natural benefits provided by forests, which in Germany e.g. lead to a fixation of 0.7 t carbon per second while at the same time 1.8 t oxygen are being released into the atmosphere. Care should be taken to differentiate between the material and energy flow of primeval forests (Fig. 8) and the forests of Germany and Europe, mostly subject to centuries of cultivation. The latter created forests where wood produced by photosynthesis with the aid of sunlight can be tended, harvested and finally supplied to the forest products industry with a minimum of operational energy (Fig. 9, 10) (4, 8).

Table 1. Forests and Wood Utilization, Germany	
Forest area	10 million ha
Growing stock	
- biomass	4,000 million m^3
- timber	2,800 million m^3
Fixed carbon	
- biomass	900 million t
- timber	600 million t
Annual increment	60 million m^3
Annual cut	42 million m^3
* Non-used increment	18 million m^3
* Increase in carbon reservoir	4 million t
* CO_2-equivalent	15 million t

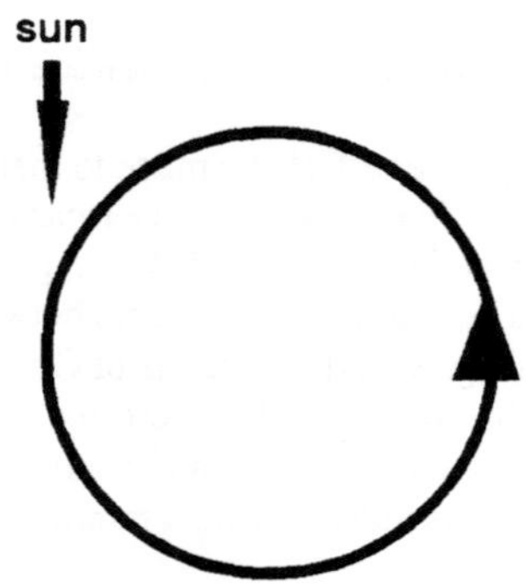

Figure 8. Natural forest cycle

In the course of the biological process by which wood is produced in the forest 51 kg carbon per 100 kg wood are being fixed, corresponding to an extraction from the atmosphere of the CO_2 equivalent of 187 kg. The use of wood in a great range of products complements and extends the eminently

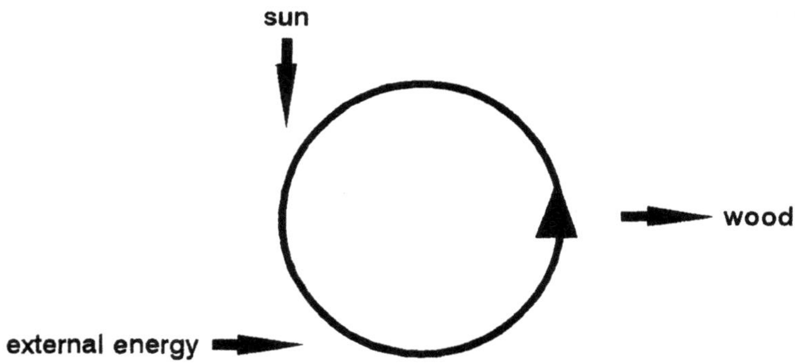

Figure 9. Commercial forest cycle

important carbon storage system forest, which leads, in Germany, to an estimated carbon fixation of approx. 1,250 million tons (Table 2) (5). Table 3 shows the production volumes, usually greatly underrated by most people, for some wood products which, according to statistics, consume about 1.1 m^3 wood (roundwood equivalents) and wood products per head and year.

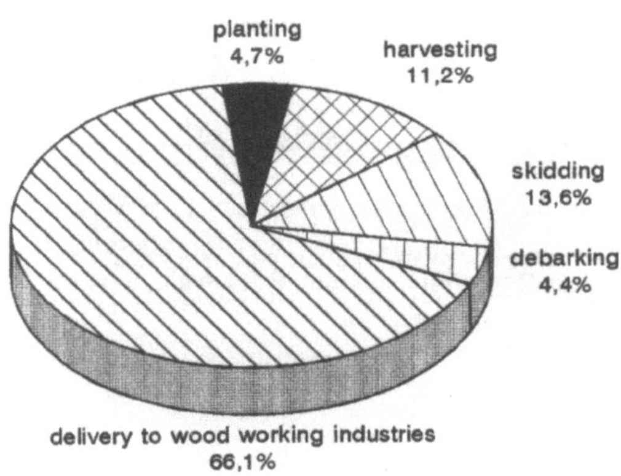

Figure 10. Energy input for roundwood production and transport

Table 2. Wood in use and resultant CO_2-mitigation potential.
1 t approx. 2 m^3 wood

35 million households x 2,000 kg furniture and wooden household objects	10 million t
17 million one and two family homes each with 25 m^3 (= 15 t) wood used in construction (framework for roofs, walls, ceilings, floors, panels)	155 million t
2.75 million residential buildings with more than 2 flats, each with approx. 40 m^3 wood used in construction	85 million t
Wood for exterior uses (fences, playgrounds, sheds etc.), sleepers and poles	80 million t
Wood as construction, fitting and furnishing material in public and industrial buildings	100 million t
Wooden packaging material	10 million t
Wood used in paper and paperboard (books, newspapers, packaging materials)	10 million t
Semifinished products pending further processing and in storage	15 million t
TOTAL	655 million t
SEQUESTERED CARBON	340 million t
CO_2 EQUIVALENT	1,250 million t

Table 3. Production volume of some finished wood products in Germany 1993

Doors	7.5 million
Windows	8.1 million
Chairs	2.8 million
Easy chairs, couches, beds	9.7 million
Wardrobes, cabinets	33.2 million
Parquet	8.8 million m^2

Apart from the storage effect the consistent use of wood constitutes a unique approach towards saving energy and reducing atmospheric CO_2 caused by emissions from fossil resources, which amount, worldwide, to about 22 billion tons per annum (Figure 11) (6).

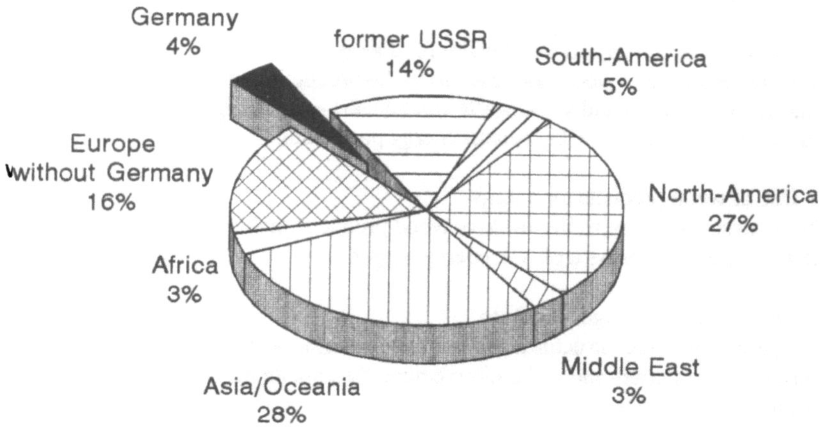

Figure 11. Energy related CO_2-emissions - country groups 1993

There are at least 10 good reasons advocating the increased use of wood, as outlined below:

1. Substitution potential
The equivalent amount of at least 1 t of atmospheric CO_2 can be saved with every cubic meter of wood (density 0.5 g/cm³) used to substitute materials which do not or have not yet fixed CO_2 (mineral oil products, steel, aluminium, concrete) or which would release surplus CO_2 in the course of production or disposal. Thus, the substitution e.g. of one ton of steel by wood would reduce atmospheric CO_2 by 4 to 5 t, in other words the substitution of 20 % of the annual steel consumption in Germany by wood would reduce atmospheric CO_2 by 40 million tons (7).

2. Extremely low energy requirements for making wood available
Since solar energy takes care of tree growth, the requisite energy input is restricted to stand establishment and tending, harvesting, transport and necessary opening-up measures, such as forest road construction. Total energy consumption in making wood available for processing amounts to merely 185 MJ/m³, i.e. in relation to the energy contained in wood (calorific value approx. 8,500 MJ/m³) the energy required to produce it amounts to only 2 % of the energy it contains (see Fig. 10) (4, 8).

3. Low energy input in the manufacture of wood products
The energy requirements for the production of finished goods is comparatively low. This is particularly true for wood as a structural material.

4. Substitution of fossil fuels
There is no doubt that neither wood nor any other renewable energy resources will be able to effectively replace fossil resources. However, wood has a large potential, especially in rural areas, for its enormous CO_2-reducing effect to be utilized on account of the use of fuelwood in the private sector.

5. Wood as an industrial energy resource
Wood and other biomass can be used to produce heat and electricity.

6. Use of by-products and wood waste for energy
In the material use of wood, too, such as in the production of wooden structural material by-products and wood waste can be used as an energy resource and thus contribute once again to CO_2-neutral energy production.

7. Use of wood products for energy
The analysis of the entire life-cycle of wood products, including thermal utilization after product use, shows a plus in energy output.

8. Energy saved through wood use
Compared with other structural and working material wood has excellent heat-conserving properties and is therefore eminently suitable for energy-saving construction methods.

9. Prolongation of utility
In the repeated use and utilization of wood products as well as in their further use and utilization the life-time of wood products and the period of CO_2-fixation can be considerably prolonged.

Moreover, wood and wood products can contribute, in very special ways, to the critical ecological and economic requirements of cyclic economies, where biological degradability and CO_2-neutral re-integration into the natural cycle are important preconditions.

10. Extension of wood utilization on a sustainable basis
If almost full use were made of the difference in volume between annual roundwood increment and actual cut in Germany, without jeopardizing sustainability, another 3.4 million tons of atmospheric CO_2 per annnum could be removed through the additional use of wood as a susbstitute for non-wood products and fossil energy resources.

3. Life-Cycle-Assessment (LCA), Instrument for the Evaluation of Ecological Benefits

While there is a vast number of economic theories, approaches and instruments designed to control economic actions by people, enterprises and economies, there are still hardly any attempts at finding corresponding ecological or economic-ecological approaches or instruments. For some years intensive efforts have been made at national (e.g. DIN in Germany, SN in Switzerland) and international (ISO) levels to define the instrument of life-cycle-assessment and to determine its applicability with a view to using it as an effective tool in ecological planning.

LCA should be applied predominantly to assess the anthropogenic effects on the environment of:

* the consumption of deplenishing or renewable and partly renewable resources such as energy, raw materials, water and land;

* global changes such as greenhouse effect, ozone hole, loss in species diversity, acidification, eutrophication etc.;

* the state of health of mankind and the animal world.

LCA can be considered merely a part of environmental management within comprehensive management systems or an all-encompassing eco-certification (Figs. 12 and 13).

Life-cycle assessments are designed to determine and assess the entire life-cycle of products from raw material extraction to disposal or re-integration into cycles as well as the impact of services on natural resources. They are thus an indispensable instrument in decision-making processes concerning suitable measures in environment protection or in selecting ecologically favourable production processes, products and services, in order to prevent, minimize or remedy damage to the environment.

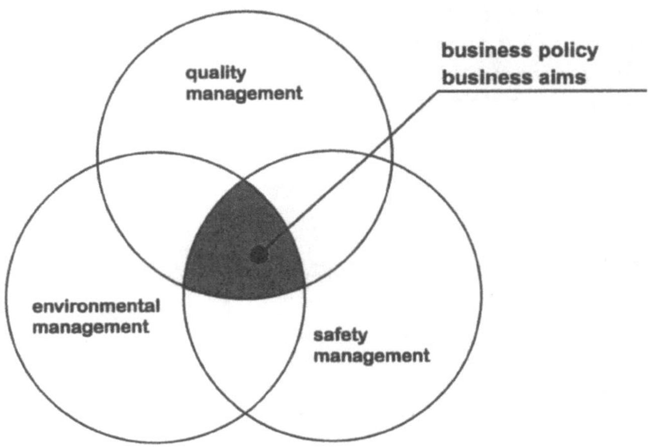

Figure 12. Management systems

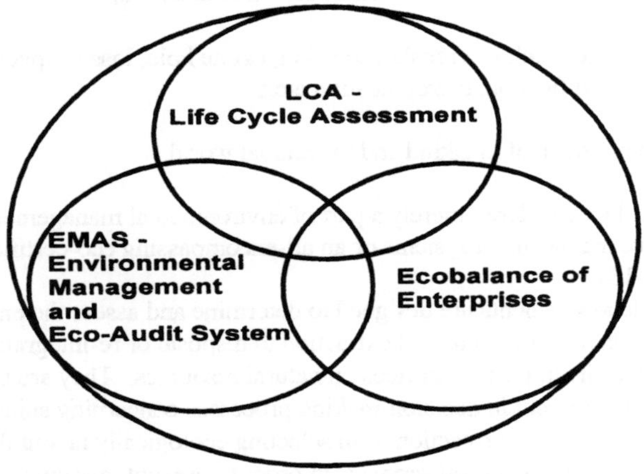

Figure 13. Elements of eco-certification

Hence, LCA is used e.g. in the implementation of the projected "Environmental-Economic-Accounting" (EEA) as a complement to data on the national gross product (Fig. 14). EEA is the attempt to quantify the state of the environment, the way it is being used and changed on account of economic operations and consumption in political economy, with the aim of creating instruments with which to exercise ecological control (9).

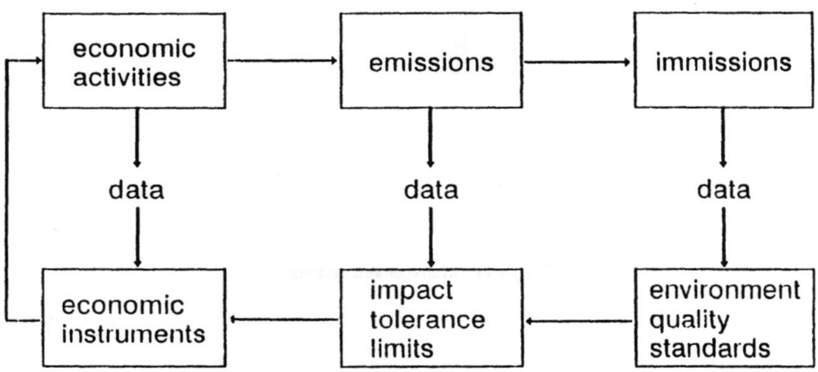

**Figure 14. Environmental-Economic-Accounting
Structures used in decision-making processes**

Other fields of LCA application are weak-point-analyses in production processes, improvement of ecological product quality as well as marketing and advertising.

LCAs are therefore equally important to producers and consumers, as well as in decision-making processes at all political and socio-political levels.

By definition LCAs are instruments of information, planning and control in product policy, as they represent ecologically evaluated material and energy balances for the entire life-cycle of products.

According to ISO draft standards 14040 to 14043, LCA consists of four essential parts:

* goal definition and scoping
* inventory analysis
* classification
* input assessment

Goal definition describes the meaning and purpose of life-cycle-analysis, its scope, limits and depth.

Inventory analysis represents the input/output assessment "from the cradle to the grave" and reviews consumption of resources, energy, water, air and landscape as well as the impact on air, water and soil.

Classification, on an interdisciplinary basis, quantifies all impacts on the environment and uses the classification categories given in Figure 15.

* Consumption of resources
* Soil impacting
* Loss of biodiversity
* Loss of ozone
* Formation of photooxidants
* Acidification
* Eutrophication
* Eco-toxicity
* Human toxicity

Figure 15. Classification categories

Last not least input assessment has to be done on the basis of political and socio-economic consideration.

The central element in this puzzle is the inventory analysis, in which all material and energy flows into and out of the system are identified and quantified.

In the comparison of the life-cycles of different materials or products it is vital that the scopes of LCA be defined (Fig. 16).

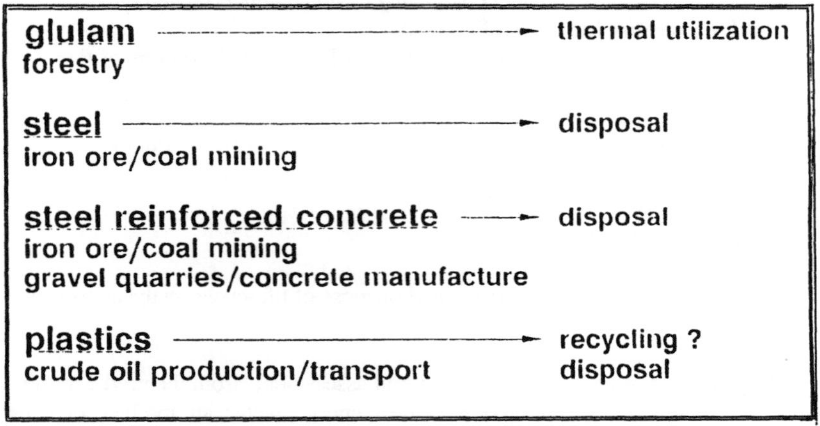

Figure 16. Life-Cycle-Assessment. Comparison of life-cycle scopes

The LCA e.g. for a glulam beam covers the entire range from the biological production of wood in the forest up to and including the thermal utilization of the structural element after use.

For steel products LCA has to include iron ore and coal mining, while that for steel reinforced concrete also comprises the production steps from gravel mining and concrete manufacture at the start of the life-cycle.

For plastic products LCA involves the calculation of the enormous ecological risks in mineral oil production and transport (e.g. tanker accidents), on the one hand, and the partly considerable technical difficulties encountered in the recycling and disposal of these products, on the other.

For a comprehensive and accurate assessment of the ecological benefits of forestry and wood utilization outlined above, the life-cycle of wood and wood products is broken down into so-called modules (Fig. 17). Great and important modules such as "Forestry" or "Sawn Timber Production" are again divided into different sub-modules where material and energy data and flows are analyzed individually and integrated into the input/output balance of the module (Figs. 18 and 19). The options in the sub-modules, in particular, permit a comparison of different forest production and management systems in different countries (10).

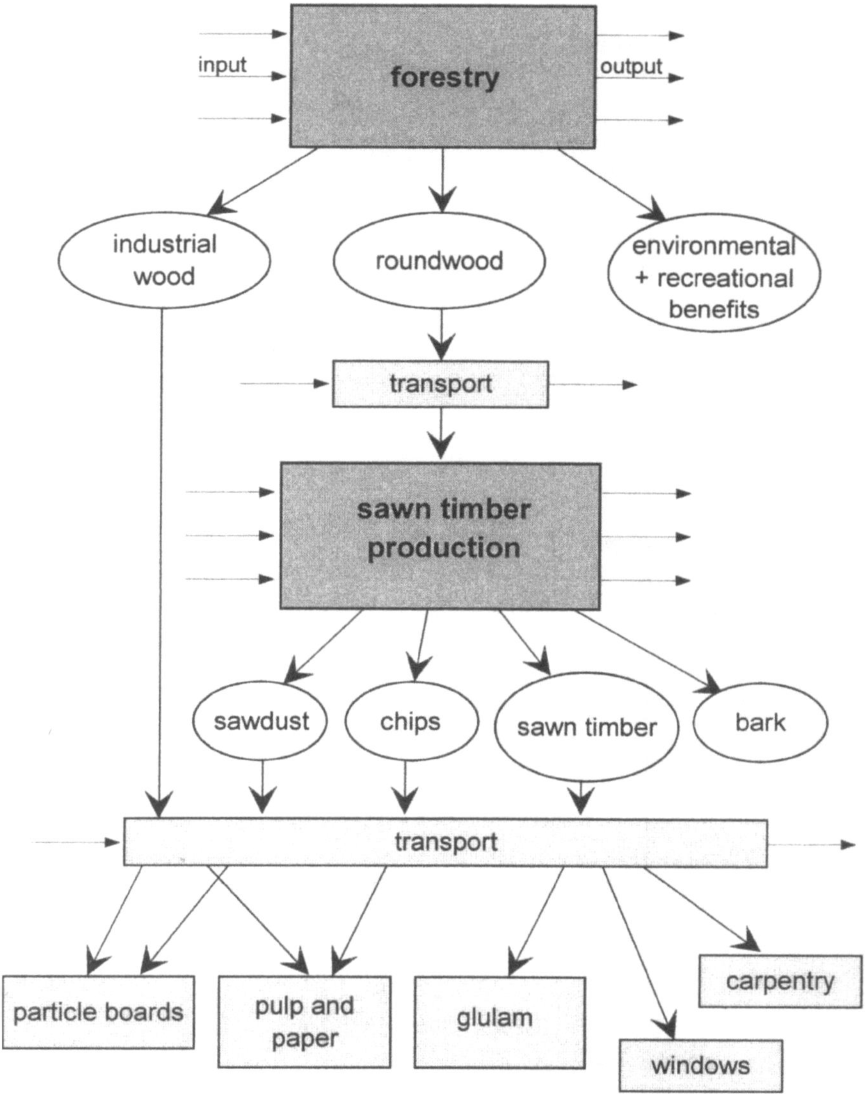

Figure 17. Material flows from forest to wood products

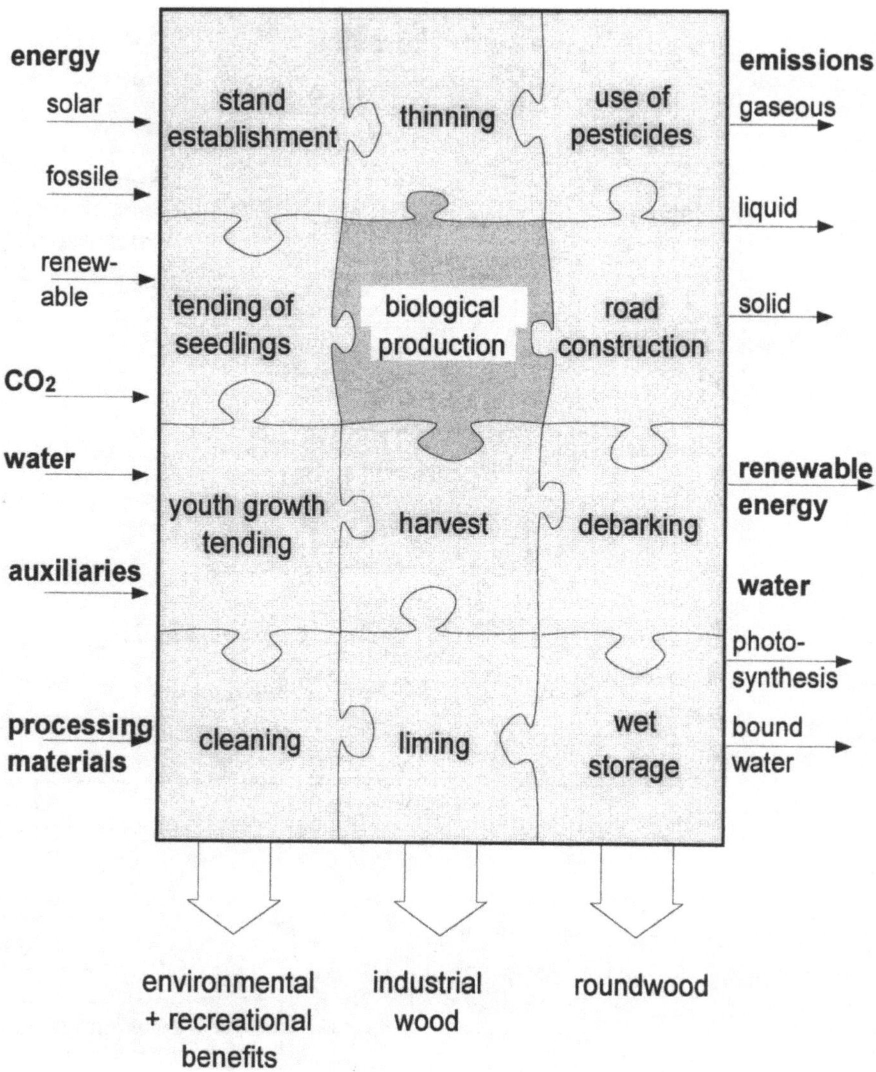

Figure 18. Module "Forestry" and its sub-modules

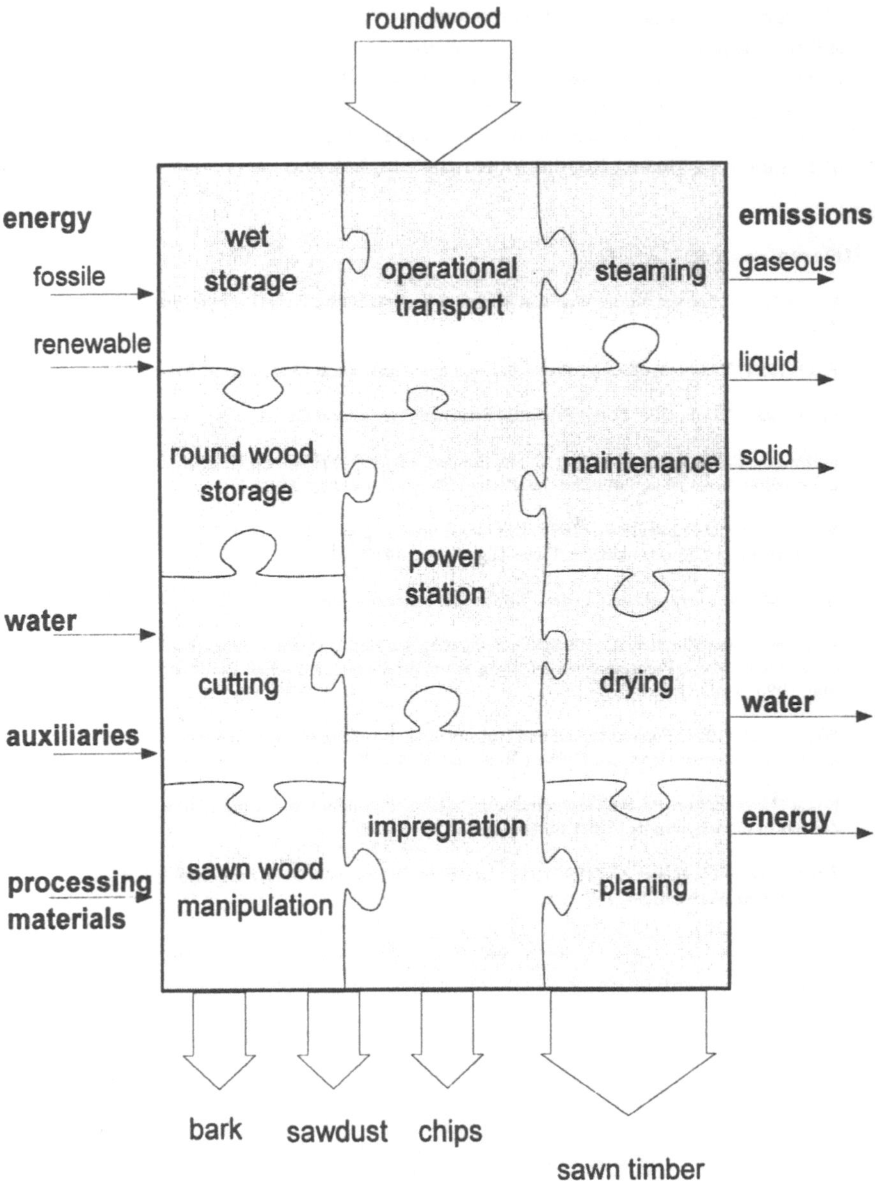

Figure 19. Module "Sawn Timber Production" and its sub-modules

4. Conclusions

In principle, the sum total of all these instruments makes it possible to quantify and interlink the positive ecological elements of forestry and sustainable wood production as well as wood utilization and to underline the great overall ecological importance of the system "Forestry and Wood."

Environmental policies based on modern, active systems with good prospects for the future have no choice but to rely on wood as a raw material, which cannot be replaced, with respect to either quantity or quality, by any other material. Wood needs to be produced effectively, on suitable sites and on a sustainable basis for the long-term securing of natural resources and in strong support of projected strategies aimed at reducing fossil resource consumption and CO_2 emissions.

5. References

1. Ott, G. (1993) Energie für die Welt von morgen, Energiewirtschaftliche Tagesfragen 43 (11) 742-749.

2. FAO (1994) Yearbook of Forest Products, Food and Agricultural Organization, Rome.

3. FAO (1991) Yearbook of Forest Products, Food and Agricultural Organization, Rome.

4. Frühwald, A., Wegener, G., Krüger, S. and Beudert, M. (1994) Holz - ein Rohstoff der Zukunft, Informationsdienst Holz, Deutsche Gesellschaft für Holzforschung, Munich.

5. Wegener, G. and Frühwald, A. (1994) Das CO_2-Minderungspotential durch Holznutzung. Holz als Energieträger, Energiewirtschaftliche Tagesfragen 44 (7) 421-425.

6. BMWi (1994) Energie-Daten. Federal Ministry of Economy, Bonn.

7. Wegener, G. and Zimmer, B. (1996) Die Bedeutung des Holzes in der ökologischen Gesamtbetrachtung, Proceedings 4. Int. Symposium für die Holzwirtschaft Biel/Switzerland, May 1996, 6 pp.

8. Wegener, G. (1994) Forstwirtschaft und Holznutzung als unverzichtbare Elemente zukunftsorientierten Handelns, Holz als Roh- und Werkstoff 52, 235-238.

9. Beirat "Umweltökonomische Gesamtrechnung" (1995) 2. Stellungnahme des Beirats UGR. Zeitschrift für angewandte Umweltforschung 8 (4) 455-476.

10. Zimmer, B. and Wegener, G. (1996) Stoff- und Energieflüsse vom Forst zum Sägewerk, Holz als Roh- und Werkstoff 54, 217-223.

SUSTAINABLE ECONOMY OF FORESTRY AND VALUE ADDED UTILIZATION OF FORESTS

The Only Chance to Save the Forests of the World

J. NATTERER
Director of the Chair of Timber Construction
Swiss Federal Institute of Technology
Lausanne, Switzerland

Abstract

The success of timber engineering and of the use of timber as load-bearing material depend on the quality criteria. The most important factor is the economical use of timber in order to increase its utilization in construction and make it competitive in comparison to other construction materials. Therefore, all kinds of wood and timber, from round wood to square timber through composite sections made of boards and square timber, as well as glue-laminated timber, plywood, etc., especially when combined with other materials, must be developed and employed with that awareness.

1. Timber Quality Wood Sampling

The modulus of rupture of massive wood, like round wood, square timber, block boards, as well as glue-laminated timber, and other wood materials have different dispersions. The influences on the resistances mostly depend upon a) raw density, and nodosity, b) section form, fiber obliqueness, etc., and c) manufacturing criteria like moisture content, use of finger joints to prevent fissuring, etc. (Figure 1).

The reliability of optical sampling methods does not correspond to the degree of accuracy presupposed by the engineer's calculations or analytical models. The assumptions of modulus of elasticity and rupture, respectively the admissible stresses, are dispersing between 100 and 200 per cent, while an exceeding stress of 3 percent measured in a static test already leads to a conflict with the expert (Figure 2). Basing admissible stresses on the five-percent fractile, i.e. on the value just above the five weakest from one hundred, leads to an uneconomical exploitation of available qualities in the very large general dispersion of timber, when compared with other construction materials.

R. M. Gutkowski and T. Winnicki (eds.), Restoration of Forests, 97–118.
© 1997 *Kluwer Academic Publishers.*

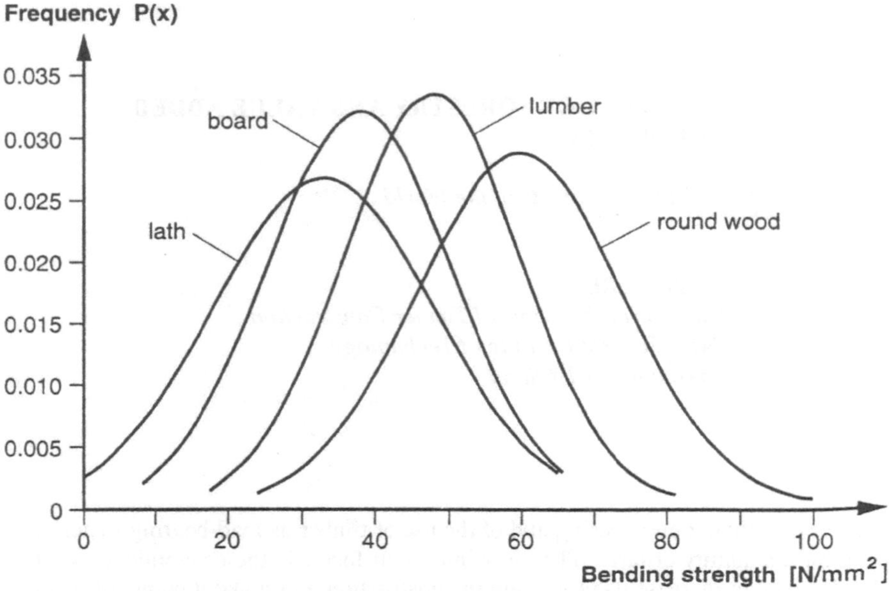

Figure 1. Qualitative variation in round timber and sawn timber according to its cross-section

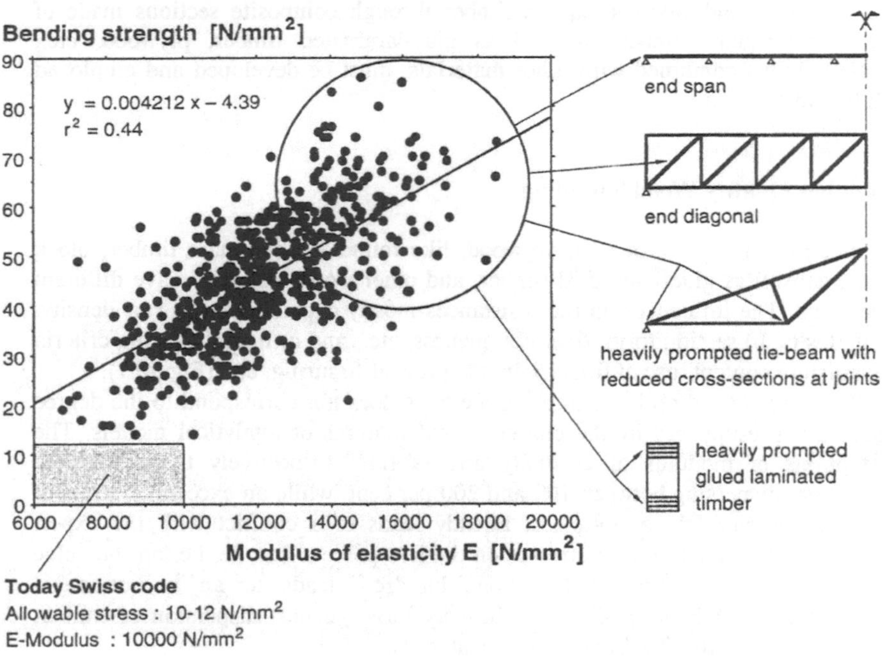

Figure 2. Correlation between bending strength and modulus of elasticity

In order to take advantage of better qualities, one needs to be able to place the best pieces of timber in the most stressed construction elements. This requires the admission of non-destructive testing equipment which can, like the Sylvatest® (Photo 1) ultrasonic method, determine single resistanceís much more precisely;

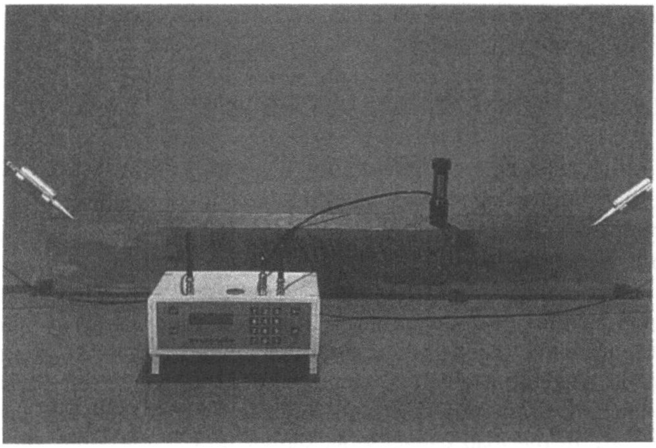

Photo 1 (See List of Captions)

including the elastic modulus as well as for the modulus of rupture. This testing equipment can be used for round wood as well as for solid sawn timber or glulam. It is also very useful for reconstruction and renovations of old buildings, as it provides the resistance values to be given to the aging of old constructions.

1.1. MATERIAL QUALITY - MATERIAL SELECTION

The manifold material values can also be expanded with the material selection. The selection between different materials like round wood, square wood, sawn wood or glued-laminated timber which has been improved through industrial methods, or different sorts of plywood, provides a range of resistance qualities. This variety enables economical and competitive construction possibilities (Photos 2-5).

1.2. MATERIAL QUALITY - PROFILES

Wood sections are a further quality to take into account besides resistance criteria when using round- or sawn wood. The treatment preceding the drying and the considering of the extension of the annual rings in different forms of sections, as well as the profiles of the wood sections are most important for an economical formation of detail in the constructive use of timber.

2. Assembling Techniques - Assembled Sections

The many types of material and section forms must be used with new assembling techniques in order to manufacture larger sections. There are many examples of an economical use of round wood, sawn round wood, square timber, profiled sections of square timber as well as assembled joists, and they will open further economical utilization of raw wood in the future.

2.1. TECHNIQUES AND MEANS OF ASSEMBLING

New, highly efficient means of assembling, i.e. connections with lower cross-section reductions and less need for steel parts, have to be developed for a highest possible degree of pre-fabrication in the workshop and in order to reduce erection time on the site as much as possible.

The use of new connecting systems like nailed tin plates, screws, lag bolts as well as connectors with wood contact allow a much higher quality of more filigree supporting systems when linked with deterministic non-destructive testing methods in order to avoid sporadic problems, which arise in highly stressed construction elements. Connectors combining fiberglass and mechanical fasteners also allow a noticeable increase in the load capacity, as shown in a recent study on fiberglass reinforced timber joints.

In order to give timber a new chance as a construction material, the different research, development and marketing programs should not aim at the quantity of material used, but at the manifold qualities of consistency of material properties, section variability, material diversity as well as facilitated construction control and faster adoption of the new techniques in timber engineering construction.

3. Composite Systems

In the history of timber construction, there have always been composite constructions – timber frameworks with glue or mortar, walls of stone and bricks – the most lasting were in timber architecture. Examples from China and Japan, to Frank and Alsatian framework construction are well known. Essential criteria are the better behavior of the whole construction during a fire, as well as excellent acoustics and vibration properties.

Today, quality criteria – fire, acoustics, vibration – are easily fulfilled through new shape applications, i.e. massive nail laminated floors and wood-concrete composite systems for wide-span and load supporting structures (Photos 6-13).

Nail laminated decks and wood-concrete decks including a load-bearing concrete slab present new advantages, especially for houses, schools and public buildings. With these techniques the steadiness and bending properties of structures with minor dead loads can be fulfilled economically. Fire resistance times of F30, F60 or F90 AB, as well as sonic insulation criteria up to 60 dB for walls and decks can be reached (Photo 14 and Figures 2 & 3).

Photos 2 & 3 (See List of Captions)

Photos 4 & 5 (See List of Captions)

Photos 6 & 7 (See List of Captions)

Photos 8 & 9 (See List of Captions)

Photos 10 & 11 (See List of Captions)

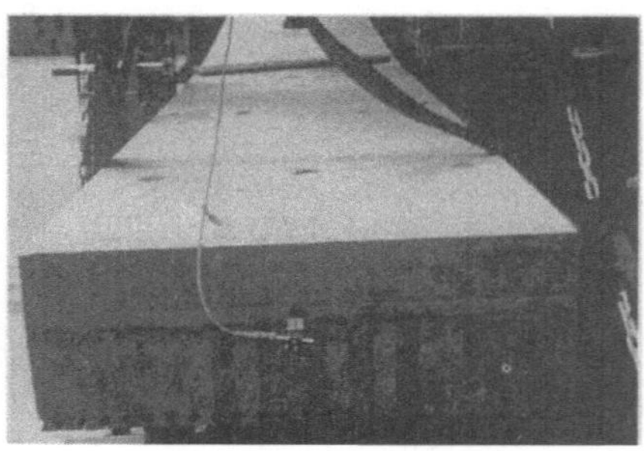

Photos 12 & 13 (See List of Captions)

Photo 14 (See List of Captions)

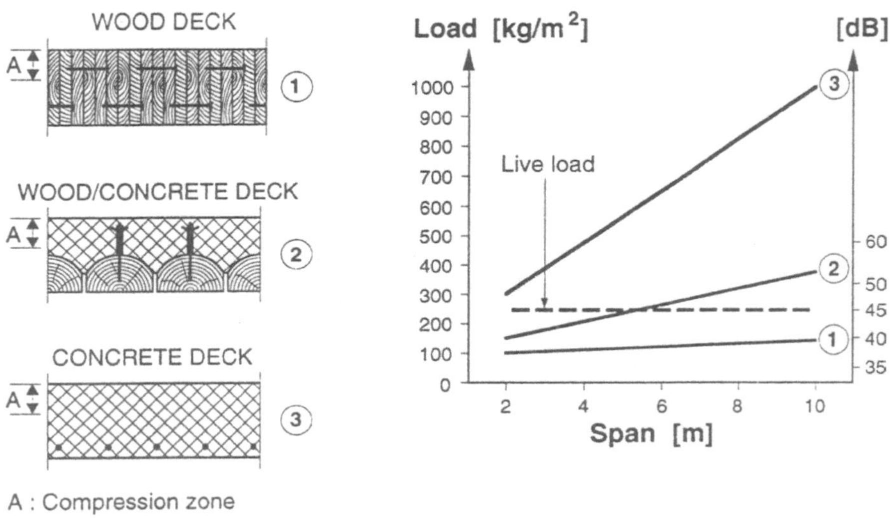

Figure 3. Comparison between live and dead load of floors

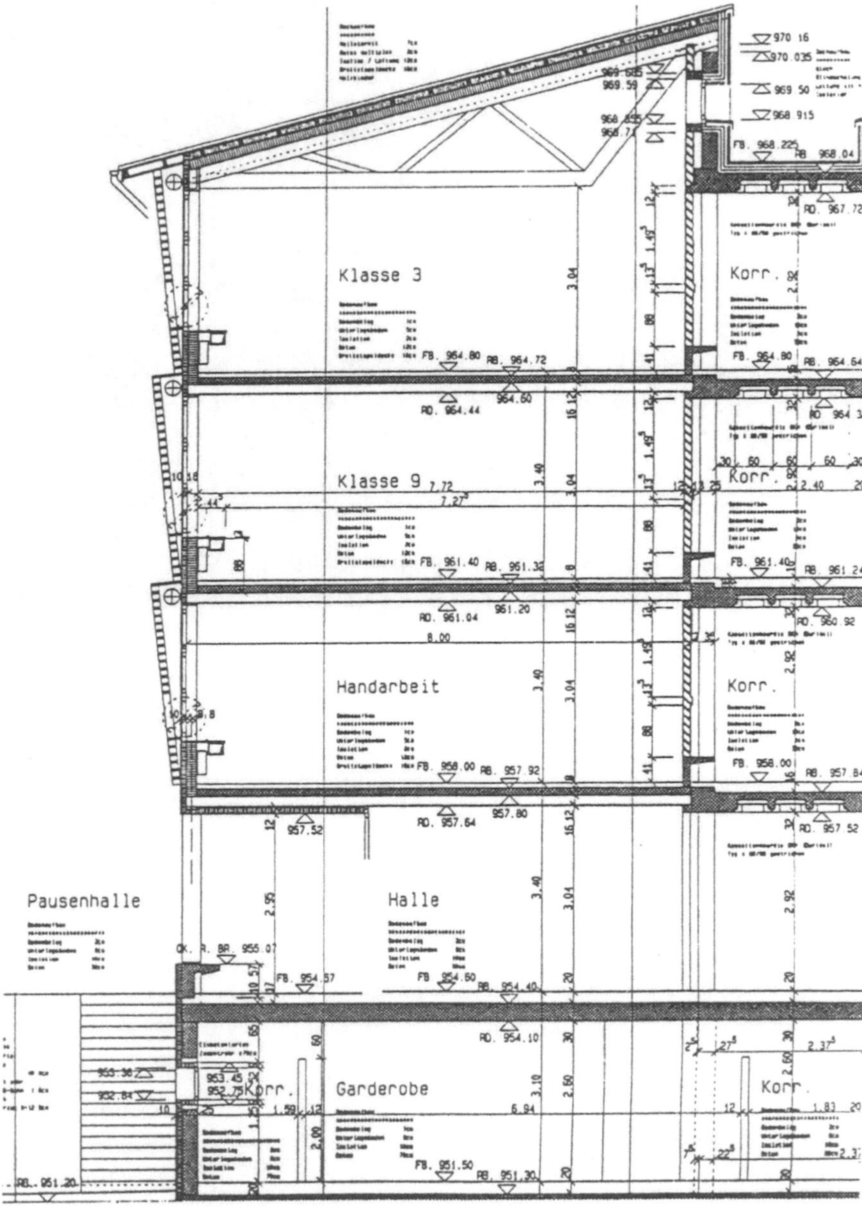

Figure 4. Timber and concrete composite ceiling with "Hilti" dowels and 7.9 span width, under test as a ceiling for a four-story school building in Triesenberg, Liechtenstein (Arch. H. Ospelt, coworker M Freund)

4. Planification Criteria

Timber as load-bearing material has a chance only if the conception of the construction can show a quality which is not only functional, technical and architectural, but which can also justify its economy. This presupposes, however, a more important planification and a better cooperation between architect and engineer, in order to make the most of the diversity of forms, structures and techniques applicable to timber.

It is essential to define clearly the quality criteria of a timber construction and to aim at reading easily the force and load fluctuations, and reducing the material through load- and detail-planification with an optimally functional adaptation to technique and construction.

5. Building Shape

The first planification criterion of a timber construction is its shape. The many design possibilities given by the easy manipulation and the low dead loads of timber, especially for houses, halls and roof shapes, are well known. The links between building shape, construction support, energy requirements and costs of maintenance are simple, but do not go without planification (Photos 15-20).

5.1. BUILDING SHAPE AND MAINTENANCE

Common perception of timber constructions being superficially planned in terms of material quality and use of material, and badly executed, leads to a mistaken image of high maintenance costs for timber constructions. That this is not the case is proven by timber structures in roofs, bridges and frameworks dating from the Middle Ages.

In spite of the good experiences made with historical constructions, planification of new timber constructions must not lead to nostalgic, history or museum oriented building or roof shapes (Photos 22 & 23).

5.2. BUILDING SHAPES AND ENERGY REQUIREMENTS

In comparison with other construction materials, using timber is already energy efficient. Further, energy saving measures like the creation of mid-temperate zones in timber-glass constructions have an important impact on the shape and allow a diversity of forms. Integration of active solar techniques can be fulfilled in a satisfactory way only with special planification.

Additional criteria concerning energy saving are lighting, ventilation and heating. They can be estimated in different ways, but must be considered first.

Photos 15 & 16 (See List of Captions)

Photos 17 & 18 (See List of Captions)

Photos 19 & 20 (See List of Captions)

6. Supporting Structures

In timber construction, each building shape can be constructed economically when the supporting structure is developed adequately in the section, whereas the structure of the inner space as well as the direction of load transmission, and main and secondary support systems have to be considered in the design. Optimization of the supporting structure through reducing flexure and developing, instead, axial forces needs special planification.

Further reduction of the material needed in timber construction can be achieved by using compressed frameworks, with the advantages of simple contact connectors. Frameworks and bracing structures can be designed in many ways, and are filling the inner space at the same time. When left visible they are part of the inner space arrangement together with the supporting structure.

A further step towards optimization, respectively minimization of the material used, is the use of statically indeterminate systems. New developments are made into this direction, whereby the dispersions of material are balanced through static uncertainties and the deformation behavior becomes critical for the dimensioning. This way, the construction of orthotropic systems made of round wood, square timber, beams or glue-laminated timber in association with concrete becomes economical (Photos 21 & 24-27).

Photo 21 (See List of Captions)

Photos 22 & 23 (See List of Captions)

Photos 24 & 25 (See List of Captions)

Photos 26 & 27 (See List of Captions)

7. Why Timber Constructions?

Given the steady decrease in reserves of fossil fuel energy and raw materials, the importance of the role of forests as suppliers of timber, respectively raw material, will grow significantly in the future. Without forest economy, society cannot cope with the costs of producing energy inefficient material, using non-renewable resources for construction material and other ecological challenges. Neglect of maintenance reduces the forest's health and increases the damages caused to forests. The use of timber is directly linked to forest conservation and the planting of new trees. The resistance of forests against wind, snow, frost, drought, insects and fungus and, of course, pollution through emissions, cannot be maintained or improved by only financial support and environmental protection actions.

Furthermore, the different functions of forests, such as protection against avalanches, landslide, erosion, flood, fire, etc., or the role of forests as life space for plants and animals, as well as relaxing areas for forest visitors and tourists – especially when in urban neighborhood – cannot be covered with the sole incomes obtained through the sale of fuelwood.

The use of timber as construction material is the only chance to save the world's forests. Two hundred years ago, forestry was financed through the use of timber, not only in rural zones, but especially in the cities where all houses – up to 10 stories – were built with timber joist floors and wooden roofs. The role the forests of the future will have to play for mankind and environment cannot be assured only through environmental protection – as little as the role of future cities can be granted through the sole protection of monuments (Figures 4-7).

8. New Techniques in Timber Construction

The increasing use of timber in the construction depends on engineering developments of timber as a load-bearing material, in order to raise the modest portion of the total construction volume from about 1 per cent to perhaps 2 or 3 per cent (Figure 8).

The criteria of development are: better evaluation of the timber quality, increase of the diversity and better treatment of material varieties, development of new time-saving assembly and construction techniques which allow the highest possible degree of pre-fabrication (Photo 28 & 29).

Techniques for thickwood and composite wood-concrete floors, walls and roofs of the dense housing and public buildings linked with other massive construction materials, as well as quality related high-tech systems, should correct the image of timber and offer a competitive alternative to other materials used in construction (Photos 30-32).

The material selection is no proof for "good architecture". It is, however, an important contribution to the environmental conservation, even if it needs more concentration in the planning phase.

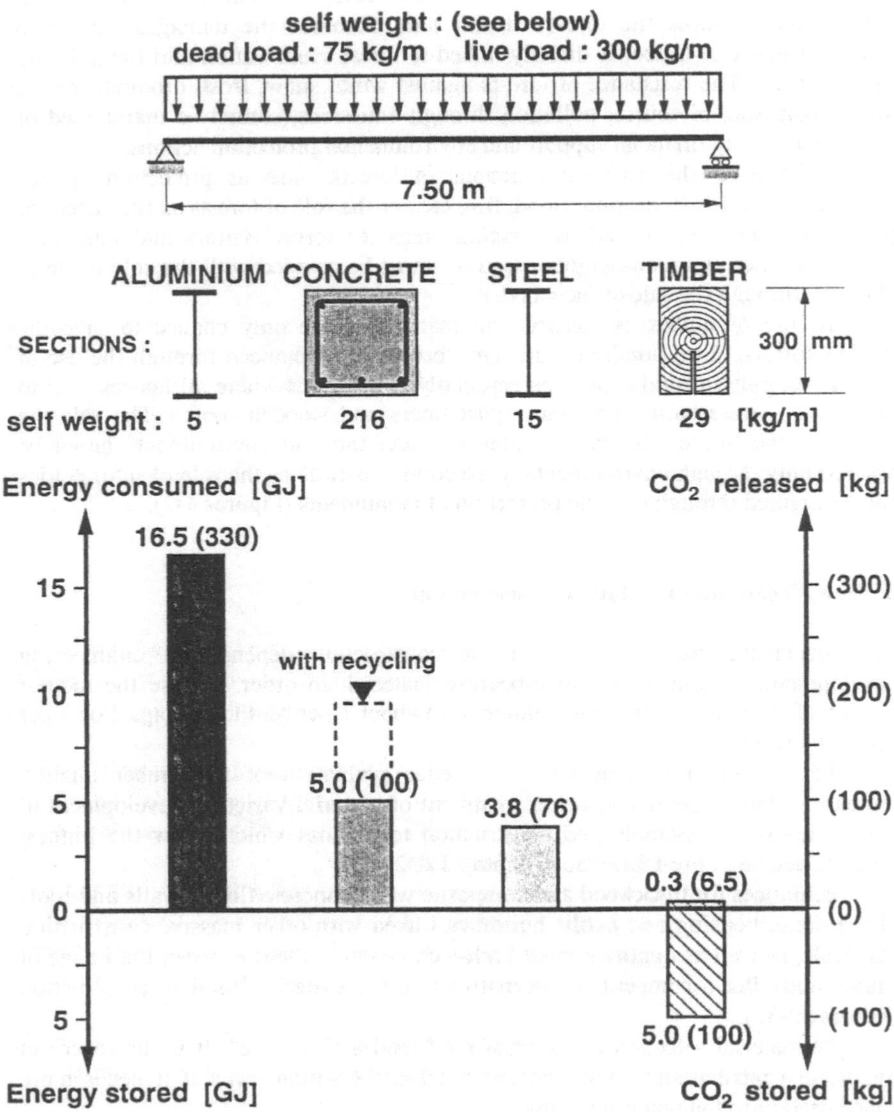

Figure 5. Energy consumed and CO_2 emitted during production of a simply supported beam

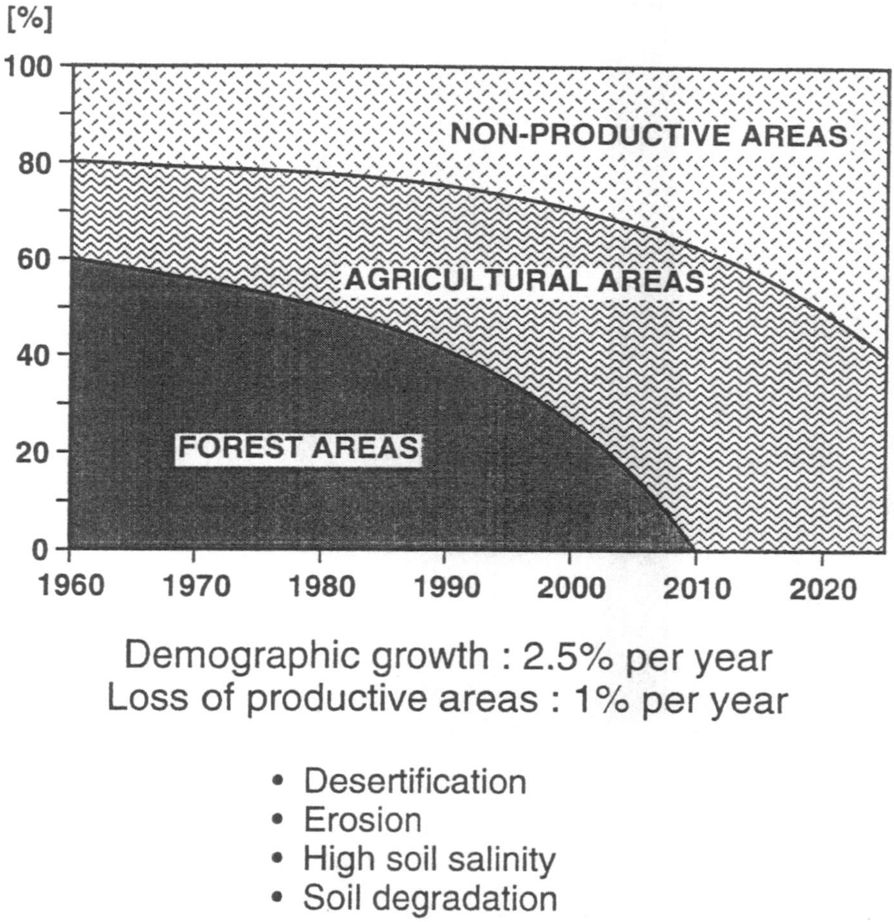

Figure 6. Decrease in the area of forests for the benefit of agricultural and non-productive areas in tropical forest regions

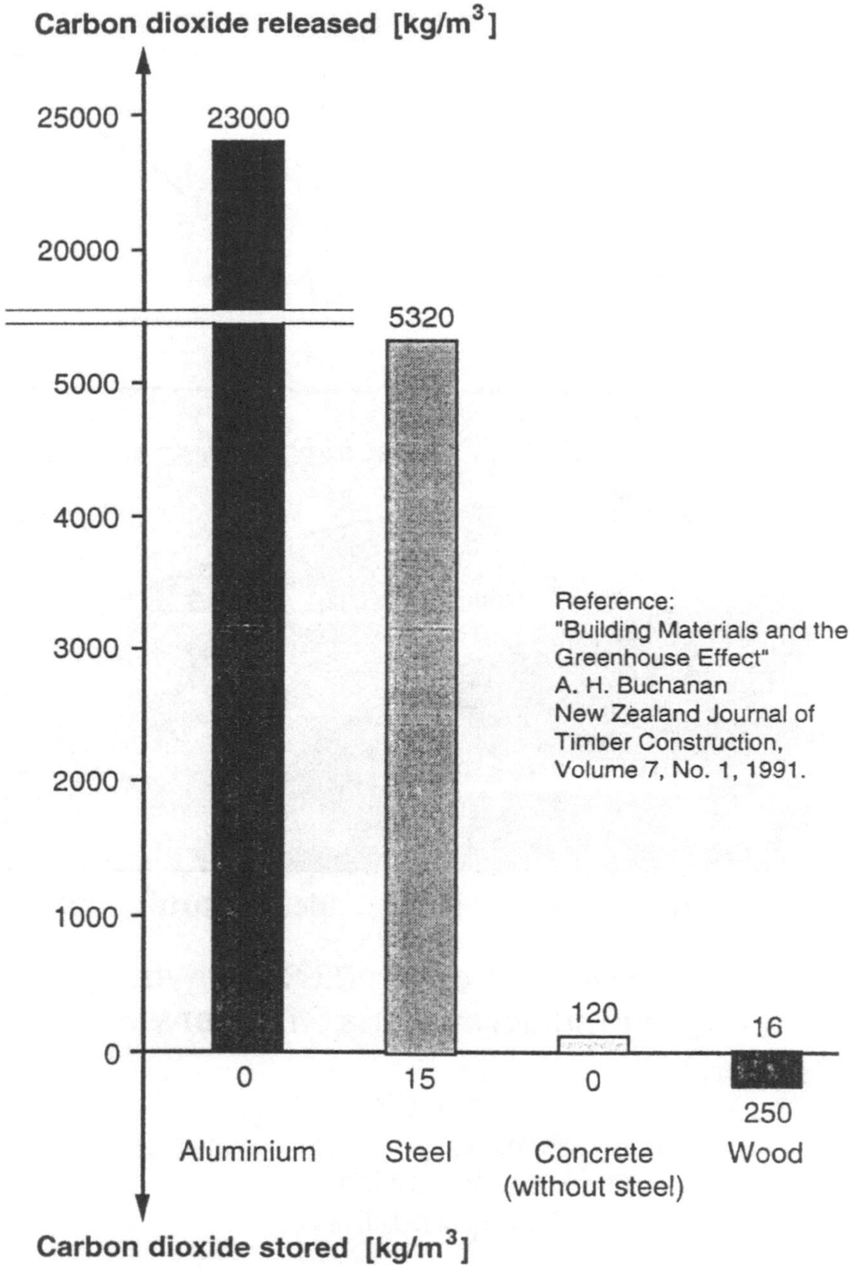

Figure 7. Carbon dioxide emissions for construction materials

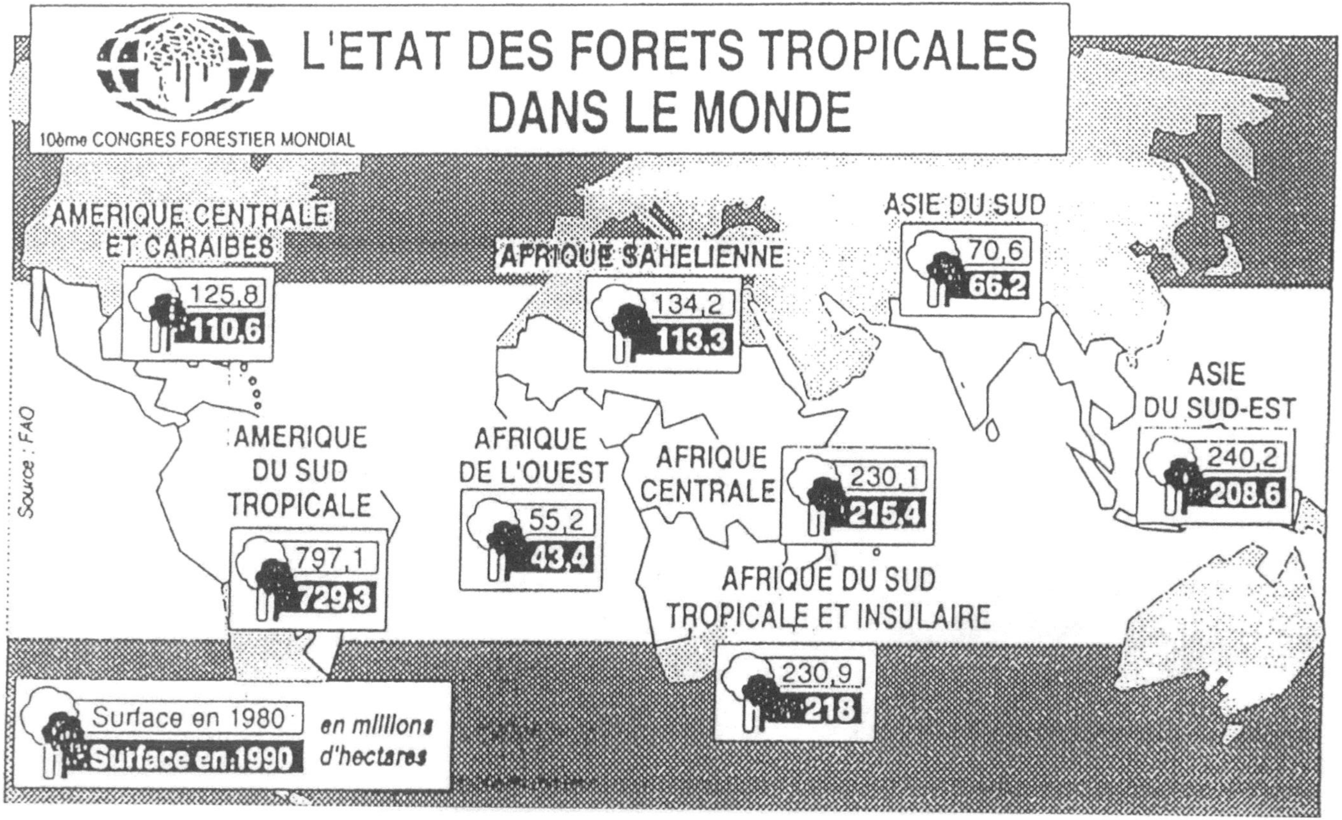

Figure 8. Global loss of tropical forests: comparison between the area of forests in 1980 and 1990 (FAO)

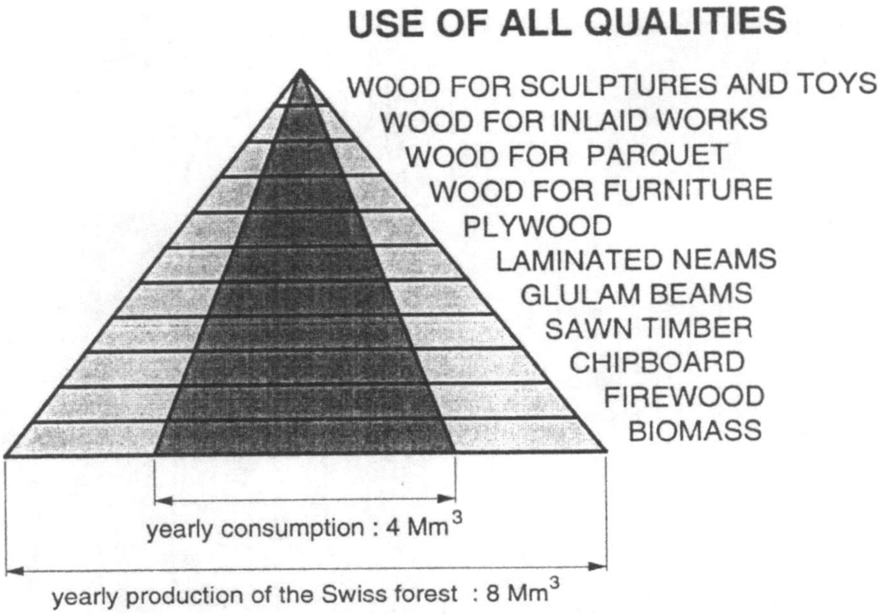

Figure 9. Yearly production and use of different qualities of wood in Switzerland

Photos 28 & 29 (See List of Captions)

Photos 30 & 31 (See List of Captions)

Photo 32 (See List of Captions)

9. Concluding Notes From NATO-ARW Discussions

It is very important to show the authorities and the public that forests can produce a construction material that is not only renewable, but can generate income and employment in regions normally with high unemployment rates in Central Europe. The private forests managed by the Thurn and Taxis family in Bavaria (Germany) are a good example of professional use of the forest resources with long term options in order to insure incomes for replantation and healthy forest management.

The sustainable use of forests is beneficial for the environment of the planet and must be put forward to the public. In order to open up new markets for forest products, it is essential to continue developing timber construction technologies. This workshop showed that some people involved with forest management are probably not well informed about the many possibilities already available for engineered structures made of wood. The transfer of know-how in timber construction must be increased in order to get more decision makers aware of these possibilities.

10. List of Captions for Photos

Photo 1. Sylvatest®. Hand-held ultrasound equipment for nondestructive determination of modulus of elasticity and strength of wood

Photo 2. Round poles timber bridge in Visp (Switzerland) *

Photo 3. Guesthouse of the Youth Training Center in Windberg, Lower Bavaria (Germany). Constructed out of glulam beams, plywood and larch cladding (Arch. T. Herzog) **

Photo 4. Several story half-timbered building made of rectangular timbers with projecting roof as weather protection (Germany) **

Photo 5. Industrial hall as a balcony-style rack made of glulam beams with rectangular wood pieces in competition with other building materials for a metalworking shop, Visp (Switzerland) *

Photo 6. Half-timbered frame of several stories with concrete joints of F-60 fire resistance for a school in Chaumont, France ***

Photo 7. Composite ceiling of nailed laminated boards. The simplest method for carrying out the improvement of old buildings in Etoy, Switzerland *

Photo 8. Combination of nailed laminated ceiling on a glulam beam support in Fribourg, Switzerland *

Photo 9. Forest University in Weihenstephan, Germany. Institute building of several stories with round timber, glulam beams and special lines, intended in particular to comply with noise protection requirements (Arch. Klein, Sänger, Scheer) **

Photos 10 & 11. Church in Nürnberg, Germany. A youth center built within an existing church with 90 minute fire-resistance. Made of composite wood and concrete **

Photo 14. Composite timber and concrete ceiling and walls of nailed laminated planks, under construction in Clarens, Switzerland (Arch. G. Bellmann) *

Photos 12 & 13. Timber and concrete composite ceiling with « Hilti » dowels and 7.9m span width, under test as a ceiling for a four-story school building in Triesenberg, Liechtenstein (Arch. H. Ospelt, coworker M. Freund)*

Photos 15 & 16. Church in Watzeneck, Austria. Made of glulam beams with existing building and adapted building shapes (Arch. W. Ritsch, R. Wäger and S. Wäger) **

Photos 17 - 19. Bridge over the river Simme, Wimmis near Thun, Switzerland, as an example of modern construction linked to the landscape, with glulam beams *

Photo 20. Building for the German Postoffice, on the Brauneck, Upper Bavaria (Arch. K. Maurer, G. Schloffer) **

Photo 21. Recycling hall, Vienna, Austria. Suspended roof made of glulam beams, about 180m diameter **

Photo 22. House in Pullach, Germany, with the integration of active and passive measures for saving energy (storage wall and transparent thermal insulation elements), (Arch. T. Herzog and M. Volz) *

Photo 23. Theater in Schwanden, Switzerland, with underspanned pressure supports, made of glulam beams under 800 kg/m^2 snow load *

Photo 24. Stick bundle system made of glulam beams for the refectory of the Technical University of Münich in Weihenstephan (Arch. Unibauamt Weihenstephan) **

Photos 25 - 27. Spherical dome constructed using the nailed laminated beam method for an auditorium over 25x25m at the Federal Institute of Technology, Lausanne, Switzerland *

Photos 28 & 29. High-Tech. Timber construction, 100% prefabricated, 3.5 hours to erect using a helicopter in La Dôle, Switzerland (Arch. V. Mangeat) *

Photos 30 & 31. Building « Boucle de Gilamont » in Vevey, Switzerland. Five-story apartment building. Wood-concrete decks and load-bearing walls made of composite structures of planks and concrete (Arch. assoc. G. Bellmann & M. Pedrolini)

Photo 32. "Waldweg" building in St-Ursen, Fribourg (Switzerland), with composite timber and concrete ceiling (Arbeitsgemeinschaft H. Binz, CH-Stimmen; D. Kurz, K. Vetter and S. Schärer, CH-Bern)

11. References

Natterer, J., Herzog, T., Volz, M. (1991) *Holzbau Atlas Zwei*. München: Arbeitsgemeinschaft Holz e.V. und Institut für internationale Architektur-Dokumentation.

Natterer, J., Herzog, T., Volz, M. (1994) *Construire en bois 2*. Lausanne: Presses polytechniques et universitaires romandes.

Engineering concept work on projects by Prof. J. Natterer in collaboration with

* Bois Consult Natterer, Etoy, Switzerland

** Planungsgesellschaft Natterer and Partner, Münich, Germany

*** I.C.S. Bois, J. Natterer, M. Flach, les Lanches, France

Position Papers

Working Group 1

"Environmental Factors in Forest Depletion"

EVALUATING THE RISK OF AIR POLLUTION TO FORESTS IN CENTRAL AND EASTERN EUROPE

D.S. ELLSWORTH
Department of Applied Science
Brookhaven National Laboratory
Upton, N.Y. 11973 U.S.A.
ellswort@bnl.gov

J. OLEKSYN
Polish Academy of Sciences
Institute of Dendrology
Parkowa 5
PL-62-035 Kórnik, Poland

Abstract

Foliar damage to trees by air pollution in Central and Eastern Europe has been a major scientific and political issue. Emissions of toxic gases such as sulfur dioxide and nitrogen oxides can have wide-ranging effects on local and regional vegetation that can be compounded by other environmental stresses to plant growth. Since uptake and physiological effects of these gases on tree leaves are largely mediated by stomata, surrogate methods for estimating pollutant conductances into leaves and forest canopies may lead to risk assessments for major vegetation types that can then be used in regional planning. Management options to ameliorate or mitigate air pollutant damage to forests and losses in productivity are likely to be more difficult to widely implement than on-the-stack emissions abatement. Informed management and policy decisions regarding Central and Eastern European forests are dependent on the development of quantitative tools and models for risk assessment of the effects of atmospheric pollutants on ecosystem health and productivity.

1. Introduction

In the latter half of the 20th century, Central and Eastern European nations have experienced an unprecedented array of environmental problems affecting the ecological integrity of forests of the region. Among these problems, the issue of declining forest health has attracted considerable scientific and public interest since forests in the region have substantial economic, ecological and social value. In the 1980's, reports of possible forest

R. M. Gutkowski and T. Winnicki (eds.), Restoration of Forests, 121–131.
© 1997 *Kluwer Academic Publishers.*

damage in Central and Eastern Europe and the implication of air pollution as a causative agent of this damage (Hinrichsen, 1986; Mazurski, 1986) heightened awareness of the effects of air pollution on forest productivity and sustainability. No specific air pollutant has been implicated as a causative agent of forest damage in the region, in part because the observed damage symptoms (such as leaf chlorosis and necrosis or crown dieback) are general for a wide variety of stress agents such as nutrient deficiencies, pathogens, severe drought and temperature anomalies, etc. (Schulze *et al.*, 1989). It is now recognized that air pollution stress most likely affects forest trees in interaction with other contributing factors such as reduced levels of plant resources for growth (e.g., low nitrogen or water availability) or unfavorable biotic conditions (e.g., presence of herbivores or pathogens). Hence air pollutants can be viewed as additional stress agents that can limit tree growth and productivity in concert with other stresses (Winner, 1994). The intrinsic variability in environmental conditions in space and time, coupled with the variable distribution of pollution levels, produces a mosaic of possible impact levels of air pollution on the landscape that can be difficult to assess. The problem becomes more accentuated when multiple pollutants are considered along with other anthropogenic pressures such as improper land management practices.

A number of detailed reviews of environmental problems associated with forest damage by air pollution have appeared recently (Kandler & Innes, 1995; Taylor *et al.*, 1994). Instead, we will here focus on topics of special interest for forest management decisions in Central and Eastern Europe with a perspective on possible strategies for mitigating effects of air pollutants on forests.

2. Nature of the Problem

Energy production technologies represent the largest sources of pollutant emissions in Central and Eastern Europe. Fossil fuel use and combustion technologies, particularly in the developing economies of Central and Eastern Europe, result in widespread emissions of sulfur dioxide (SO_2), nitrogen oxides (NO_x) and carbon dioxide (CO_2), carbon monoxide and volatile organic carbon compounds. For instance, in Poland 77% of the primary energy needs are met by coal-burning, which accounted for 64% of the estimated SO_2 emissions from that nation (approximately 3 million tons of SO_2; Figure 1A) and over 80% of CO_2 emissions (457 million tons) in 1991 (Anonymous, 1992). Much of the coal used for electricity generation in Poland and other Central and Eastern European nations is high-sulfur, "brown" (lignite) coal. In the so-called "Black Triangle" region from southern Poland to the tri-nation border area with Germany and the Czech Republic, lignite coal extraction at a rate of 200 tons year^{-1} over scores of years, and the operation of 17 major regional power plants have resulted in high levels of SO_2 pollution believed to have contributed to forest damage in these nations (Oleksyn & Reich, 1994). Mean annual SO_2 concentrations in Upper Silesia in excess of 35 ppb (100 μg SO_2 m^{-3} air) were documented for 15 towns in the region in 1990-91 despite a severe economic crisis accompanied by reductions in SO_2 emissions from industry at that time (Anonymous, 1992). Similar annual SO_2 levels were reported in the Saxony region of former East Germany in the 1980's (Slovik *et al.*, 1996).

Of 105 European cities of more than 500,000 inhabitants reviewed by the

European Environmental Agency (EEA), over half of those in Central and Eastern Europe are estimated to exceed WHO Air Quality Guidelines (AQG) for pollutant ozone (O_3) in the summer months and two-thirds of the cities exceed the AQG for SO_2 and particulates in the winter (Stanners & Bourdeau, 1995). However, the problem is far greater than local in the sense that long-range air pollutant transport results in considerable pollutant deposition across regional and national boundaries and into rural areas. Because forests display very large surface areas for pollutant deposition or uptake into leaves, they have the capacity to sustain large doses of pollutants and large impacts on tree function resulting from long-term exposure to air pollutant emissions. Considering that much of the projected future forest products need in Europe (to the year 2020) will be supplied by forests in Central and particularly Eastern Europe (Kuusela, 1994), it is critical to assess risks to forest productivity from air pollutants and develop managment strategies to maintain sustainable yields in stands in the region.

3. Risks to Central and Eastern European Forests by Air Pollution

Elevated levels of the trace gases SO_2, NO_x, CO_2, and O_3 resulting from industrial emissions have well-documented physiological effects on trees (Keller, 1984; Ceulemans & Mousseau, 1994; Matyssek *et al.*, 1995). With the exception of CO_2, these trace gases are toxic to plants and cause reductions in tree growth which for coniferous trees can range between approximately 5 and 25% at realistic exposure levels, e.g. chronic levels that forests experience in industrialized areas (Matyssek *et al.*, 1995). These gases enter leaves almost exclusively through the stomata and impact the internal physiology of the tree through effects on mesophyll tissue. The resultant reductions in photosynthetic processes in tree leaves are the primary mechanism of direct effects of these pollutants on tree growth and productivity. Indirect effects of these pollutants arise from the chemical changes that emitted pollutants undergo in the atmosphere resulting in O_3 production and wet and dry deposition of sulfur and nitrogen compounds with subsequent cascade effects on various components of the forest ecosystem (Schulze *et al.*, 1989; Taylor *et al.*, 1994).

Surveys of forest damage attributable to environmental change via industrialization have been conducted in most European nations under the auspices of the EEA and the United Nations Economic Commission for Europe (UNECE). In Germany, damage level to Picea abies canopies during 1988-1991 was strongly correlated with annual SO_2 dose (Slovik *et al.*, 1996). Poland's survey in cooperation with the Commission of European Communities in 1989 showed 32% of trees were considered "damaged" in terms of putative loss of foliage from trees (Figure 1B), and its own national forest health survey indicated at least moderate acute damage symptoms on forests covering 270,000 ha in that year. In southern Poland alone, 46,000 ha of moderate to severe forest damage to Picea abies stands was noted in the 1980's in the Sudety Mountains resulting in large volumes of standing dead timber (Mazurski, 1986; Oleksyn & Reich, 1994). Poor management has contributed to this forest decline, since forests in much of the region were harvested in the 19th century and replaced with Picea abies in even-aged stands. While air pollution was certainly not the unique cause of this phenomenon, there is much evidence suggesting that chronic pollutant loads in the region were a major contributing factor in this forest decline.

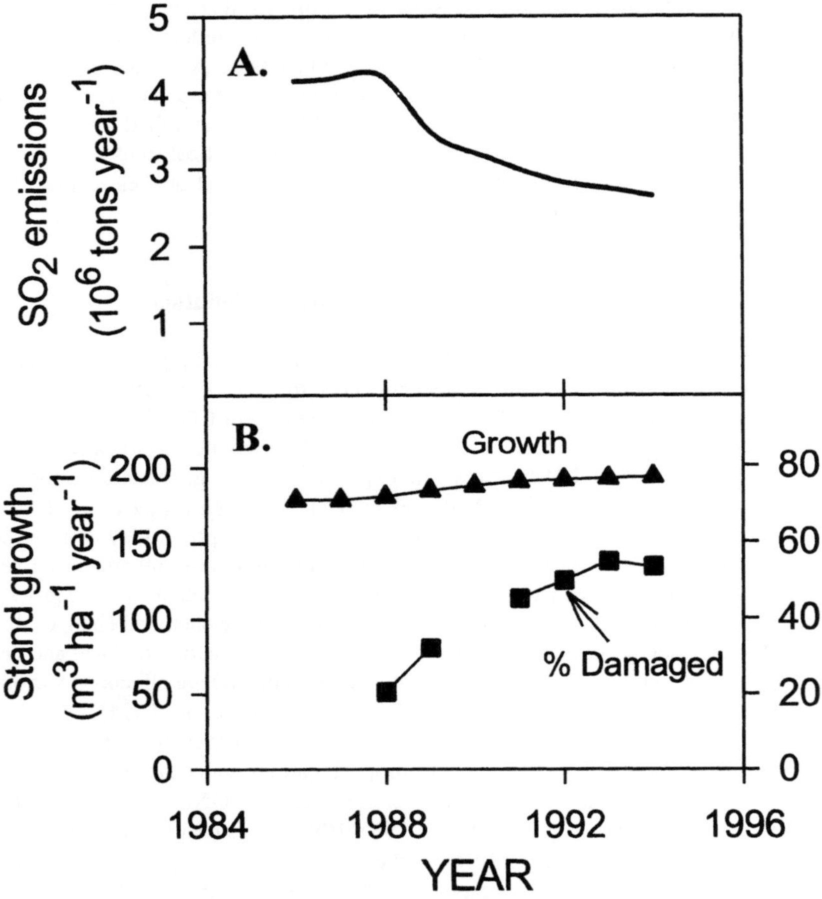

Figure 1A. Time trend in annual emissions of SO_2 from Poland calculated from industrial parameters and from sulfur content of combustion fuels (Anonymous, 1995; Anonymous, 1992; and previous statistical yearbooks for Poland). Note the decline in emissions beginning in the late 1980's as a result of economic recession. *Figure 1B.* Time trend of stand volume growth (all species) in state-owned forests in Poland, along with the trend of the proportion of trees surveyed that exhibited putative damage, defined as foliage loss > 10% from visual assessment. Data from a network of over 2000 plots throughout the country from the Institute of Environmental Protection, Forest Research Institute, and the Ministry of Environmental Protection, Natural Resources and Forestry, Warsaw, Poland.

In the 1992 European survey, 24% of trees on approximately 184 million ha in Europe were considered "damaged" (e.g., loss of foliage assessed at more than 25% of the crown; Commission of European Communities, 1993). The damage levels were significantly higher than average for many nations in Central and Eastern Europe. While large-scale multinational surveys involving measuring very large numbers of trees will reduce error significantly, there can be bias in site selection and in the subjective evaluation of foliage and crown conditions in field surveys (Wilmot *et al.*, 1995). Also, it is difficult to translate results from such surveys into more relevant terms such as losses in volume growth or biomass production. As illustrated by the data in Figure 1 for Poland, a simplistic approach to estimating pollution effects on forest production is not possible, since at a national level the data exhibit a dramatic decline in SO_2 emissions in the late 1980's despite an increase in the proportion of stands exhibiting damage symptoms. While the volume growth data and the forest damage surveys in Figure 1B were not conducted at the same time or on the same stands, it is clear that a closely coordinated, quantitative and less subjective approach is needed to be able to evaluate the risk of air pollution to forests in Poland and among Central and Eastern European nations.

A more quantitative approach to assess the risk of impacts of air pollutants to forests in Central and Eastern Europe may be possible using the 'water-vapor surrogate' method whereby atmospheric pollutant uptake is calculated from measurements of ambient air pollutant concentrations, canopy conductance to water vapor, and substitution of the water flux out of a forest canopy by the pollutant flux into the canopy using biophysical principles (Matyssek *et al.*, 1995). Since data on ambient air pollutants and climate and forest water balance exist for many areas, there is a potential for the development of analytical tools for predicting vegetation risk to exposure to air pollutants that can then be used in management and regional planning efforts. While the dose-response characteristics of many tree species have not been evaluated, the water vapor surrogate approach may be applicable to species with high dominance in forested landscapes once sufficient data become available. Moreover this approach will be most valuable when linked with forest growth models and stand productivity estimates, and has the advantage that extensive field measurements at many sites are possible. A variant of the water vapor surrogate approach was applied by Slovik *et al.* (1996) to Norway spruce in Germany, demonstrating a parallel decrease in ambient SO_2 dose and crown damage related to mandatory regulation of SO_2 emissions from industry in the late 1980's. Hence this approach shows promise for widespread application, and it is currently under evaluation by the European Monitoring and Evaluation Program (EMEP/ECE) for spatial characterization of regional O_3 risk to vegetation.

Despite much indirect evidence of the impact of air pollutants on forest condition and physiological function, recent assessments suggest that roundwood production in Europe has increased over the past decade (Kuusela, 1994). This appears to be largely a result of afforestation and to a lesser extent more efficient harvesting technology, the maturing age of European forests, and possible growth enhancements by atmospheric nitrogen deposition and increasing atmospheric CO_2 concentration. While the latter two conditions may to some degree mitigate the effects of other environmental pollutants by stimulating tree growth, these enhancements may be transient until the onset of critical limitations by other resources such as water or mineral nutrients (Hinrichsen, 1986; Schulze, 1989; Ceulemans & Mousseau, 1994). Although there may be some technical difficulties in assessing the economic impacts of air pollution on forest production (since 'damaged' and unaffected stands are rarely located in otherwise completely comparable environments), recognition by governments and industry of quantifiable impacts on forests is an important precursor to policy or management strategies for maintaining forest health and productivity.

4. Options for Mitigating Air Pollution Losses

4.1. FOREST FERTILIZATION

Fertilization of forest stands has a long history as a management alternative. While this strategy has been applied most successfully in intensively-managed, short-rotation forests used in industrial pulpwood production, it has also been widely suggested as an appropriate ameliorative strategy for forests undergoing growth declines (Hüttl, 1989). More specifically, forest liming or base cation fertilization have been proposed to ameliorate symptoms of nutrient deficiency in forests where acidic deposition and elevated ozone may have promoted this condition through elevated base cation leaching from foliar surfaces or from surface soil horizons (Zöttl &Hüttl, 1986; Evers & Hüttl, 1990). This approach would also be useful in areas where soil fertility has been degraded by logging and soil erosion, or by toxic metal contaminants from mining and smelters. In Germany, a national program in the 1980's undertook the large-scale liming and fertilization of over 1 million hectares of damaged forests at a rate of 2-3 tons ha^{-1} lime or amended lime (Hüttl & Zöttl, 1993). However, long-term data from liming trials have yielded mixed results indicating that this strategy may not be a long-term solution (Hüttl & Zöttl, 1993; Wilmot *et al.*, 1996). Frequently, heavy applications of lime (1 ton ha^{-1} or greater) are required on acid soils to sufficiently alter soil pH such that nutrient availability can be elevated, or to overcome adsorption onto soil colloidal surfaces (Wilmot *et al.*, 1996). Forest liming at such rates may have detrimental effects on the rhizosphere (Persson & Ahlstrom, 1990), soil microbial processes and groundwater quality (Kreutzer *et al.*, 1989; Marschner *et al.*, 1992). In many cases, the effects of chemical fertilizers can be short-lived because much of the fertilizer is leached out of the system, or causes an initial growth stimulation that may dilute foliar nutrients and thus exacerbate the initial condition once soil nutrients return to their initial level.

Primary operational factors limiting the utilization of fertilization as a management alternative include the prohibitive cost of purchasing and applying fertilizers in forests, the potential need for application on vast land areas, the lack of diagnostic guidelines for the type of fertilizer mixture needed, the limited access to some forest areas for fertilizer application, and lack of knowledge of the efficacy of single or multiple fertilizer applications to a variety of sites. Because of these considerations, the use of fertilizers in European forestry is rare, except in limited areas of the U.K. and Germany. Economic considerations alone have limited the use of fertilization in many forest systems, and are likely to be paramount in dictating whether this approach will be widely applied in Central and Eastern European forests.

4.2. REFORESTATION

Extensive land conversion from forests to agriculture has occurred in Central and Eastern Europe as a result of development and expansion of agriculture, particularly in the last 100 years. The increasing fragmentation and over-exploitation of forests, coupled with forest damage and loss of productivity by anthropogenic pollutants has severely degraded the biotic resources of the region. Reforestation or natural afforestation of marginal agricultural lands can help restore the ecological integrity of former forest sites. Because of the nature of anthropogenic disturbances in many locations, some candidate sites for reforestation may need to be planted with species or genotypes that are tolerant of environmental stresses such as drought, acid soils, soil contamination by toxic metals, and elevated atmospheric pollutant loads in order to assure successful plantings.

Reforestation efforts most commonly involve coniferous trees in Europe. Some provenances and genotypes of Scots pine (Pinus sylvestris) have been found to be especially susceptable to air pollution damage and hence may not be appropriate for reforestation in some regions (Białobok *et al.*, 1984; Oleksyn, 1988). From a number of comparative studies it appears that more productive clones/provenances have greater susceptibility to air pollution, probably due to greater physiological exposure dose via pollutant uptake through stomata. Hence in Scots pine, more northern provenances appear to be less susceptible to chronic pollution in central Poland than are local provenances (Oleksyn, 1988). In some areas subject to acute pollution exposure, Austrian pine (Pinus nigra) has been adopted for reforestation efforts because of its low inherent physiological uptake for air pollutants and hence reduced incidence of air pollution damage symptoms (Oleksyn *et al.*, 1987). While caution must be used in selecting species and provenances for reforestation efforts, it appears that native species but not necessarily native genotypes are to be encouraged for reforestation on marginal sites or in areas of chronic air pollution. Ultimately, the management goals for reforestation along with site conditions (including atmospheric and soil pollutant exposure) must dictate what kind of plant material is used.

According to many authors, genetic variation in tree sensitivity to pollution is not a sufficient argument for resistance breeding. There are several reasons for this, such as the forseeable exhaustion of the known resources of lignite and hard coal (the main source of pollution in Central and Eastern Europe), the lack of stability of the chemical composition of emissions through time, the existence of ecological variability that affects tree resistance to pollutants (e.g., variability in soils). Perhaps most important of all are the ethical considerations of plant breeding for air pollution resistance. Resistance breeding cannot be justified as an alternative to emission control, but it is a component of a comprehensive pollution control strategy that can be recommended with caution under very special circumstances (Gerhold, 1977).

4.3. ATMOSPHERIC ENHANCEMENT OF FORESTS

While it is generally not considered to be an 'option' for the future, the possibility of atmospheric enhancements of the condition and growth of forests is of major interest to forest managers and wood technologists. There is little question that a sharp rise in global atmospheric CO_2 concentration as a result of fossil fuel combustion and deforestation has been documented in this century and is expected to continue into the next one (Houghton & Meira-Filho, 1995). Seven European nations including Poland, Russia and the Ukraine are among the 15 countries that emit the largest quantity of CO_2 annually as of 1995. The increase in atmospheric CO_2 concentration has important implications for tree function although the long-term sustainability of the 'CO_2 fertilization effect', the CO_2-induced stimulation of photosynthesis in trees and other plants, is still the subject of much investigation. There is evidence of an increase in subalpine tree growth in the Austrian Alps beyond the simple constraints of climatic variation in this century (Nicolussi *et al.*, 1995), suggesting that some sensitive forest types already exhibit this growth stimulation. Considerable scientific debate has focused on whether CO_2-induced growth stimulation of trees in the field can be sustained over a tree's lifespan without increased biological requirements for other resources such as nutrients (Ceulemans & Mousseau, 1994). Also, interactions between CO_2 and other components of the atmospheric environment of forests in Central and Eastern Europe have not yet been evaluated. It is often suggested that rising atmospheric CO_2 concentration into the future could offset stress-induced losses in tree growth and net production, or losses due to air pollutants such as SO_2 and O_3. Evidence from some experimental studies suggests the inhibitory effects of O_3 on photosynthetic capacity in trees may be exacerbated rather than ameliorated by elevated CO_2 (Kull *et al.*, 1996). It is unlikely that gains in growth resulting from increasing atmospheric CO_2 in future decades will ameliorate effects of other atmospheric pollutants on plant function. Since increasing atmospheric CO_2 is a global prospect for the forseeable future, interactive effects on forests between elevated CO_2 and other air pollutants such as elevated O_3 or SO_2 concentrations will probably still be perceptible in 50 years as growth differences between stands under unpolluted versus polluted conditions in otherwise similar environments. Moreover, effects of elevated CO_2 on other components of forest ecosystems such as carbon and nutrient cycling between trees and the soil may further alter ecological processes in forests that are subjected to local and regional air pollution.

5. Conclusions and Policy Options

Many of the current environmental difficulties facing Central and Eastern Europe in the 1990's can be traced to an emphasis on industrial production in the previous three decades in the region and to poor governmental organization and lack of enforcement of environmental protection laws. With the large-scale governmental changes associated with changes in the ruling structure of many Central and Eastern European nations, there is a strong need for legislative mandates for clean air and environmental protection to reverse previous mismanagement. While there are still a number of pressing research needs regarding air pollutant effects on vegetation, enforceable controls should be implemented in the near future to 'err on the side of environmental protection', given the anti-environment tendency of the past regimes in Eastern Europe in this century. Financial incentives for environmental protection must be created to assure modernization of environmental controls and implement ecologically-sound development. These options are far more tractable to implement than are management options to mitigate chronic and acute air pollution impacts on forest lands. Once it is clearly understood that environmental pollution can lead to losses in forest (and agricultural) productivity and threatens the economic vitality of forest management, then it will become possible to develop regional plans for maintaining the ecological as well as economic vigor of forests in Central and Eastern Europe.

6. Pertinent Observations from the ARW

Based on the presentations at the ARW in this volume, it is clear that air pollution is considered to be the major environmental factor affecting forests in Central and Eastern Europe. There appears to be strong consensus on this issue among scientists from Central and Eastern Europe who are familiar with forestry in the region, almost surprisingly so compared to the North American perspective on the problem.. It was said at the meeting that "We can live from the compound interest on our forests (e.g., growth), not the capital". The fact that air pollution is diminishing this 'interest', and its implications for the future growth of forests throughout Central and Eastern Europe is a perspective that should not be lost on any inhabitant of the region with interests in forestry and forests.

Finally, for one who has been concerned with research to understand the basic physiological and ecological processes of trees that may be altered by exposure to atmospheric pollutants and increasing atmospheric CO_2 concentration, the concept of carbon sequestration by wood utilization offers possibilities to stabilize a component of global atmospheric change. This option appears to have not been fully considered by the UNEP's Intergovernmental Panel on Climate Change (IPCC) in their recommendations to policy makers, and merits serious consideration for implementation in nations with sufficient forest resources and high per-capita emissions of so-called 'greenhouse' gases.

Many participants at the ARW had the sense that the populations of highly developed nations were becoming increasingly marginalized from forests and forestry. This is largely a phenomenon of the increasing urbanization and suburbanization of Europe's population, which results in forestry activities being increasingly carried out on private land and away from the view of the general population. Media attention in Europe has become attracted only to examples of mismanagement of forests, as is also true in North America. The emerging view from the ARW of the importance of education the general public regarding the possible benefits of forestry and forest managment to current and the future populations in the region cannot be overemphasized. A common theme at the ARW was one of optimism regarding forest resources in the face of past mismanagement and environmental problems in Central and Eastern Europe. In the present atmosphere of emerging nations and international cooperation in Europe, we forsee even greater promise for the region's forests as long as those who make management decisions do not overlook environmental priorities.

7. References

Anonymous. (1992) *Rocznik Statystyczny 1992.* Głowny Urzad Statystyczny, Rok LII, Warszawa, Poland. [*Statistical Yearbook 1992.* State Main Statistical Office, Volume 52, Warsaw, Poland].

Anonymous. (1995) *Rocznik Statystyczny 1995.* Głowny Urzad Statystyczny, Rok LV, Warszawa, Poland. [*Statistical Yearbook 1995.* State Main Statistical Office, Volume 55, Warsaw, Poland].

Białobok, S., Oleksyn, J. & Rachwal, L. (1984) Selection of trees and shrubs for forest restructuring in industrial regions. In Grodziński, W.; Weiner, J. & Maycock, P.F. (eds.), *Forest Ecosystems in Industrial Regions*, Springer, Berlin, pp. 239-244.

Commission of European Communities/UNECE. (1993) *Forest Condition in Europe*. Report of the International Cooperative Program on Assessment and Monitoring of Air Pollution Effects on Forests. Results of the 1992 Survey. Commission of European Communities, Direction Générale VI, Brussels, Belgium.

Ceulemans, R., & Mousseau, M. (1994) Effects of elevated CO_2 on woody plants. *New Phytologist* 127, 425-446.

Gerhold, H.D. (1977) Effects of air pollution on Pinus strobus L. and genetic resistance. A literature review. U.S. Environmental Protection Agency, EPA 600/3-77-002, Corvallis, Oregon, 44 pp.

Evers, F.H., & Hüttl, R.F. (1990) A new fertilization strategy in declining forests. *Water, Air, and Soil Pollution* 54, 495-508.

Hinrichsen, D. (1986) Multiple pollutants and forest decline. *Ambio* 15, 258-265.

Houghton, J.T., & Meira-Filho, L.G., editors (1995) *The IPCC Report on Radiative Forcing of Climate Change*. Intergovernmental Panel on Climate Change (IPCC), Cambridge University Press, Cambridge.

Hüttl, R.F.(1989) Liming and fertilization as mitigation tools in declining forest ecosystems. *Water, Air, and Soil Pollution* 44, 93-118.

Hüttl, R.F., & Zöttl, H.W. (1993) Liming as a mitigation tool in Germany's declining forests-reviewing results from former and recent trials. *Water, Air, and Soil Pollution* 61, 325-338.

Kandler, O. & Innes, J.L. (1995) Air pollution and forest decline in Central Europe. *Environmental Pollution* 90, 171-180.

Keller, T. (1984) Direct effects of sulfur dioxide on trees. *Philosophical Transactions of the Royal Society of London Series B* 305, 317-326.

Kreutzer, K., Reiter, H., Schierl, R., & Göttlein, A. (1989) Effects of acid irrigation and liming in a Norway spruce stand (Picea abies L. (Karst.)). *Water, Air, and Soil Pollution* 48, 111-125.

Kull, O., Sober, A., Coleman, M.D., Dickson, R.E., Isebrands, J.G., Gagnon, Z., & Karnosky, D.F. (1996) Photosynthetic responses of aspen clones to simultaneous exposures of ozone and CO_2. *Canadian Journal of Forest Research* 26, 639-648.

Kuusela, K. (1994) *Forest Resources in Europe*, Cambridge University Press, Cambridge, U.K.

Marschner, H., Stahr, K., & Renger, M. (1992) Lime effects on pine forest floor leachate chemistry and element fluxes. *Journal of Environmental Quality* 21, 410-419.

Mazurski, K.R. (1986) The destruction of forests in the Polish Sudetes Mountains by industrial emissions. *Forest Ecology and Management* 17, 303-315.

Matyssek, R., Reich, P.B., Oren, R., & Winner, W.E. (1995) Response mechanisms of conifers to air pollutants. In Smith, W.K. & Hinckley, T.M. (eds.), *Ecophysiology of Coniferous Forests*, Academic Press, San Diego, pp. 255-308.

Nicolussi, K., Bortenshlager, S., & Körner, C. (1995) Increase in tree-ring width in subalpine Pinus cembra from the central Alps that may be CO_2-related. *Trees* 9, 181-189.

Oleksyn, J., Oleksynowa, K., Kozłowska, E., & Rachwal, L. (1987) Mineral content and the sensitivity of black pine (Pinus nigra) of various provenances to industrial air pollution. *Forest Ecology and Management* 21, 237-247.

Oleksyn, J. (1988) Height growth of different European Scots pine (Pinus sylvestris L.) provenances in a heavily polluted and a control environment. *Environmental Pollution* 55, 289-299.

Oleksyn, J., & Reich, P.B. (1994) Pollution, habitat destruction and biodiversity in Poland. *Conservation Biology* 8, 943-960.

Oren, R., & Schulze, E.-D. (1989) Nutritional disharmony and forest decline: a conceptual model. In Schulze, E.-D., Lange, O.L., & Oren, R. (editors), *Forest decline and air pollution: a study of spruce (Picea abies) on acid soils,* Springer-Verlag, Berlin Heidelberg, pp. 425-443.

Persson, H., & Ahlstrom, K. (1990) The effects of forest liming and fertilization on fine-root growth. *Water, Air, and Soil Pollution* 54, 365-375.

Schulze, E.-D. (1989) Air pollution and forest decline in a spruce (Picea abies) forest. *Science* 244, 776-783.

Schulze, E.-D., Lange, O.L., & Oren, R., editors. (1989) *Forest Decline and Air Pollution.* Ecological Studies 77. Springer-Verlag, Berlin.

Slovik, S., Siegmund, A., Führer, H.-W., & Heber, U. (1996) Stomatal uptake of SO_2, NO_x and O_3 by spruce crowns (Picea abies) and canopy damage in Central Europe. *New Phytologist* 132, 661-676.

Stanners, D. & Bourdeau, P., editors. (1995) *Europe's Environment: The Dobříš Assessment.* European Environment Agency, Copenhagen.

Taylor, G.E., Johnson, D.W., & Andersen, C.P. (1994) Air pollution and forest ecosystems: A regional to global perspective. *Ecological Applications* 4, 662-689.

Wilmot, T.R., Ellsworth, D.S., & Tyree, M.T. (1995) Relationships among crown condition, growth, and stand nutrition in seven northern Vermont sugarbushes. *Canadian Journal of Forest Research* 25, 386-397.

Wilmot, T.R., Ellsworth, D.S., & Tyree, M.T. (1996) Base cation fertilization and liming affects nutrition and growth of Vermont sugar maple stands. *Forest Ecology and Management,* in press.

Winner, W.E. (1994) Mechanistic analysis of plant responses to air pollution. *Ecological Applications* 4, 651-661.

Zöttl, H.W. & Hüttl, R.F. (1986) Nutrient supply and forest decline in southwest Germany. *Water, Air, and Soil Pollution* 31, 449-462.

FORESTS AND AIR POLLUTION IN THE CZECH REPUBLIC

V. HENZLIK
Lesprojekt, Forest Management Planning Institute
Nábrezní 1326
250 44 Brandýs n. L.
Czech Republic

1. Introduction

One of the main problems in Czech forestry is its decline due to the effect of air pollution. Although it is a problem that is quite common in Europe, the damage has different history and characteristics in this part of the continent.

Talking about the history, situation, and forecasts of the forest deterioration it is useful to recall some basic data on the Czech forests.

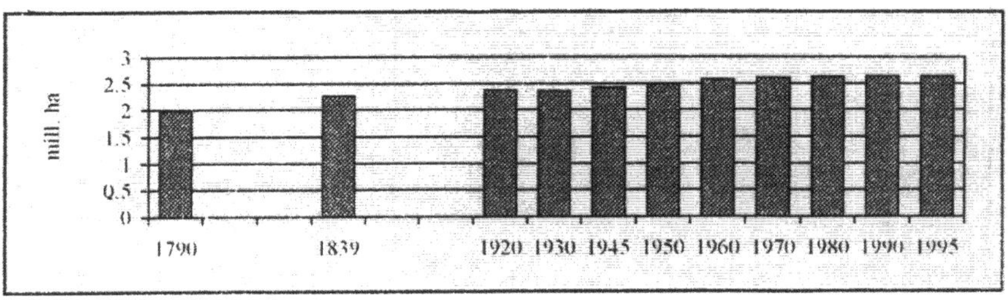

Figure 1. Forest Land Area (1790 - 1995)

About 25 % in the 18th century, the forest coverage has increased up to today's 33.3 %. Over the last three decades, the increment in size of forest land is no longer significant.

R. M. Gutkowski and T. Winnicki (eds.), Restoration of Forests, 133–149.
© 1997 *Kluwer Academic Publishers.*

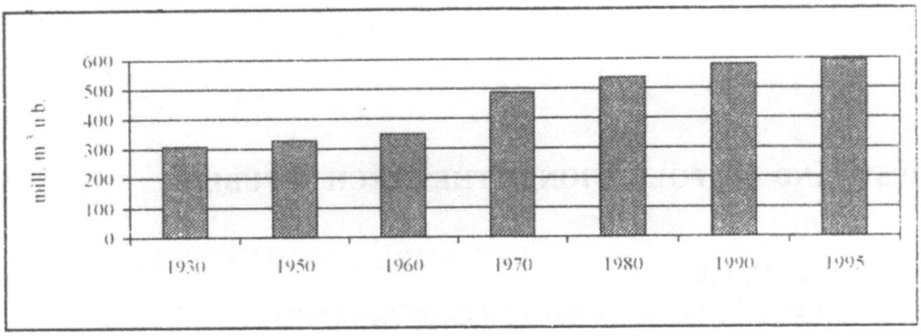

Figure 2. Growing Stock Volume (1930 - 1995)

Rapid growth of registered stock volume is mainly an effect of the following three factors: (i) the current increment starts culminating in large size forests restocked after bark beetle and nun moth catastrophic outbreaks that happened at the beginning of this century and the increment itself is also increasing, (ii) a mean age of the forest stands has increased to 62 years, and (iii) the methods of stock volume assessment are more accurate.

Species	Area		Site index*	Mean age
	ha	%		years
Norway spruce	1 398 498	54.1	3.3	62
fir	24 153	0.9	3.3	76
Scots pine	452 993	17.5	3.9	68
larch	90 043	3.5	2.4	54
other conifers	24 582	1.0	-	-
oak	157 753	6.1	4.4	66
beech	145 557	5.6	3.4	74
birch	76 729	3.0	2.3	43
other non-conif.	174 684	6.8	-	-
conifers	1 990 260	77.0	-	63
non-conifers	554 723	21.5	-	60
total stocked	2 544 992	98.7	-	62

* Mainly Schwappach's yield tables.

Figure 3. Species Composition (1995)

In the course of the last four decades, the percentage of conifers (except of larch) was decreasing.

The share of Norway spruce on the forest composition is a result of forestry policy in the past: the main aim of the forestry was to produce the maximum volume of wood. Although it is a species very sensitive to pollution, spruce will continue being planted, because of site conditions of the Czech forests. Within the vegetation zones 5 to 9, spruce is a natural component of forests. In lower zones, some spruce ecotypes have probably been developed in refuges of the species since the glacial period.

Ecolog. category	Forest vegetation zone										
	0	1	2	3	4	5	6	7	8	9	Total
	% of forest stands area										
extreme	**0.3**	**0.4**	**0.1**	**0.1**	**0.0**	**0.2**	**0.2**	**0.1**	**0.3**	**0.1**	**1.8**
acid	**2.9**	**1.3**	**3.8**	**7.8**	**2.3**	**10.2**	**5.6**	**2.4**	**0.8**	**0.1**	**37.2**
fertile	**0.3**	**1.8**	**4.6**	**10.7**	**3.7**	**12.2**	**2.9**	**0.6**	**0.2**	**0.0**	**37.0**
enriched	**0.0**	**1.6**	**0.9**	**2.2**	**0.6**	**2.3**	**1.2**	**0.2**	**0.0**	**0.0**	**9.0**
gleyic	**0.4**	**1.3**	**0.8**	**1.5**	**3.0**	**2.5**	**2.0**	**0.4**	**0.1**	**0.0**	**12.0**
water-logged	**0.5**	**0.2**	**0.0**	**0.0**	**0.2**	**0.3**	**0.5**	**0.7**	**0.5**	**0.1**	**3.0**
Total	**4.4**	**6.6**	**10.2**	**22.3**	**9.8**	**27.7**	**12.4**	**4.4**	**1.9**	**0.3**	**100.0**

Figure 4. Site Conditions (except military forests)

As shown in Fig. 5, the rotation period has also increased in the Czech forests.

Year								
1920	1930	1940	1950	1960	1970	1980	1990	1995
Rotation period lenght								
93.4	**92.5**	**95.4**	**101.1**	**101.2**	**102.6**	**108.1**	**112.4**	**114.6**

Figure 5. Average Rotation Period (1920 - 1995)

Average rotation period in the production, protection, and special purpose forest is 111.28, 156.28, and 115.34 years, respectively.

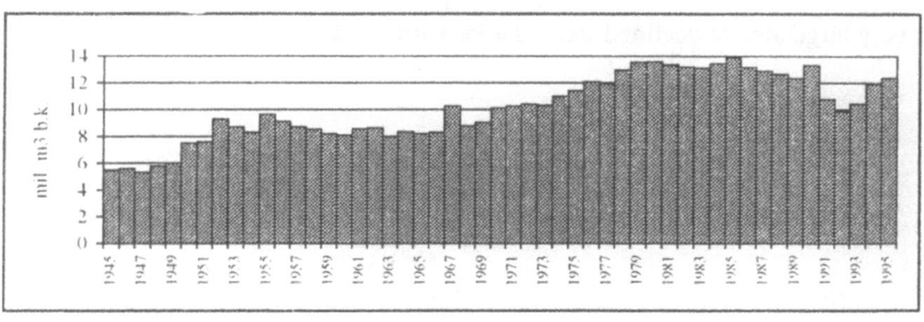

Figure 6. Total Fellings (1945 - 1995)

Volume of fellings is adequate to the cutting possibilities. The high portion of unplanned salvage fellings is however a great obstacle to the stewardship and planned management of forests.

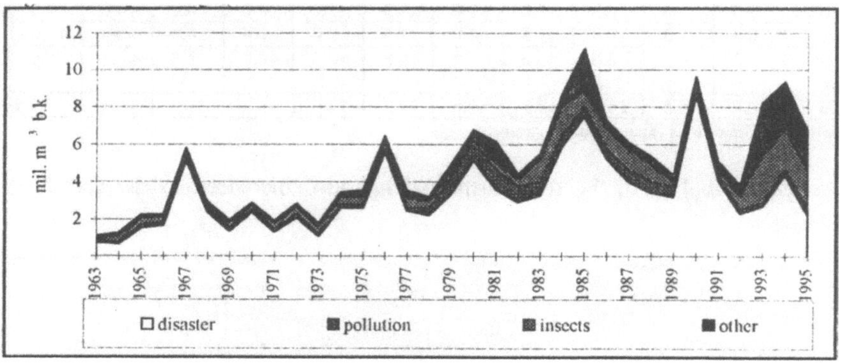

Figure 7. Salvage fellings (1963 - 1995)

A large portion of Norway spruce in the Czech forests composition enabled pollution to kill about 47,300 ha of forest stands since 1958.

Pollution causes losses not only to public forests but also to the other ownership categories. In the course of the national economy transformation, the ownership of the main part of private forests has been restituted. One case is known, the owner of a large forest refused restitution of his holding because of a very large area of declined forest stands within his original property.

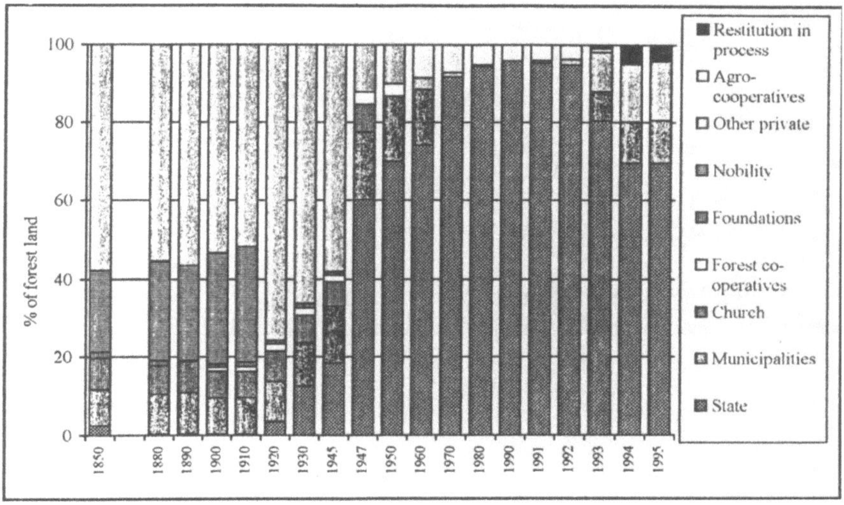

Figure 8. Forest Ownership (1850 - 1995)

According to a very recent agreement of the Czech political parties, an ownership restitution of church forests will be allowed in the near future.

In a big holding, as the state forest in the Czech Republic is, the elevated costs and losses in pollution damaged forests could be covered by an income coming from less damaged parts of the holdings. It is impossible to reach any permanent profit from small lots of forest land. The mean size of a private forest holding in the Czech Republic is about 3 ha. Pollution caused damage creates even more difficulties.

Owner		Number of cases			Restituted, '000ha			Mean size of one holding, ha		
		1993	1994	1995	1993	1994	1995	1993	1994	1995
Municipalities		2637	3 572	3 820	234.9	277.0	289.6	89.1	77.5	75.8
Individuals	< 200 ha	97020	11 9806	127 690	210.4	239.2	241.4	2.2	2.0	1.9
	> 200 ha	171	185	200	135.1	156.4	156.3	790.1	845.4	781.5
	total	97191	11 9991	127 890	345.5	395.6	397.7	3.6	3.3	3.1

Figure 9. Size Structure of Restituted Holdings (1993 - 1995)

A large portion of the Czech forests should respond to demands for non-wood goods and benefits. A percentage of protection and special purpose forests in the Czech Republic has increased (but a great part of originally

protection forests is now overlapped by the pollution damaged forests, a sub-category of the special purpose forests).

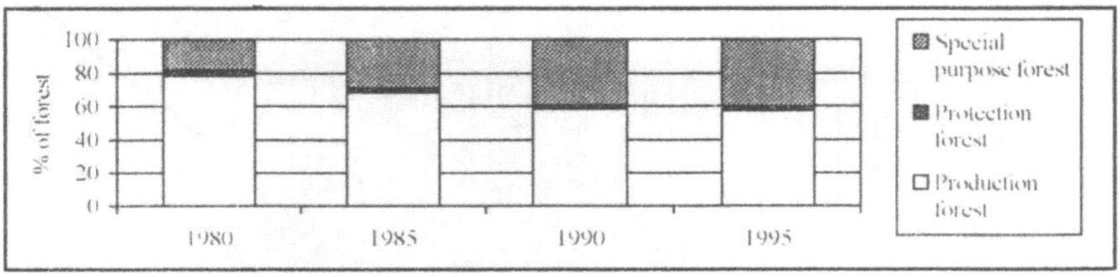

Figure 10. Forest Categories (1980 - 1995)

The following table gives an illustration example how the protected forests and Protected Landscape Regions are endangered by the pollution.

Name	Forest area	Forest coverage	Pollution threat zones			
			A	B	C	D
	ha	%				
CHKO Beskydy	**85 931**	**74.1**	-	**7.6**	**50.8**	**41.6**
CHKO České Středohoří	**31 826**	**29.7**	-	**20.7**	**79.1**	**0.2**
CHKO Šumava	**52 909**	**56.0**	-	-	-	**11.3**
CHKO Jeseníky	**57 914**	**78.3**	-	**16.0**	**68.0**	**16.0**
CHKO Žďárské vrchy	**33 219**	**46.5**	-	-	**43.6**	**56.4**
CHKO Bílé Karpaty	**32 365**	**45.3**	-	-	**1.6**	**9.4**
CHKO Třeboňsko	**33 280**	**47.5**	-	-	-	**30.0**
Šumavský národní park	**55 501**	**81.0**	-	-	-	**8.8**
CHKO Slavkovský les	**34 966**	**54.6**	-	**17.6**	**69.1**	**13.3**
CHKO Křivoklátsko	**39 165**	**62.2**	-	-	**7.8**	**8.9**
CHKO Broumovsko	**16 348**	**39.9**	-	**8.2**	**91.8**	-
Krkonošský národní park	**31 955**	**83.1**	-	**21.0**	**33.4**	**45.6**
CHKO Jizerské hory	**27 968**	**79.9**	**6.1**	**45.4**	**43.3**	**0.2**
CHKO Lužické hory	**17 667**	**50.5**	**0.4**	**5.7**	**82.5**	**11.4**
CHKO Labské pískovce	**23 346**	**77.8**	**3.6**	**18.7**	**61.0**	**16.7**
CHKO Železní hory	**12 511**	**43.6**	-	-	**20.0**	**80.0**
CHKO Kokořínsko	**13 526**	**50.1**	-	**10.0**	**88.0**	**2.0**
CHKO Blanský les	**12 190**	**57.4**	-	-	-	**1.4**
CHKO Orlické hory	**15 607**	**78.0**	-	**38.5**	**61.5**	-
CHKO Český kras	**5 106**	**38.7**	-	**15.0**	**50.0**	**35.0**
CHKO Český ráj	**4 429**	**35.4**	-	-	-	**100.0**
CHKO Litovelské Pomoraví	**5 380**	**56.0**	-	-	-	**100.0**
CHKO Moravský kras	**5 436**	**59.1**	-	-	-	**26.2**
CHKO Poodří	**786**	**9.6**	-	**9.6**	**19.5**	**70.9**
CHKO Pálava	**2 638**	**37.7**	-	-	-	-
Národní park Podyjí	**2 896**	**45.9**	-	-	-	-
CHKO Blaník	**1 265**	**31.6**	-	-	-	**100.0**
Celkem	**656 130**	**57.2**	**0.4**	**9.6**	**37.1**	**22.4**

Figure 11. Landscape Protected Areas and National Parks (1994)

The European and the additional national indicators give an overall idea about sustainability in the Czech forests.

		unit	1980	1990
1.1	Forest land area	´000ha	**2 623**	**2 629**
1.2 a)	Total growing stock	mill. m^3 u.b.	**536**	**564**
1.2 b)	Average growing stock	m^3 u.b. per ha	**204**	**214**
1.3	Carbon storage in forest stands	´000kg per ha	**76**	**79**

2.1	Total deposition of main pollutants on permanent plots	**			
2.2	Defoliation changes - percentage of sample trees (ICP-Forest)	unit	1992	1993	1994
	class 2	%	**51.8**	**47.2**	**53.8**
	--classes 3 + 4	%	**4.6**	**5.8**	**5.9**
	total classes 2 to 4	%	**56.4**	**53.0**	**59.7**

		unit	1980	1990
2.3	Damage to forest	unit	1980	1990
	serious damage by insects and diseases	´000ha/year	**116**	**80**
	serious damage by fire	´000ha/year	**0,16**	**0,98**
	serious damage by wind	´000ha/year	*	*
	serious damage by wind	´000m^3/year	**9 370**	**9 219**
	serious damage to regenerated stands by game	´000ha/year	*	**21**

3.1	Balance between increment and logging	unit	1980	1990
	total increment	mill. m^3 u.b./year	**17.1**	**17.0**
	total logging	mill. m^3 u.b./year	**13.6**	**13.3**
	balance	mill. m^3 u.b./year	**+ 3.5**	**+ 3.7**
3.2	Share of forest managed according to a management plan	%	**100**	**100**
3.3	Non-wood goods and benefits bag of red deer	´000 pieces	**10.9**	**20.8**

4.1	Forests within regions and localities of nature protection (total area)	unit	1980	1990
	national parks	'000ha	38	111
	protected landscape regions	'000ha	999	1 000
	small protected areas	'000ha	41	67
4.2	Change of the endangered species number		**	
4.3	Forests managed for rescue and use of the forest trees gene resources	unit	1980	1990
	total area of domestic species	'000ha	144	260
	total area of introduced species	'000ha	0.49	0.51
4.4	Share of mixed stands from total forest land area (assessment)	%	20	20,5
4.5	Share of natural regeneration from total forest regeneration	%	3.6	2.6

		unit	1980	1990
5.1	Total soil protection forest area	'000ha	102.1	113.6
5.2	Forests within water protection zones (1st degree)	'000ha	11.4	17.3

		unit	1980	1990
6.1	Share of the forestry on the GNP	%	1,0	0,8
6.2	Area of forests accessible for recreation per capita	ha	0,2	0,2
6.3	Number of employees in forestry	individuals	51 300	57 700

* Data missing. **Not recorded.

Figure 12. European Indicators of the Sustainable Management of Forests (CZ)

Annual share of salvage felling on the total fellings	%	63.5

Share of broadleaves on afforestation and reforestation carried out over the reporting year (1995)	%	27.4

Areal share of broadleaves in the age classes								
Age class/ age in years								
1 1-10	2 11-20	3 21-30	4 31-40	5 41-50	6 51-60	7 61-70	8 71-80	9 81-90
17.6	21.1	24.7	30.1	31.6	26.9	21.5	18.0	14.4
Age class/ age in years								
10 91-100	11 101-110	12 111-120	13 121-130	14 131-140	15 141-150	16 151-160	17 161 +	
14.9	17.3	19.8	22.9	26.0	33.9	42.1	41.9	

Figure 13. National Indicators of Sustainable Forest Management

2. Air Pollution and Deposition of Pollutants

Except for some pollutants from mobile sources (transportation), a drop in the pollution ammount has been recorded in recent years resulting from temporary reduction of industrial activities during the transformation of the national economy and, in the same time, being an effect of emission cleaning.

Total Emission	1990	1991	1992	1993	1994	1995*
'000 metric tons / year						
Dust	**631.4**	**591.6**	**500.7**	**441.3**	**346.7**	**319.2**
SO_2	**1 875.7**	**1 775.6**	**1 538.2**	**1 418.7**	**1 270.1**	**1 284.1**
No_x	**531.6**	**514.6**	**487.2**	**364.1**	**235.4**	**239.3**
CO	**680.5**	**891.7**	**835.1**	**756.2**	**741.0**	**768.5**
C_xH_x	**152.7**	**154.7**	**132.7**	**131.7**	**127.8**	**124.7**

* Assessment

Figure 14. Emission of Main Pollutants from Stationary Sources (1990 - 1995)

Total Emission	1990	1991	1992	1993	1994	1995*
'000 metric tons / year						
Dust	**8.5**	**8.2**	**8.4**	**8.5**	**8.0**	**6.8**
SO_2	**5.7**	**5.6**	**6.2**	**6.8**	**7.5**	**7.5**
No_x	**183.9**	**173.3**	**182.7**	**190.7**	**192.9**	**192.9**
CO	**318.9**	**287.0**	**298.0**	**296.8**	**285.4**	**266.4**
C_xH_x	**23.8**	**22.5**	**23.9**	**25.2**	**25.8**	**24.9**

* Assessment

Figure 15. Emission of Main Pollutants from Mobile Sources in the Transportation Sector

Heavily pollution loaded regions did not change, it is still the northern part of the Bohemia (Ore Mountains - *Krusne hory, Erzgebirge*) where the highest concentration of pollutants is registered. Local incidental concentrations continue to be unacceptable in some locations. Following are some examples of a monitoring results on the pollution concentration and deposition.

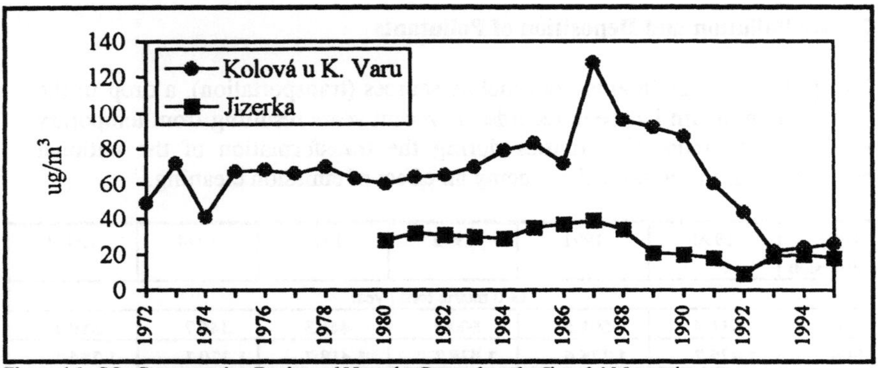

Figure 16. SO_2 Concentration Registered Near the Ore and on the Jizerské Mountains

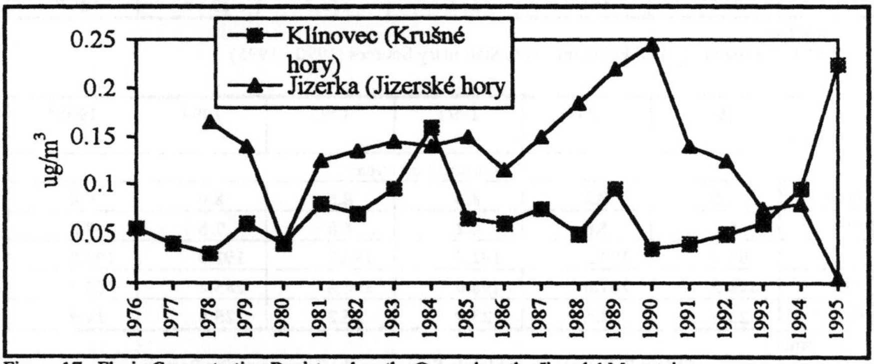

Figure 17. Florin Concentration Registered on the Ore and on the Jizerské Mountains

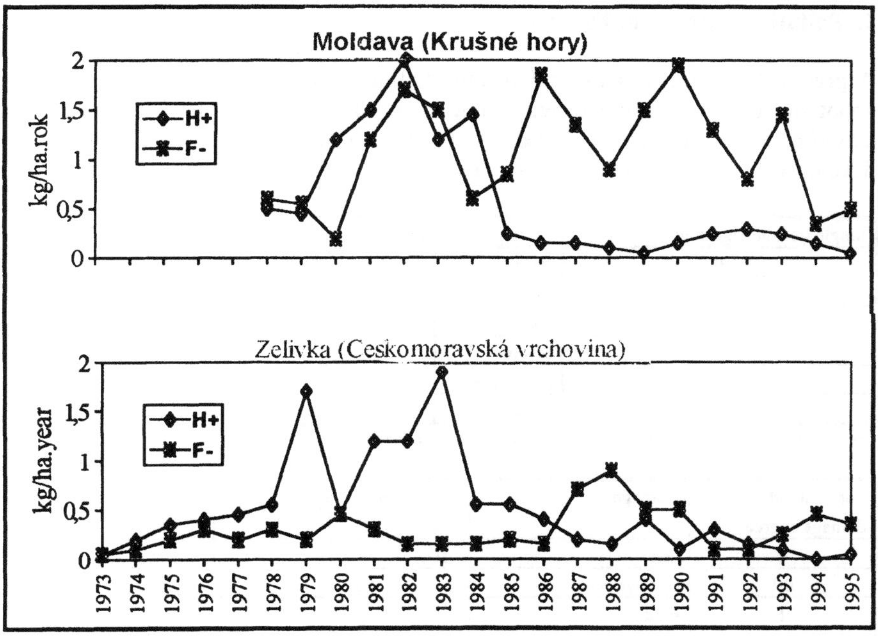

Figure 18. Bulk Deposition on the Ore and Ceskomoravská vrchovina Mountains

3. Pollution Damage to Forests

A practical forester needs objective information about the damage status and about the expected future development of it. To meet this, an appropriate classification systems is needed. The Czech classification system was developed in 1958 and - with minor modifications - it is still in practical use.

Single tree damage degree	Damage	Defoliation in %
0	none	0
1	light	1 - 25
2	moderate	26 - 50
3	severe	51 - 75
4	dying tree	76 - 100
5	dead tree	100

Figure 19. Czech Classification of Forest Tree Damage Degree (Norway Spruce)

Forest stand	Damage	Maximum % of the single tree damage degrees				
damage degree		0	1	2+	3+	5
0	none	100	0	0	0	0
0/I	first symptoms		20	0	0	0
I	light			32 or	5	
II	moderate			84 or	30	
IIIa	severe				50	
IIIb	very severe				70	
Iva	dying and dead				100	
Ivb	depreciated timber in dead stands					100

Figure 20. Czech Classification of Forest Stand Damage Degree (Norway Spruce)

The damage caused by pollution has been recorded since the last century and registered just between the World Wars. Local decline faced in the 1960's became severe after a deep drop of temperature in December 31, 1978. Since that, the damage dramatically spread within all of the mountain ridges located in the northern part of the Czech Republic.

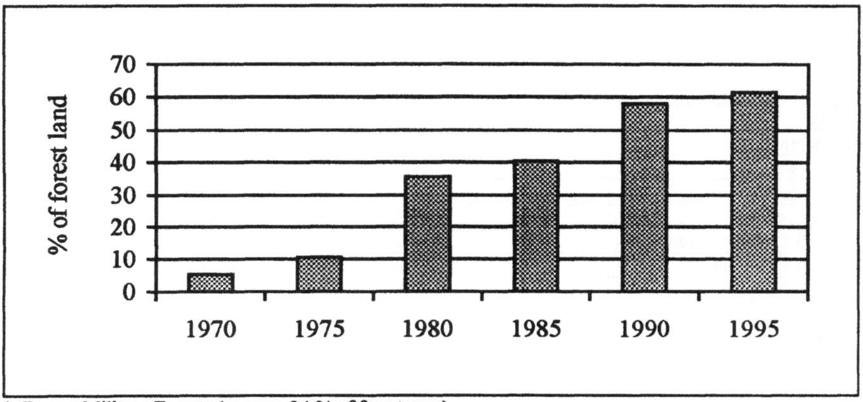

* Except Military Forests (approx. 94 % of forest area)
Figure 21. Air Pollution Damage to the Czech Forests* (1970 - 1995).

Year	Damage degree (Czech scale)				Harvested since 1958
	0/I + I	II	IIIa + b	IVa + b	
	% of the forest stands area (civil forests only)				
1981	**32.7**	**2.6**	**0.9**	**0.3**	
1985	**33.2**	**5.7**	**1.4**	**0.2**	**1.4**
1990	**51.3**	**4.5**	**1.2**	**0.1**	**1.9**
1991	**51.5**	**4.4**	**1.1**	**0.1**	**2.0**
1992	**54.6**	**4.2**	**0.9**	**0.1**	**2.0**
1993	**56.0**	**4.0**	**0.8**	**0.1**	**2.0**
1994	**58.3**	**3.8**	**0.8**	**0.1**	**2.0**
1995	**57.4**	**3.4**	**0.7**	**0.1**	**2.0**

Only those forest stand areas are mentioned that were stocked in the year of damage inventory.
Figure 22. Air Pollution Damage to Coniferous Forests by Damage Degree (1981 - 1995)

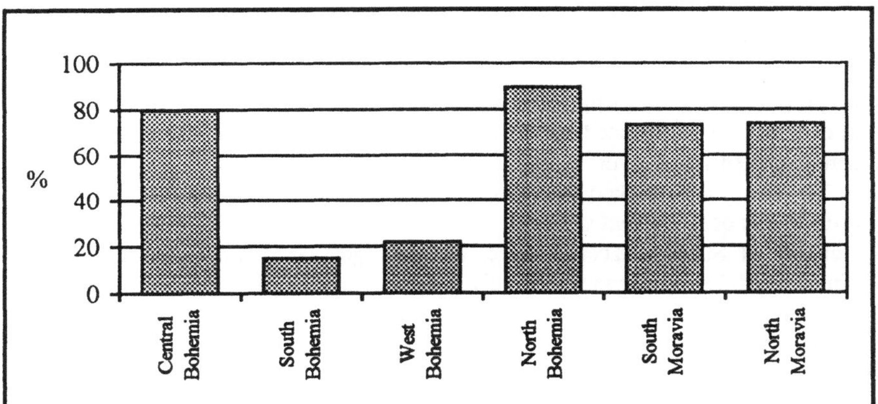

Military forests not included
Figure 23. Pollution Damage to Forests by the Regions (June 30, 1995)

Damage degree	Area		Growing stock	
	ha	%	tis. m^3	%
Conifers total	**1 142 264**	**47.6**	**297 771**	**55.6**
0/I with first symptoms	**739 486**	**30.9**	**177 004**	**33.1**
I light	**306 641**	**12.8**	**95 255**	**17.8**
II moderate	**77 650**	**3.2**	**22 076**	**4.1**
IIIa severe	**15 033**	**0.6**	**2 941**	**0.5**
IIIb very severe	**2 461**	**0.1**	**329**	**0.1**
IVa dying	**1 101**	**0.0**	**134**	**0.0**
IVb dead and destroyed	**492**	**0.0**	**32**	**0.0**
Non-conifers total	**334 790**	**14.0**	**56 010**	**10.5**
0/I with first symptoms	**281 413**	**11.8**	**47 646**	**8.9**
I slight	**48 283**	**2.0**	**6 972**	**1.3**
II moderate	**4 886**	**0.2**	**1 344**	**0.3**
IIIa severe	**158**	**0.0**	**39**	**0.0**
IIIb very severe	**40**	**0.0**	**7**	**0.0**
IVa dying	**10**	**0.0**	**2**	**0.0**
IVb dead and destroyed	**0**	**-**	**0**	**-**
Conifers and non-conifers total	**1 477 654**	**61.6**	**353 781**	**66.1**

Figure 24. Air Pollution Damage by June 30, 1995

Damage dynamics are expressed by means of the so called pollution threat zones.

Pollution threat zone	Area of forests	Mean period of the damage growth by 1 degree	Life expectancy* since the pollution affect beginning
	ha	years	
A	**28 994**	**5**	**< 20**
B	**136 352**	**6 - 10**	**20 - 40**
C	**626 317**	**11 - 15**	**40 - 60**
D	**787 249**	**16 - 20**	**60 - 80**

* A mature stand of Norway spruce

Figure 25. Czech Classification of a Pollution Threat to Forests (1991)

In 1994, about 681,800 ha of forests (with the growing stock of 138 millions m^3 u. b.) were categorized as the sub-category 3e, forests affected by the pollution and requiring a change of forest management methods. 12,023 ha of temporary unstocked stands were found and other 5,442 ha of clearings were expected yearly over the next decade in the sub-category.

The zones are checked this year on a base of a damage classification using satellite imagery. Created by the Lesprojekt, Brandýs n. L., and a private company of Stoklasa Tech.,Prague, the damage classification based on the Landsat TM-5 images has a relatively long tradition in the Czech Republic. In 1995, an improvement of the damage state has been recorded. However promising, that break of the pollution damage trend is not sure yet. After the first stage of my field surveys this year, the damage has increased in the southern (less polluted) regions of the Czech Republic.

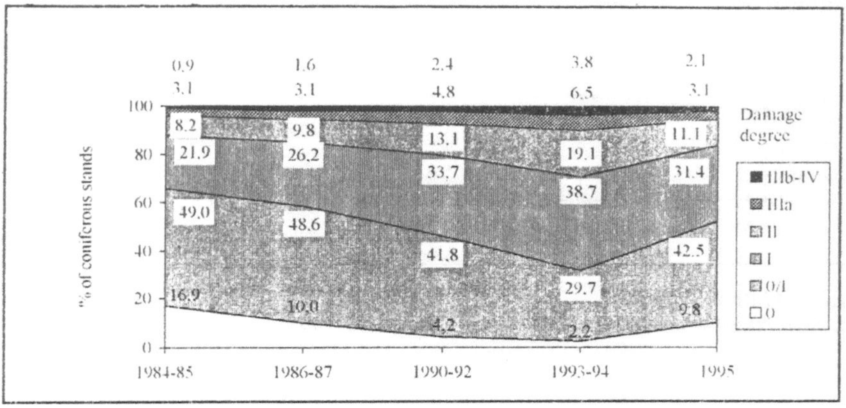

Figure 26. Damage Degree Changes of the Coniferous Forest Stands (1984 - 1995)

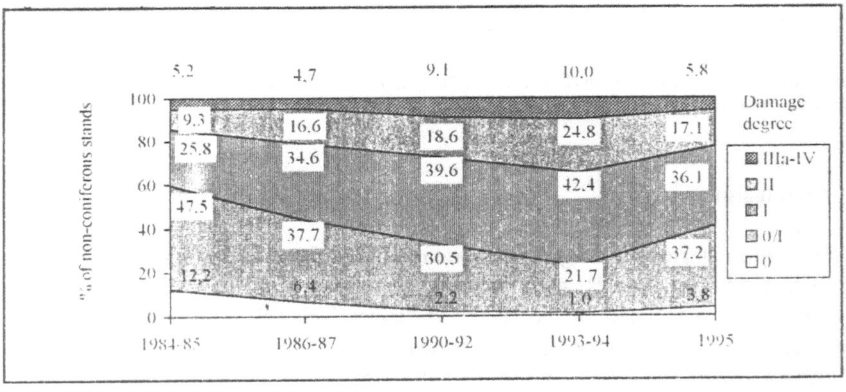

Figure 27. Vitality Changes of the Non-coniferous Forest Stands (1984 - 1995)

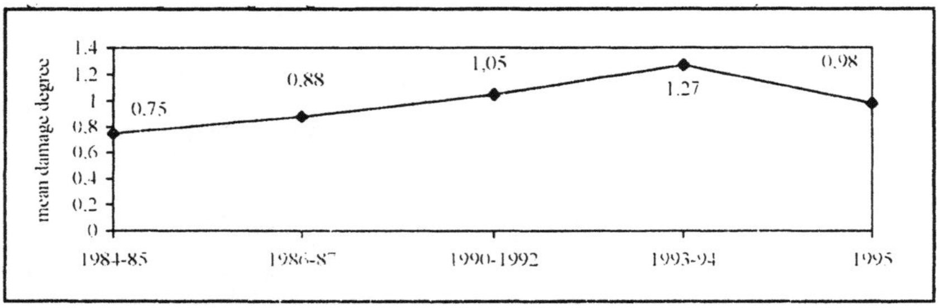

Source: STOKLASA Tech.
Fig. 28: Average Damage Degree of the Coniferous Stands (1984 - 1995)

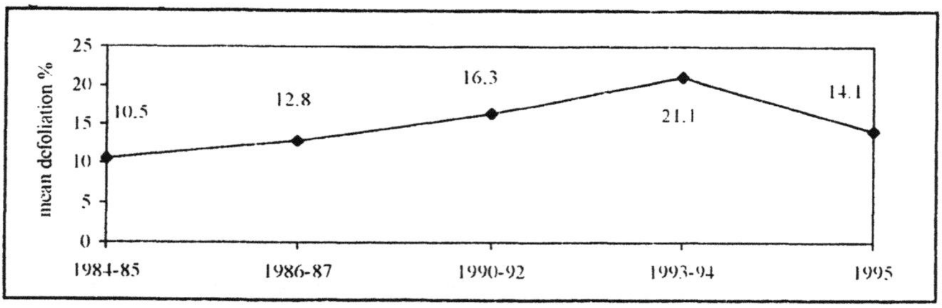

Source: STOKLASA Tech. Note: In the above four graphs, the 1995 classification has not been performed for the entire area of the CR because of the lack of convenient satellite images.
Figure 29. Average Defoliation of the Coniferous Stands (1984 - 1995)

4. Conclusions

Air pollution started causing severe damage to the Czech forests since the late 1970's. Emission sources, mainly the power plants using brown coal of very low quality, are situated in the so-called "Black Triangle" along the borders between the Czech Republic, East Germany and South Poland.

About 2% of the Czech forests (mainly Norway spruce stands located on the tops of mountain ridges) have been thoroughly ruined. The negative effect has been multiplied by the environmental importance of the destroyed forest ecosystems. Deforested forest land has been reforested by temporary stands composed of birch, mountain ash, and introduced spruces and pines considered capable to withstand the pollution affect. These should be replaced by local ecotypes of spruce and beech after pollution load drops.

The acute damage disappeared since about 1990. The chronic damage symptoms probably started diminishing in 1995. That break of the damage development trend is uncertain yet.

The present chronic damage is even more dangerous than the acute one because of the very large areas of damaged forest ecosystems. Chronically damaged forests are susceptible to diseases and create an environment for outbreaks of insects and fungi. Weakened forest stands dramatically react in any extreme climatic situation that is unpredictable and can happen any time.

5. References

Zpráva o stavu lesního hospodáøství Èeské republiky 1996. Report on Forestry of the Czech Republic 1996. Ministry of Agriculture of the Czech Republic, in print.

ENVIRONMENTAL FACTORS AND STATE OF UKRAINIAN FORESTS: MONITORING AND IMPACT ASSESSMENT

M.M. KALETNIK
Ministry of Forestry of Ukraine
Kreshchatik 5
252005 Kiev
Ukraine

I.F. BUKSHA,
O.M. RADCHENKO,
O.S. SIDROV
Ukrainian Research Institute of Forestry and Forest Agromelioration, Kharkov
Ukraine

Abstract

Ukrainian forested lands cover 10 million hectares and about 14.3% of the total territory. Ukrainian forests are considered mostly as a valuable environment forming factor and thus having a limited value for industrial utilization. Now Ukrainian forests are affected by environmental pollution, intensive forestry practice and recreational uses. Almost a third of the forested areas is radio-actively contaminated due to the Chernobyl accident. Therefore, there is a strong need to provide forest monitoring in Ukraine and since 1989 Ukraine has participated in the International Co-operative Programme on Assessment and Monitoring of Air Pollution Effects on Forests (ICP-Forests).

The first (extensive) level of monitoring is designed for evaluation of the dynamics and distribution of forest damage in the country. The main estimation criteria are defoliation and discoloration of forest stands. From 1990 until 1995 the mean defoliation increased from 10% to 23%, part of moderately and severely defoliated trees increased from 4% to 27%. Recent decrease of air pollution concentrations, however, has not been reflected by the lowering grades of the defoliation or discoloration. It may be explained by the accumulative effect of air pollutants. Defoliation and discoloration can also be affected by other factors, e.g. pests, fungi or extreme weather conditions.

Actually, the actual source of changes in forests state is difficult to indicate without extensive level of monitoring. In summer 1995 forest monitoring activities in Ukraine were extended under US Forest Health Monitoring Programme (FHM) that provides additional information to extensive monitoring data. Research on climate change impacts on Ukrainian forests was carried out under the US Country Studies Program for forest vulnerability and adaptations assessment. Estimation of a potential distribution of forests, distribution of main tree species, and species biodiversity was accomplished with use of a computer Geographic Information System (GIS). Methodology used in the project is based on a forest types classification developed in Ukraine. It is valuable for interpretation of forest monitoring data, evaluation of forests

R. M. Gutkowski and T. Winnicki (eds.), Restoration of Forests, 151–160.
© 1997 *Kluwer Academic Publishers.*

dynamic under environmental impacts including global climate change.

1. Introduction

Ukraine is located in the south-eastern part of Europe. Its territory extends from 52°22'54" to 44°23'18" of northern latitude (about 900 km) and from 22°08'42" to 40°13'05" of northern longitude (about 1300 km). The total area is 603.7 square kilometers. Population is about 52 million, of which 68% of population are living in cities.

Ukrainian forested lands cover 10 million hectares which is about 14.3% of the total territory. Forest distribution pattern, growth and state are related to climate conditions of temperate continental type, except for a narrow strip of subtropical coastal forest in South Crimea. In all there are 5 natural zones where forest growth conditions are essentially different.

Approximately 54% conifer species and 46% deciduous species comprise the Ukrainian forests. The average annual increment is about 4 cubic meters per year. About half of the forest land is in managed plantations. Main forest forming coniferous species ate Pinus L. (about 35% of forest stands), Picea A. Dietr. (16%) and Abies Mill. (13%). Dominating hardwood species are Quercus L. (22%), Fagus L. (13%), Carpinus L. (2%). Softwood species are Betula L., Populus L., Alnus Mill., Tilia L.

Ukrainian forests, as indicated in the national legislation, are considered mostly ad a valuable environment forming factor and presently it is under a large anthropogenic load. Among former Soviet Union countries, Ukraine has had the highest level of air pollution. Overall industrial emissions of pollutants to atmosphere were about 17 million tons per year. the main sources are ferrous metallurgical plants and coal industry. It is estimated that motor transport adds about 40% of the overall air pollution.

The primary objective of the Ukrainian national forests monitoring program is to describe spatial distribution and degree of forests' anthropogenic damage. Forest monitoring has been ongoing since 1989 in the frame on International Co-operative Programme on Assessment and Monitoring of Air Pollution Effects on Forests (ICP-Forests). Net of plots (grid size 16x16 kilometers) have been established in 14 of 24 administrative districts and the Crimea republic. Until now 32 plots have been established and data base on them has been gathered. In summer 1995 Ukrainian forest monitoring has been extended with support from the US Forest Health Monitoring Programme (FHM). Until now 32 plots have been established and a data base on them has been collected (Figure 1).

2. The Regional Assessment of the Forest Contamination by the Industry Air Pollutants and Chernobyl Radioactive Deposits

For assessment of air pollution intensity used generalized data on industrial emissions of pollutants in administrative regions and big industrial cities for the period from 1985 to 1993.

Figure 1. ICP Forests and FHM plots location

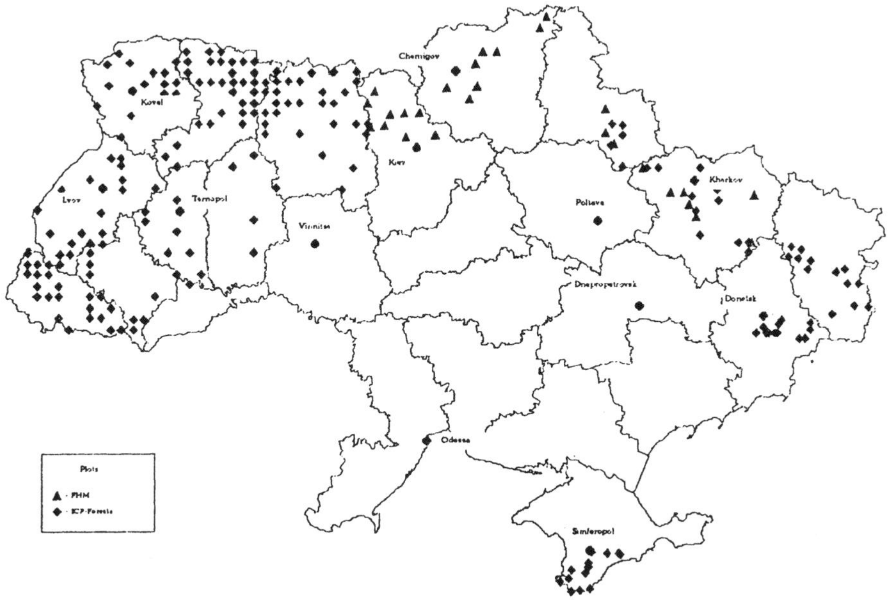

The total amount of the pollutant emissions from industrial sources has decreased from 1985 to 1993 by 52.8%. Air pollution level is still decreasing along with the industry recession. Intensity of the pollutant emissions has been decreased from 35% to 28%.

All the administrative regions can be divided into 2 groups that differ by total intensity and composition of emission. The first group includes Donetskaya region, partially Luganskaya, Zaporozhkaya, and Dnepropetrovskaya region. The second group includes the rest of the units. The first group has higher intensity density level of pollution and impact of stationary sources than the second group. (Figure 2, Figure 3).

Due to the accident at Chernobyl'skaya Power Station in 1986, about 3.5 million hectares of forests in Ukraine were radio-actively contaminated. Density of the contamination with ^{137}Cs at 1.5 million hectares of the forests is greater than 1 Ku/km^2 (Table 1).

Figure 2. Emissions by regions with intensive air pollution (Group 1)

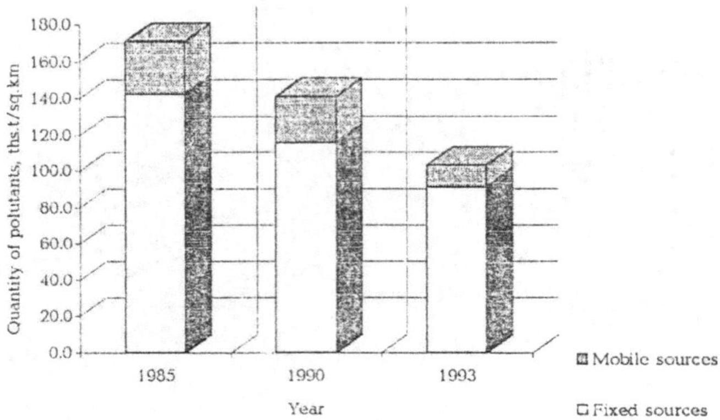

Figure 3. Emissions by regions with low air pollution (Group 2)

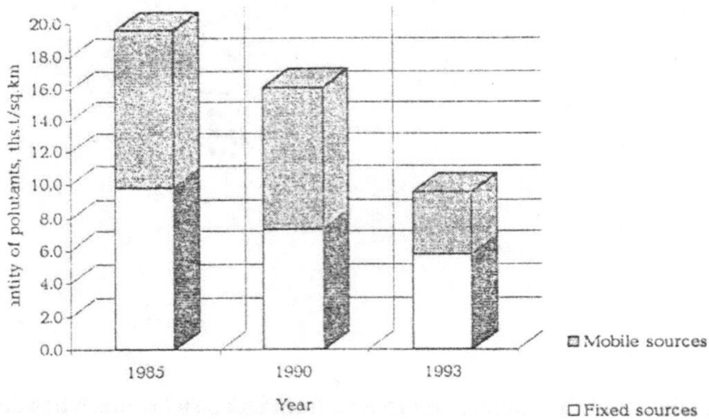

Table 1. Forest areas that were contaminated with ^{137}Cs, dated 01.01.93

Areas' distribution by contamination density (Ku/km^2)	Area, ths ha	%%
< 1.0	1955.1	61.4
1.1-2.0	692.1	21.6
2.1-5.0	396.9	12.5
5.1-10	78.4	2.5
10.1-15	23.1	0.7
15.1-40	31.3	1.0
40.1-80	7.4	0.2
> 80.0	2.1	0.1
Total in Ukraine	3186.4	

The radioactively contaminated forest territories are predominating the management of forests. They require uninterrupted forestry care to save the forests' ability to keep the contamination inside and prevent its further spreading.

Distribution of radioactive elements inside forests eco-system, effects of the ionizing radiation on main forest-forming tree species, migration of radio-active elements with surface and ground water are studied to maintain forestry and utilization at the contaminated lands. Research that was carried out on forest plots in 1992-1995 show that main portion of radio-active elements is resting in forest litter (dead organic layer over ground) and in 2 cm upper ground layer (approx. 90%). Content of ^{137}Cs in forest litter depends on type of forest habitat: where conditions are more fertile the content of ^{137}Cs in forest litter is less. It was also found that ^{90}Sr in 2-20 cm upper ground layer and accordingly ^{90}Sr has a greater coefficient of quasi-diffusion.

Comparison of ^{137}Cs migration processes in coniferous and deciduous forests shows that there are certain differences. Forest litter in a deciduous forest contains about 50% of overall ^{137}Cs and movement of ^{137}Cs to 2-5 cm ground layer is more intensive (Figure 4).

Figure 4. Comparison of ^{137}Cs distribution in coniferous and deciduous forests

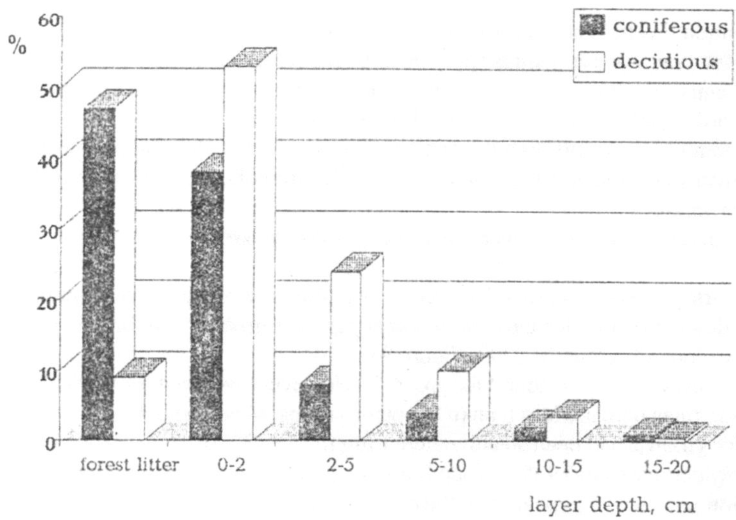

3. Dynamics and Distribution of Forest Damage in Ukraine

There is an established network of 215 forest monitoring plots in Ukraine at this time. The network was formed under ICP-Forests Programme and presently cover 15 of 25 districts. Over five thousand species are in the observation. The monitoring is designed to identify zones of forest decline and to study cause-consequence relationships between forest state change and atmospheric pollution. The degree of crown defoliation in the stand is a common indicator of monitoring which characterizes forest condition.

Figures 5-7 show the state of defoliation in Ukraine's forests. The fraction of trees that defoliated greater than by 25% has been increasing (from 4% in 1990 to 27% in 1995). Average year defoliation of the forest stands has been increased during this time from 9.5% to 22.6%. Maximum of the defoliation up to 44% was found in Luganskaya region in 1995 and minimum is about 5%, which was found in Ternopol'skaya region in 1990. Averaged for the whole period defoliation was highest in Donetskaya region (36%) and in Luganskaya region (28%). Lowest levels of average defoliation for the period were in Ternopol'skaya region (8%) and L'vovskaya region (9%). The 1994 results show the greatest defoliation.

Most of the plots are in the Western part of Ukraine and a few in the East. The map below shows the maximum defoliation levels were found in the Luganskaya and Donetskaya regions. High levels exist in the Kharkovskaya and Rovenskaya regions because of their highly developed industry (Figure 8).

The estimates are showing growth of tree defoliation during the observational period from 1989 to 1995. Comparative analysis of tree species was accomplished through indication of differences in levels of defoliation by years. Species set divided to classes in such a way that within a class there is no species with reliable differences in defoliation levels. The following grouping was developed:

1. Betula verrucosa, Ulmus spp., Quercus robur, Alnus incana, Pinus sylvestris;
2. Fagus sylvatica, Quercus petrea, Acer Tataricus, Acer pseudoplatanus, Tilia cordata, Juniperus communis, Abies alba, Fraxinus exelsior;
3. Betulus pubescens, Fagus orientalis, Caprinus orientalis, Caprinus betularius, Acer platanoides, Alnus glutinosa, Populus tremula, Pinus pallasiana, Padus racemosa;
4. Quercus pubescens, Picea abies, Juniperus oxycedrus

The first tree species group has relatively uniform defoliation growth rate. Its average defoliation rate for the observational period is greater than for any of the other groups (except 4th group in 1993, Figure 9).

The same pattern was found for group 3. However, averaged for the whole period, defoliation rates here are less than in the other groups. Average defoliation rate for the period for group 2 had been steadily increasing from 1989 to 1994. Then, it was noticeably decreased in 1995. Group 3 differs from the others with high peak of defoliation in 1993. It appears that the cause is particularly of climate conditions in 1993 and decreasing of defoliation in 1994 and 1995.

The pattern of changes in defoliation of the forests in Ukraine shows that the photosynthetic apparatus continues to decline. Defoliation of the first species group, that consists of the main forest forming species Quercus robur and Pinus sylvestris, has a steady tendency to increase. Finding out the causes of defoliation change pattern requires longer time series of observation and more adequate and complete set of measurements.

The expected decrease of general defoliation intensity has not been found in spite of the considerable decrease of air pollutants emission since 1992. The possible causes of this phenomenon are the following:

- the affect of accumulation of pollutants in soil for previous years;
- global climate changes;
- general increasing of anthropogenic load on forest ecosystems.

However, data of forest pathology observation show the most significant desiccation is in the Central part of the country (region of Dnieper drainage-basin), especially in the Kherson and Kiev regions, due to the water regime of soil disturbance.

Figure 5. Defoliation of conifers from 1989-1995

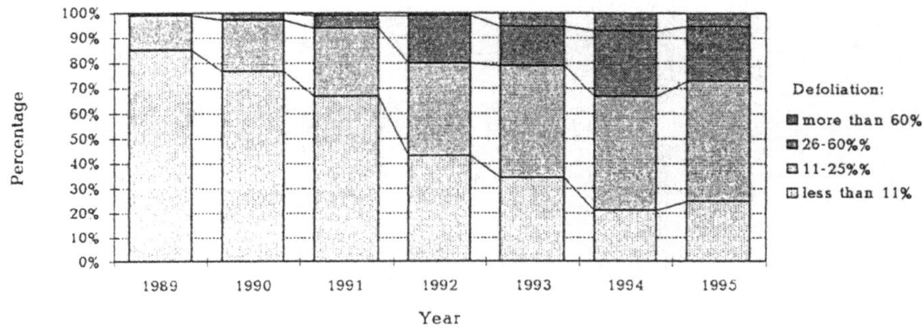

Figure 6. Defoliation of broadleaves from 1989-1995

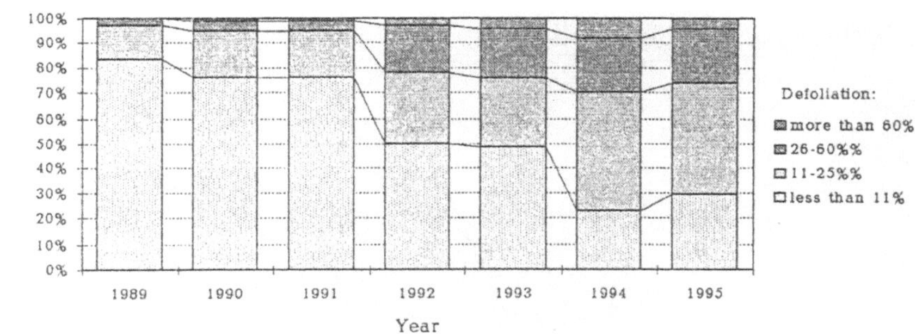

Figure 7. Defoliation of all species from 1989-1995

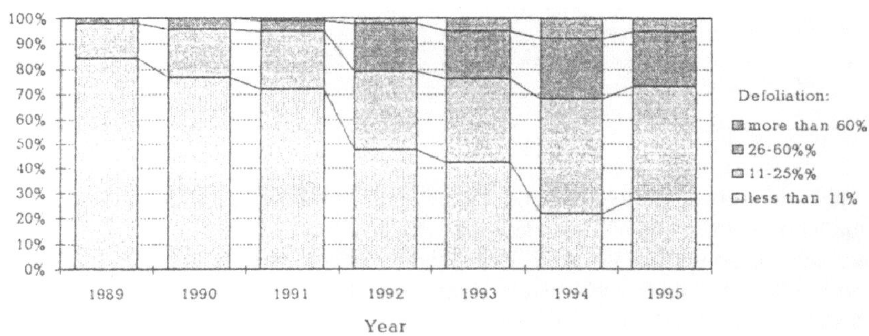

Figure 8. Ukrianie. Mean defoliation by regions 1989-1995

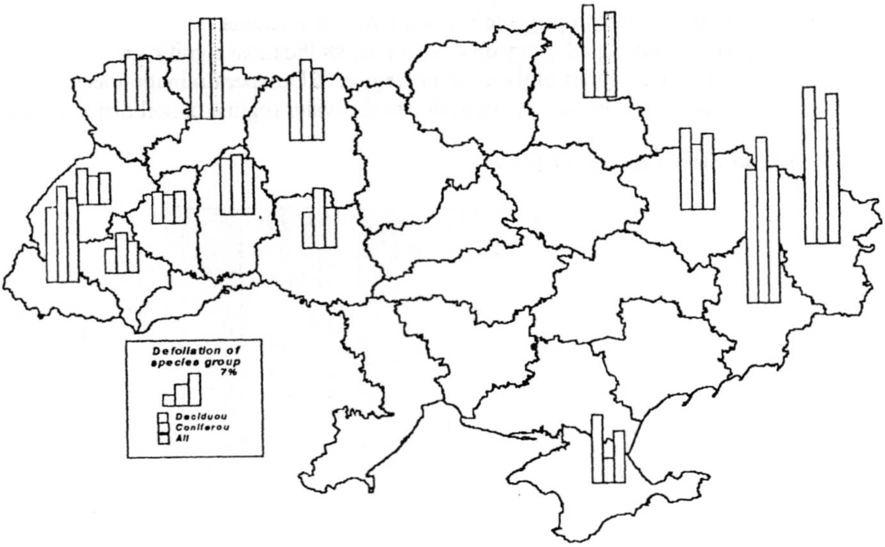

Figure 9. Defoliation of species groups from 1989-1995

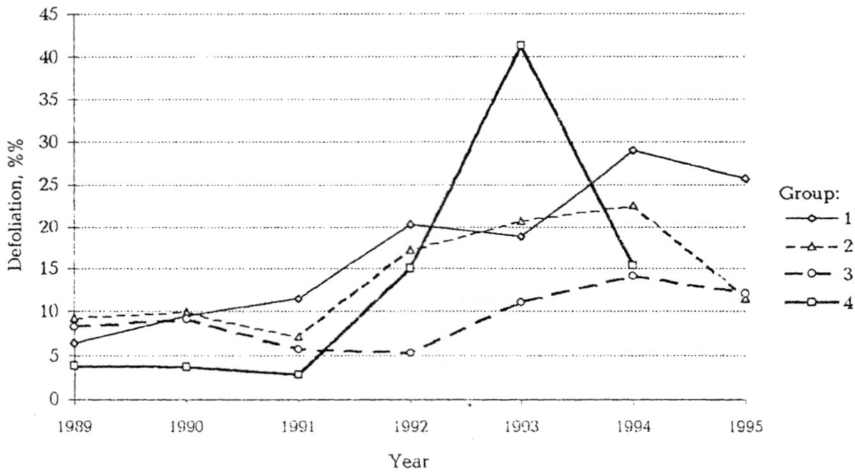

At the same time, in 1989, 82.9% of coniferous trees was not defoliated, 15.9% was slightly defoliated and 1.4% - moderately defoliated. In 1995 the proportion of non-defoliated trees fell (24.7%), the parts of slightly and moderately defoliated trees increased (48.7% and 24.6% respectively). As to deciduous, in 1989, 84% of them were not defoliated, 12.9% were slightly, 3.0% moderately, and 0.1% severely defoliated. In 1995 non-defoliated trees were at least (19.8%), and parts of slightly, moderately and severely defoliated trees increased (49.3%, 30.0% and 0.9% respectively).

Among coniferous species, Pinus and Picea (Juniperus in Crimea) were defoliated the most. Those were Quercus, Fagus, Carpinus and Betula among deciduous.

Also, the observed defoliation and discoloration may have been induced by pests, or diseases, air pollution or weather anomalies.

4. Research on Vulnerability of Ukrainian Forests to Global Climate Change Impacts

Our approach to research on vulnerability of Ukrainian Forestry Research Institute has examined vulnerability of forests to global climate change under the US Countries Studies Programme. The primary approach to assess impact on forests is mapping via GIS of potential vegetation land cover with the Holdridge life zone classification model and forest-oriented Ukrainian classification model of climate zone. Climate change scenarios from general circulation models (GCM) are used for projection with the classification models. Maps of potential forest distribution allow one to assess the general pattern of climate change impacts. Change of biological productivity of forest stands in regional scale is estimated with climate-related productivity index (Paterson). Climatic species ranges were modeled for currently dominating tree species as a first approach to describe future forest composition. Incorporation of forest ecosystems models (gap-type models) allows one to reproduce more detailed predictions for most vulnerable forest sites. However, there is lack of adequate gap-models for Ukraine and serious problems in use of gap models have developed in other regions: Ukrainian forestry uses particular assessment indices for plant response to climate change such as those required by the gap models.

There is a definite methodological gap in the described assessment approach. The gap is between large-scale level of climate conditions classification scheme and "too local" gap-model that is difficult to spatially extrapolate. A method has been proposed for mid-scale impacts assessment on the basis of Ukrainian school of forests topology (Aleksyev, Pogrebniak, Vorobjov, Lavrinenko). The idea is to synthesize forest-topological maps from mid-scale maps of soils, hydrology and relief (1 : 100 000 - 1 : 1 000 000).

There are a number of forest-topological classification schemes developed by the Ukrainian school. They mainly differ by forest lands classification criteria that are used to indicate soil humidity and trophic conditions. The current research concerns improvement of quantitative indices for the criteria and extension of climatic and edaphic grid models onto representation of more trophically poor, more warm and more dry sites. To make a mid-scale map of forest growth conditions (forest land maps, forest edaphic maps) according to the forest topology scheme required synthesis of two maps: a forest trophic conditions map and a map of soil moisture availability. The forest trophic conditions map may be derived from available soil maps: with tables of correspondence between soil types and trophic index compiled from empirical data. However, the preferable way for trophic index evaluation (as recent research indicated) is to estimate contents of phosphorus and potassium in the root inhabited layer of soil. The territorial unit of the resulting mid-scale edaphic map describes edaphic conditions with some grade of generalization. Apparently, it depends on scale and accuracy of source maps. However, the forest-topological meaning of such an edatope is subject to study. Actually, adjustments of forest estimation methods to the "mid-scale" edatope are required.

Apparently, the described methodology is applicable to broad scale of the impacts assessment: for issues of global climate change assessment as well as for assessment of damage caused by air pollution.

5. Conclusion

Forest monitoring in Ukraine reflects the forest damage pattern affected by unfavorable environmental factors. These factors are air pollution, pests, fungi diseases, unfavorable weather conditions and intensification of anthropogenic pressure on forest ecosystems. Assessment of relative changes of trees assimilation apparatus gives evidence of progressive deterioration of crown condition from 1989 to 1994 and reveals a slight improvement in 1995. The highest level of defoliation among main conifers was in Pinus spp. and Picea spp. and among main broadleaves - in Quercus spp., Fagus spp., Carpinus spp and Betula spp. Taking into account the extent of Ukrainian forest in environmental protection and stabilization of migration processes on territories that are radio-actively contaminated, it is necessary to develop forest monitoring programs aimed at design of assessment system with several levels of observation including extensive and intensive levels. Also, it is necessary to use approaches that have been developed on the basis of forest topology classification of forest growth conditions for proper interpretation of monitoring results and for adequate assessment of fores condition dynamics affected by various environmental factors (including global climate changes).

AIR POLLUTION IMPACTS ON FOREST TREES AND BIOLOGICAL ASPECTS OF THE ENVIRONMENT OPTIMIZATION IN BELARUS

S.A.SERGEICHIK
Central Botanic Garden
Belarusian Academy of Sciences
Surganova 2a Minsk, 220012
BELARUS

Abstract

This paper discusses the effect of air pollution on forest trees and biological aspects of the environment optimization in Belarus. Data on gas-tolerance and air pollution accumulation capabilities of some trees species are presented. Some recommendations for creating green plantations in areas with air pollution by SO_2, H_2S, CS_2, NO_2, NH_3 are given. This study indicated that vegetation in Belarus is greatly endangered by sulfurous and nitrogenous air pollutants. Sulfur accumulation in the Scots pine (*Pinus sylvestris* L.) needles from 0.11 to 0.17% negatively impacts on fundamental physiological processes, primery photosynthetical reactions, pigment and protein pools concentrations, chemical elements N : S : P ratios. These changes are factors inthe forest ecosystems destabilization in Belarus.

1. Introduction

The problem of forest biospheric function preservation and of the environment optimization is quite urgent for Belarus which is a densely populated region of Europe having highly-developed power engineering and chemical industry. The environmental pollution is one of the main causes of deteriorating living conditions for Belarusian population. Many other serious ecological effects also take place in Belarus, such as the decline of forest and green plantations in some areas. As a result of the Chernobyl disaster 1.685 thousand hectares of forest in the republic are radioactively polluted. Every year total emission of air pollutants consist about 2.5 million tons in Belarus. From their sulfur dioxide (SO_2) - 318 thousand tons and nitrogen dioxide (NO_2) - 188 thousand tones [1]. Air pollution (mainly SO_2) is considered one of the main causes of forest decline in eastern Europe [2]. About 30% of the country's area is under severe environment influence. Forest and greenery belts in some parts of Belarus are under strong air pollution stress [3,4]. Although many studies have been carried out in Europe and America on the effect of air pollution on

R. M. Gutkowski and T. Winnicki (eds.), Restoration of Forests, 161–176.
© 1997 *Kluwer Academic Publishers.*

forests there are still major gaps in our understanding of physiological and biochemical mechanisms leading to forest decline. Physiological and biochemical changes associated with stressors can occur before there are any visible symptoms that are normally associated with decline in forest health and may provide early warning signals to protect many forests.

Reduction of air pollution and improvement of conditions of the forest in Belarus and other countries will have both local and global benefits. The results of international air pollution and forest health monitoring efforts will aid policy makers in implementing necessary air pollution standards in the region. Cleaner air means improving living conditions of people in that part of Europe and also improvement of the global air pollution situation. Reduction of air pollution deposition and better understanding of processes in forest ecosystems will help to prevent further forest decline and to rebuild the damaged forests of the region. This is especially important as many countries in the region are looking to timber and forest products as a means to stabilize their economies. Healthier and more productive forests provide a healthier and more aesthetic environment for living and recreation, and they prevent the accelerating deterioration of the world's biological diversity.

Botanic gardens as centers of introduction of trees and shrubs and storages of valuable native species carry out an important function in the preservation and improvement of the environment, in the rational use and reproduction of natural resources.

Areas with polluted atmosphere require stable plant communities, which should have large capacity of absorbing harmful admixtures. This problem may be solved by selecting highly gas - resistant plant species with ultimately exhibited gas - absorbing ability which posses appreciable biomass of the leaves and are characterized by different leafage terms. Biological accumulation and neutralization of poisonous gases should be considered as a means of smog control, an addition to conventional engineering methods, and as factor of atmosphere aftercleaning.

2. Forest Wealth of Belarus

Belarus is a republic of many forests. The total area of the Belarus state forest fund is 8055 thousand hectares including 7018 thousand hectares of forest covered land. Forests occupy 33.9% of the republic's total area.

Depending on their national economy and commercial value and their location, the republic's forest fall into two groups. The first group of forest having water protection,sanitary and other protective functions takes up 36.8%, while the second one, which is of commercial value, constitutes 63.2% of forest.

No felling is permitted within the forests belonging to the first group. The forest belts along river banks, lake shores occupy 426.5 thousand hectares, the protective strip along railways and highways - 936 thousand hectares, the green zones - 1008.5 thousand hectares. The area of the first and second belts in the sanitary protection zones is 52.8 thousand hectares, that of nature reserves is 63.5 thousand hectares , that of nature reserves in 63.5 thousand hectares.

The mature plantations take up 146.6 thousand hectares of forests, including 79.8 thousand hectares of coniferous trees. Young plantations prevail in the republic's forest fund, occupying 2817.5 thousand hectares (48.9%), fully-grown trees occupy a little less territory while ripening forests constitute 10.5% .

The species composition of Belarusian forests is as follows: coniferous trees - 68.7% of the total area, including pine - 57.6%, fir -11.1, hard-wood broad-leaved - 4.2, oak - 3.8, hornbeam - 0.2, alder - 7.4, grey alder - 0.1, other species - 0.1%. The total wood reserves in the Belarusian forest fund are 921,300,000 m^3. The mean annual growth is 21,700,000 m^3 [5].

In 1988, 12,140,000 m^3 were obtained through all types of forest felling (56% of mean annual growth). Annually, 10,000,000 to 11,000,000 m^3 of merchantable wood are produced in the republic, including 6,200,000 m^3 by clear felling, 4 to 5 million m^3 by intermediate felling. Clear felling has been done over 30,000 hectares while gradual and selective felling - over 376 hectatrs. Annual improvement and sanitary felling is done over 300,000 hectares, which yields almost 4,500,000 m^3 of timber. The output of mercantable wood is 90.8%, that of industrial wood - 37.6%. Improvement of young plantations is done over 90,000 hectares. The annual forest felling area is about 30,000 hectares, i.e. 0.4% of the forest area which is subject to regeneration using valuable and ecologically strong species within 1 or 2 years.

Forest regeneration is carried out annually over an area of 30,000 hectares, including 27,000 hectares of forest planting and seeding. Forest planting in ravines and gullies, and other inconvenient areas is done annually over 1,100 hectares. Ample seedling and nursery reserves have been created in the republic providing all planting material required for forest regeneration. There are 158 forest nurseries occupying 1464 hectares, which supply forestries and other organizations in the republic with nearly 250 million seedlings and saplings. Seed plantations occupy 1450 hectares. Extensive research is carried out on selection and creation of new hybrid species. Nearly 200 tons of seeds are collected annually in forests and plantations, of which 25 tons of seeds are sold outside the republic. Homogeneous plantations in Belarus occupy over one and a half million hectares, a quarter of all forest-covered area [5].

Under the scientific guidance of experts from the Belarusian Academy of Sciences, Institute of Forest Research, the Belarusian forest management enterprise and Republic's forest protection and monitoring station under the Belarusian Ministry of Forestry are completing the creation of a network of forest monitoring in Belarus.

3. Effect of Air Pollution on Sulfur Accumulation and Physiological Tests in the Scots Pine (*Pinus sylvestris* L.) Needles in Belarus

There is an urgent need to prevent further deterioration of forest health and to help in rebuilding forests in Belarus and other East and Central European countries. This can be accomplished only with well coordinated activities in the areas of air pollution and forest heath monitoring, investigation of the effects of air pollution on forest functioning, and selection and implementation of proper response strategies. Belarus is under air pollution stress from industrial sources. Because of this, the Central Botanical Garden of the Belarus Academy of Sciences began research in 1974 problems of air pollution impact on vegetation. Scots pine (*Pinus sylvestris* L.) needles were chosen as the best bioindicator of pollutant impact on vegetation, because Scots pine grows in everywhere in the country and pollutant accumulation in the needles occurs. Scots pine is the main tree species of Belarusian forests. It occupy by 57.6% of the total area.

The method development research was conducted in the Berezinski Biosphere Reserve, representing the cleanest place in Belarus, and in Forest District near Novopolotsk, a location with the highest level of pollution in the country. Studies have been performed in this areas for several years. It has been found that the best time to collect samples of pine needles are the winter months. In 1992 , a survey of the entire territory of Belarus was initiated. The entire country - a map of Belarus (scale 1:1000000) was divided into 72 main squares (60 x 60 km). Each of the main squares was subdivided into 100 base squares (6 x 6 km). In every base square five collection squares were randomly selected where the pine needles for chemical determinations were collected. A total of 316 collection squares were selected for the entire country. The sampling sites were always located at least 2 km away from direct emission sources such as industrial plants or settlements, and at least 300 m away from transportation routes. At each of the collection sites, needle samples were collected from eight trees of uniform age (about 30-40-year-old) from the second to the fourth whorls, counting from the tree top. Only 2-year-old needles were sampled. Samples were collected between January and March when the highest pollution of air by SO_2 occurs in Belarus, resulting in the highest concentrations of S in needles. During this time the largest differences in S concentrations of needles collected from areas of various degrees of environmental contamination also occured [6 - 9].

Concentrations of S were determined with a LECO - SC - 132 instrument. Physiological and biochemical investigations were determined according to [10]. On the base of our investigations four zones of pollution with sulfur were selected in Belarus [11].

Sulfur concentrations of 0.065% in second year pine needles is a minimum for growth; below that amount , there can be a deficiency. Therefore, 0.065% can be considered a "normal" level of sulfur, as presented as 100% of its content. Elevated content up to 0.13% can be considered a "luxury" concentration, i.e. not needed but also not toxic. Levels greater that 0.13% can be considereded phyto-toxic [12]. A sulfur level of 0.3% for pine needles in the second year of growth is critical, as above that level the tissue dies. All pine trees with sulfur content greater than 0.13% may exhibit symptoms of toxicity, but 0.3% generally results in mortality. Of course, some individual needles can be found wich are more resistant or more susceptible to toxicity than others. Factors such as ecological conditions, water supply, availability of other nutrients, and extreme temperatures can influence greatly the survivability or death of trees.

The distribution of different zones of pollution in Belarus on the basis of determination sulfur content in pine needles shows that only a sample part of Belarus (about 10% of territory) has air relatively high level in sulfur more than 0.13% (i.e. more than 215% of the normal level of this element). About of 20% Belarussian territory has pine forest where sulfur content excees 0.11 - 0.13% (200% of the normal level). About 70% territory of Belarus has pine forest, where sulfur content there is in the bound less than 0.08 - 0.010% (about 133 - 166% of the normal level of 0.06%). The highest level sulfur content in pine needles in Belarus is 0.17 - 0.2% dry matter and that is near the large industrial centers of republic and along the border with Poland on the reason influence of transboundary air pollution with SO_2. Sulfur deposition in Belarus exceeds 2 tons of sulfur on km^2 per year. This is the reason of middle acidiation of atmospheric precipitations.

As compared with Poland Scots pine needles in Belarus have a lower level of sulfur accumulation. Four zones of pollution with sulfur were selected in Poland too [9]. The least polluted was zone 1, in which S concentrations in needles did not exceed 0.09% (e.g. about 150% of the normal level of 0.06%) and was found mostly in eastern Poland , with some smaller areas found also in northern and central parts of the country. About 5-7% (18 000 km^2) of the country's territory was in this zone of pollution. In zone 2 , the S concentrations in needles were between 0.091 and 0.12%, i.e. up to 200% of the normal level of this element. This zone was located mostly in northen and eastern Poland and comprised 47.4% (148 000 km^2) of the countrry's territory, with some small areas within the more polluted zone 3. In zone 3 , the total S concentrations were between 0.121 and 0.15% (200-250% of the normal level of this element). This zone was found mostly in the southwestern part of the country, in some areas near Warsaw and Ostroleka within zone 2, and also within zone 4 in Kotlina Klodzka near the Czech border. In zone 4, the S concentrations were above 0.15% (more than 250% of normal). About 14.7% (46 000 km^2) of the country's territory, mainly in southwestern Poland and some areas near Torun and Wloclawek, was in this zone of pollution.

Attempts to develop comparison scales, in which S concentrations in Scots pine needles are correlated with average ambient concentrations of SO_2, have been made by four different groups [6], in which measurements of SO_2 concentrations were made with modern electronic monitoring instruments. Their scale is as follows:

Zone 1: lower 20 mkg SO_2 m^{-3} (< 0.09% S dry wt)

Zone 2: 21-29 mkg SO_2 m^{-3} (0.091 - 0.12% S dry wt)

Zone 3: 30-36 mkg SO_2 m^{-3} (0.121-0.15% S dry wt)

Zone 4: more 37 mkg SO_2 m^{-3} (> 0.15% S dry wt)

Attempts to develop a 'reliable' method of converting S concentrations in pine needles into ambient SO_2 concentrations have been caused by a need for an inexpensive method of evaluating environmental pollution by this compound [9]. On the other hand, predicting the status of forest health on the basis of S concentrations in needles alone is not possible because no linear relationship between S concentrations in air and plant foliage and degree of forest damage exists. For Scots pine, the visible symptoms of injury in old forest stands start to occur after several years exposure at 25-30 mkg

SO_2 m^{-3}. In harsher climate conditions, the threshold value may be lower (15-25 mkg m^{-3}). Death of trees starts to occur at concentrations of 70-80 mkg m^{-3} [6]. Eastern white pine (*Pinus strobus* L.) trees near Sudbury, Canada, experienced chronic injury at an annual average SO_2 concentrations of higher than 21 mkg m^{-3}, whereas levels higher than 41 mkg m^{-3} caused severe injury of trees in the same stands [12].

According to results of our investigations air pollution with SO_2 in Belarus (sulfur concentrations in the Scots pine needles from 0.117 to 0.157%) negatively effects on physiological and biochemical parameters of trees (see Table 1).

TABLE 1. Effect of air pollution on physiological parameters of *Pinus sylvestris* L. needles

Physiological parameters of needles (average from 20 dimensions)	Areas of air pollution with SO_2, based on S concentrations (%) in the needles			
	Control area site	High pollution site	Middle pollution site	Low pollution site
Sulfur contents, % dry wt	0.108	0.157*	0.135*	0.117*
Chlorophyll A, mg/g dry wt	0.175	0.565*	0.645*	0.604*
Chlorophyll B, mg/g dry wt	0.363	0.323*	0.333*	0.316*
Chlorophyll A+B, mg/g dry wt	1.108	0.889*	0.978*	0.920*
Carotenoids, mg/g dry wt	0.283	0.203*	0.220*	0.218*
Chlorophyll A/Chlorophyll B	2.052	1.749*	1.937*	1.911*
Chlorophyll A+B/Carotenoids	3.915	4.379*	4.445*	4.220*
Chlorophyll (A+B)+Carotenoids	1.391	1.092*	1.197*	1.269*
Peroxidase enzyme's activity, D/g raw wt/s	3.567	6.935*	5.962*	5.155*
Cell's sap acidity, pH	5.275	4.745*	4.663*	5.127
Photochemical activity of chloroplasts, mkmol of recovered $K_3Fe(CN)_6$/mg Chlorophyll/h	283.1	237.3*	257.9*	263.4*
Noncyclical photophosphorilation of chloroplasts, mkmol P/ mg Chlorophyll / h	103.4	83.6*	90.2*	92.1*
Cyclical photophosphorilation of chloroplasts, mkmol P/ mg Chlorophyll / h	161.1	153.0*	157.8*	161.3*
Speed of electron movement in the electron-transport chain of chloroplasts, P/2e	1.03	0.85*	0.92*	0.96
Protein contents mg / g dry wt	180.5	100.8*	131.0*	139.4*
Ratio nitrogen/sulfur contents	20.046	15.779*	17.537*	19.743
Ratio phosphorus/sulphur contents	1.407	0.713*	0.851*	0.932*
Ratio nitrogen /phosphorus contents	14.243	21.696*	20.614*	21.190*

* Statistically different from control values; $P<0.05$

It has been found that plants metabolism is significantly affected by toxic ingredients of industrial emissions. Before visible symptoms of injury appearance it is possible to observe changes in concentration of pigments, primary photosynthetical reactions activity, peroxidase's enzyme activity, protein contents and chemical elements ratio (see Table 1). In the zone of high pollution (SO_2 with sulfur accumulation 0.157%) in the Scots pine needles concentration of chlorophyll A was lower as control of 24.2%, chlorophyll B - of 17.1% and carotenoids - of 28.3%. Carotenoids are more sensitive compared with chlorophyll A, and chlorophyll A is more sensitive compared with chlorophyll B. The present study indicate on violation primary photosynthetical reactions in the needles in high pollution site: Hill's reaction activity was lower as control of 16.2%, cyclical photophosphorilation - of 5.1% and noncyclical phosphorilation - of 19.2%. Speed of electron movement in the electron-transport chain is lower compared with control of 17.5%, protein contents - of 45%. Air pollution changes optimal ratios of chemical elements: N/S, P/S, N/P and cell's sap acidity. We have established increase of peroxidase enzyme activity of 94.4% compared with control zone. Increase of peroxidase activity as compared to reference samples is related to a significant intensification of free-radical oxidizing chain reactions and the formation of organic peroxides. Peroxidase activation may also be regarded as substrate one, resulting directly from atmospheric pollutants influence on the electron transport chain of respiration. The principal function of peroxidase is to protect the organism against the adverse effects of peroxides, although it plays an important role in the total cellular metabolism, oxidizing many important intermediates.

Large amounts of sulfur dioxide and nitrogen oxides are continously released into the atmosphere. When they react with water, these gases can form sulfurous acid, nitrous acid and nitric acid. The weak acids are easily oxidized to the strong acids sulfuric acid and nitric acid. Photodissociation of nitrogen dioxide in the atmosphere produces oxygen radicals wich give rise to the formation of ozone. SO_2, NO_x, O_3 major gaseous air pollutants. There is no doubt that they are potentially toxic to man, animals and plants. Air pollution negatively impact on photosynthesis, transpiration, respiration,metabolism and ultrastructure of plants [3]. Another form of cellular damage by air pollutants is oxidation [13,14]. In the light, oxigen radicals are involved in the oxidation of the sulfite wich has been formed from SO_2. If not trapped by cellular quenching systems, the radicals can also oxidize cellular constituents. In addition to attacking the unsaturated fatty acides of biomembranes, phytooxidants rapidly oxidize SH-groups. In many enzymes, SH-groups form part of the catalytic center. The active conformation of other enzymes depends on the reductive splitting of S-S bridges. Oxidation will inactivate these enzymes. Clearly, cells must be capable of trapping oxidants before they can act, or reverse oxidation after it has occurred. Oxidized components which have been damaged beyond repair must be replaced. Survival will depend on the capacity of trapping and on repair mechanisms. It is reported that under the influence of sulfur dioxide the intensity of transport of assimilates is inhibited and this inhibition is greater than would be indicated by the lowering of the photosynthetic rate [15,16].

Chronic atmospheric air pollution by toxic compounds of industrial emissions leads to the violation of assimilative organs Scots pine (*Pinus sylvestris* L.) in Belarus. This violation consists of an increase in sulfur accumulation level, in considerable reduction of carotenoids, in changing of optimum correlation of green and yellow pigment pools, in acidiation of cellular contents and reduction of cytoplasm buffering, in activation of peroxidase tissue activity, in inhibition of primary photosynthetical reactions (Hill's reaction activity, cyclical and noncyclical phosphorilation) until appearance of visible symptoms of injury. The revealing changes of metabolism are the factors of plant stability reduction in industrial environment and may be used for ecological monitoring of forest ecosystems.

4. Tolerance and Air Pollution Accumulation Capabilities of Some Tree Species in Belarus

In the industrial regions there is the necessity of the good protection zones around the factories and plants of power engineering. Such zones should consist of resistant trees and shrubs, able to accumulate great amounts of toxic substances from the polluted atmosphere. However, we still do not have enough good plant material for the creation of such zones [17]. Determination of the chemical composition of plants is one of the most frequently used methods of monitoring environmental pollution [9].

Air pollution (mainly SO_2, NO_x) is considered one of the main causes of forest decline and green plantations damage under conditions of urban environment and in the vicinity of chemical plants and industrial centers.

Various strategies exist for controlling atmospheric pollution. The most desirable is to eliminate atmospheric pollution. The most desirable is to eliminate, or a least reduce the emission at the source. This is not always possible, and in any case, is expensive. A second and sometimes necessary approach is to grow pollutant-tolerant plants [18].

In Europe pollutant tolerance has sometimes been incorporated as an important criterion in reforestation programmes and environmental optimization by planting of greenery (green-belt setting) [3,18,19].

Gas-resistance of different plant species is not the same because of their anatomical, morphological, physiological and biochemical characteristics. Some species can tolerate 5- to 10- fold concentrations of a harmful gas without any noticeable damage as compared with the others [20].

Selection of the species assortment is a key point in producing gas-resistant plant communities that may efficiently clean the air of poisonous admixtures. It is well known that most species have a wide range of tolerance to different stressors and that certain genotipes may be more susceptible to damage than other. In order to increase the effects of using trees and shrubs, an official list was prepared of plants which may be used for optimization of the environment and creation of verdure in the urbanized and industrial landscape in Belarus.

4.1. COMPARATIVE TOLERANCE AND GAS-ACCUMULATION CAPABILITIES OF SOME TREE SPECIES IN SO_2, H_2S, CS_2 - POLLUTED ATMOSPHERE

Many authors have studied the range of variability resistance to industrial pollution between various species [16-20].

According to results of the investigations of some plant species in the vicinity of the chemical factory in Svetlogorsk , air pollution by SO_2, H_2S,CS_2 in most cases evokes chlorosis and necrosis foliage. Trees and shrubs affected by phytotoxic air pollutant concentrations react to this stress with decreased radial and volume growth.

In zone relative low level of air pollution (average-day concentration of SO_2 - 0.021 mg m^{-3}, H_2S - 0.034 mg m^{-3}, CS_2 - 0.045 mg m^{-3}) we can identify two group of species. One of them has gaseous necrosis of leaves: (*Betula pendula* Roth., *Ulmus pinnato-ramosa* Dieck., *Picea pungens* Engelm., *Caragana arborescens* Lam., *Cotoneaster lucidus* Spaech., *Acer saccarinum* L., *Aesculus hippocastanum* L., *Tilia cordata* Mill., *Juglans mandshurica* Maxim., *Syringa vulgaris* L., *Pinus sylvestris* L., *Populus simonii* Carr., *Malus domestica* L.), and another hasn't any visual damage of leaves during all vegetation period: (*Chaenomeles maulei* (Mast.) C.K.Schneid., *Amorpha fruticosa* L., *Ligustrum vulgare* L., *Pyrus communis* L., *Swida alba* L., *Guercus borealis* Michx.f., *Lonicera tatarica* L., *Salix alba* L., *Acer platanoides* L., *Physocarpus opulifolius* (L.) Maxim., *Ribes nigrum* L., *Phyladelphus coronarius* L., *Fraxinus excelsior* L.

In the zone of high level of air pollution (average-day concentration SO_2 0.045 mg m^{-3}, H_2S - 0.05 mg m^{-3}, CS_2 - 0.06 mg m^{-3}) all species had damage from air pollution, but in different degrees. We can identify 3 group of plant species: 1 - species with considerable gaseous necrosis, which surpass 40% surface of their foliage (*Betula pendula* L., *Ulmus pinnato-ramosa* Dieck., *Caragana arborescens* Lam., *Aesculus hippocastanum* L., *Tilia cordata* Mill., *Syringa vulgaris* L., *Populus simonii* Carr.,); 2 - species which have injured assimilation organs in limits from 20% to 40% of their surface: (*Cratargus oxyacanta*L., *Pirus communis* L., *Lonicera tatarica* L., *Salix alba* L., *Acer platanoides* L., *Populus canadensis* Moench., *Juglans mandshurica* Maxim., *Malus domestica* L., *Fraxinus excelsior* L.); 3 - species which have weak gaseous necrosis in limits from 1% to 19% their surface: (*Ligustrum vulgare* L., *Swida alba L.*, *Elaeagnus angustifolia* L., *Physocarpus opulifolius* (L.) Maxim., *Robinia pseudoacacia* L., *Rosa rugosa* Tunb., *Phyladelphus coronarius* L.).

It has been established that trees and shrubs take up from air and neutralize in their tissues a considerable amount of sulfur-containing poisonous ingredients of industrial fume wastes. Gas-resistant plants with a great capacity for absorbing harmful admixtures perform filtering and protective functions more successfully.

In the atmosphere without noticeable air pollution each of the plant species has its own level of sulfur accumulation that varies from 0.02 to 0.9% [20]. In an atmosphere contaminated by gaseous necrosis sulfur-containing organs sharply increases [3,16,18,19,20].

Sulfur is taken up by plants as sulfates and is partially bound with organic based matter, being transferred into a reduced form. The specificity of sulfur accumulation may by attributed to biological characteristics of the species, different gas uptake and unequal rates of the outflow and redistribution of the substances in plant organs [21].

As has been demonstrated by present data, in SO_2, H_2S, CS_2 - polluted air sulfur contents in leaves is already higher than in the control at the initial vegetation (May) and continues to increase while the plants grow and develop. By autumn (September) some species are found to contain lower sulfur contents which might be explained by a decay of the sulfur-containing compounds, their migration to the stems and roots and washing out from the organs with the rain water. The higher the air pollution level, the more pronounced is the phytoindicating function of the species.

In lowly polluted regions, a maximum level of sulfur accumulation (as the result uptake S from air polluted atmosphere) was observed in leaves of *Ligustrum vulgare* L., *Swida alba* L., *Quercus borealis* Michx.f., *Lonicera tatarica* L., *Salis alba* L., *Elaeagnus angustifolia* L., *Ribes nigrum* L., *Populus simonii* Carr. - 0.416 - 0.736 % S. A minimum level of sulfur accumulation was in leaves of *Amorpha fruticosa* L., *Betula pendula* Roth., *Cotoneaster lucidus* Spaech., - 0.124 - 0.188% S. Intermediate position engaged species such as *Chaenomeles maulei* (Mast.) C.K. Schneid., *Ulmus pinnato-ramosa* Dieck., *Pyrus communis L.*, *Acer saccarinum L.*, *Aesculus hippocastanum* L., *Physocarpus opulifolia* L., *Syringa vulgarus* L., *Phyladelphus coronarius* L., *Malus domestica* L. - 0.224 - 0.396% (see Table 2).

TABLE 2. Sulfur accumulation in leaves of some tree and shrub species in the result of SO_2, H_2S, CS_2 absoprtion (zone of low air pollution)

Species	Sulfur accumulation, % dry wt.		
	May	July	September
Chaenomeles maulei (Mast.) C.K.Schneid.	0.256	0.320	0.329
Amorpha fruticosa L.	0.045	0.124	0.096
Betula pendula Roth.	0.128	0.160	0.188
Ligustrum vulgare L.	0.237	0.640	0.544
Ulmus pinnato-ramosa Dieck.	0.224	0.352	0.320
Pyrus communis L.	0.128	0.160	0.288
Swida alba L.	0.480	0.736	0.516
Quercus borealis Michx.f.	0.288	0.416	0.392
Lonicera tatarica L.	0.352	0.480	0.448
Salix alba L.	0.448	0.448	0.296
Cotoneaster lucidus Spaech.	0.092	0.128	0.098
Acer saccharinum L.	0.045	0.128	0.224
Aesculus hippocastanum L.	0.320	0.340	0.320
Elaeagnus angustifolia L.	0.320	0.479	0.510
Juglans mandshurica Maxim.	0.222	0.256	0.448
Physocarpus opulifolius (L.) Maxim.	0.360	0.384	0.320

TABLE 3. **Sulfur accumulation in leaves of some tree and shrub species in the result of SO_2, H_2S, CS_2 absorption from air (zone of high air pollution)**

Species	Sulfur accumulation, % dry wt		
	May	July	September
Ligustrum vulgare L.	0.525	0.699	0.448
Crataegus oxyacantha L.	0.320	0.352	0.384
Ulmus pinnato-ramosa Dieck.	0.257	0.352	0.310
Pyrus communis L.	0.224	0.288	0.210
Swida alba L.	0.512	0.896	0.554
Lonicera tatarica L.	0.352	0.704	0.640
Salix alba L.	0.416	0.448	0.421
Caragana arborescens Lam.	0.256	0.512	0.384
Acer platanoides L.	0.256	0.512	0.704
Aesculus hippocastanum L.	0.288	0.568	0.490
Tilia cordata MillL.	0.320	0.732	0.576
Elaeagnus angustifolia L.	0.416	0.576	0.824
Juglans mandshurica Maxim.	0.384	0.416	0.380
Physocarpus opulifolius (L.).Maxim.	0.192	0.416	0.390
Robinia pseudoacacia L.	0.237	0.448	0.480
Rosa rugosa L.	0.205	0.320	0.340
Syringa vulgaris L.	0.416	0.597	0.503
Populus tremula L.	0.232	0.672	0.384
Populus canadensis Moench.	0.320	0.668	0.446
Populus simonii Carr.	0.288	0.512	0.352
Philadelphus coronarius L.	0.392	0.480	0.352
Malus domestica L.	0.120	0.384	0.512
Fraxinus excelsior L.	0.350	0.490	0.430

In highly polluted regions species such as *Ligustrum vulgare* L., *Swida alba* L., *Lonicera tatarica* L., *Acer platanoides* L., *Tilia cordata* Mill., *Elaeagnus angustifolia* L., *Populus tremula* L. and *Populus canadensis* Moench. have exhibited the highest gas-accumulative ability: 0.668-0.896% S. *Pyrus communis* L., *Crataegus oxyacanta* L., *Ulmus pinnato-ramosa* Dieck., *Rosa rugosa* Thunb. are found to have the lowest air pollution accumulation capabilities: 0.288 - 0.384% S. Intermediate capability was observed in *Salix alba* L., *Caragana arborescens* Lam., *Aesculus hippocastanum* L.,

Juglans mandshurica Maxim., *Physocarpus opulifolius* (L.) Maxim., *Robinia pseudoacacia* L., *Syringa vulgaris* L., *Populus simonii* Carr., *Phyladelphus coronarius* L., *Malus domestica* L., *Fraxinus excelsior* L.: 0.416-0.597% S, which was uptaken from air (see Table 3).

The present data show good comparative results on the ability of different plant species to absorb and neutralize gaseous sulfur poisonous ingredients. The actual gas uptake of trees and shrubs is however considerably higher, since sulfur is absorbed by shoots, migrates to other organs and is periodically removed from the leaves by rain water. Experimental washing out of sulfur from leaves with distilled water has indicated that from 8 to 40% of the sulfur taken up from air is washed out with distilled water, therefore the role of the plants in cleaning the air from harmful substances becomes many times more important. In annual shoots 2- to 6-fold less sulfur is accumulated as compared with the leaves. However in view of their large biomass, the shoots may also be considered an important contributor to sulfur accumulation.

Perspective species for creation of verdure in the industrial and urbanized landscape in zones air polluted by SO_2 H_2S, CS_2 are: *Ligustrum vulgare* L., *Crataegus oxyaccanta* L., *Salix alba* L., *Acer platanoides* L., *Elaeagnus angustifolia* L., *Juglans mandshurica* Maxim., *Physocarpus opulifolius* (L.) Maxim., *Robinia pseudoacacia* L., *Rosa rugosa* Tunb., *Fraxinus excelsior* L. They combine gas-tolerance and high level air pollution accumulation capabilities.

Caragana arborescens Lam., *Aesculus hippocastanum* L., *Tilia cordata* Mill., *Syringa vulgaris* L., *Populus simonii* Carr., *Populus tremula* L., *Picea pungens* Engelm., *Pinus sylvestris* L. cannot be recommended for creation of green plantations in the areas with high concentrations of SO_2, H_2S,CS_2 due to their gas-sensibility.

4.2. COMPARATIVE TOLERANCE AND GAS-ACCUMULATION CAPABILITIES OF SOME TREE SPECIES IN NO_2, NH_3 POLLUTED ATMOSPHERE

On the basis of comparative tolerance a number of aboriginal and introduced woody plant species from results of investigations in the areas of industrial air pollution by NO_2 (0.042 mg m^{-3}) and NH_3 (0.04 mg m^{-3}) near Grodno; 3 groups of species were identified.

The first group is comprised of 37 of gas-tolerance to NO_2, and NH_3 such as *Berberis vulgaris* L., *B.v. 'Atropurpurea'*, *Berberis heteropoda* Schneid., *Berberis thunbergii* D.C., *Sambucus racemosa* L., *Sambucus nigra* L., *Vitis amurensis* Rupr., *Vitis vulpina* L., *Vitis riparia* Michx., *Vitis rupestris* Scheele., *Gleditsia triacanthos* L., *Partenocissus quinquifolia* L., *Lonicera caerula* L., *Caragana arborescens* Lam., *Cotoneaster lucidus* Spaech., *Acer platanoides* L., *Acer saccarinum* L., *Acer negundo* L., *A.n. 'Auratum'*, *Aesculus hippocastanum* L., *Schizandra chinensis* (Turcz.) Baill., *Elaeagnus angustifolia* L., *Hippophae rhamnoides* L., *Robinia pseudoacacia* L., *Rosa rugosa* Thunb., *Rhus aromatica* Ait., *Rhus typhyna* L., *Populus bolleana* Lauche., *Populus trichocarpa* Torr. et Gray, *Populus balsamifera* L., *Populus alba* L., *Populus berolinensis* Dipp., *Populus nigra 'Italica'*, *Forsythia ovata* Nakai, *Forsythia europaea* deg. et Bald., *Sheferdia argentea* Nutt.

The second group is comprised by 79 species with middle gas-tolerance such as coniferous trees: *Larix decidua* Mill., *Larix sibirica* Ldb., *Larix leptolepis* Gard., *Thuja occidentalis* L., *T.o. ' Globosa'*, *T.o. 'Columna'*, *T.o. 'Elwangeriana aurea'* and leaf-bearing trees: *Actinidia kolomicta* Max.., *Amorpha fruticosa* L., *Aronia melanocarpa Elliot.*, *Euonimus sacrosanta* Koidz., *Euonymus latifolius* (L.) Mill., *Betula pendula Roth.*, *Betula pubescens* Ehrh., *Ligustrum vulgare* L., *Crataegus oxyacantha* L., *Crataegus sanguinea* Pall., *Crataegus monogyna* Jacq., *Fagus silvatica* L., *F.s. 'Atropunicea'*, *Cerasus avium* (L.) Moench., *Ulmus laevis* Pall., *Ulmus pinnato-ramosa* Dieck., *Hamamelis virginiana* L., *Hydrangea arborescens* L., *Carpinus betulus* L., *Pyrus ussuriensis* Maxim., *Quercus borealis* Michx.f., *Quercus robur* L., *Celastris orbiculata* Thunb., *Lonicera tatarica* L., *Lonicera caprifolium* L., *Lonicera xyloterum* L., *Lonicera caerula* L., *Salix alba* L., *S.a. 'Vitellina pendula'*, *S.a. 'Aurea'*, *Salix fragilis* L., *Salix caprea* L., *Caragana turcestanica* Kom., *Acer ginnala* Maxim., *Acer pseudoplatanus* L., *Acer tataricum* L., *Acer turcestanicum* Pax., *Dasiphora fruticosa* (L.) Rydb., *Coryllus avellana* L., *C.a. 'Atropurpurea'*, *Tilia cordata* Mill., *Tilia europea 'Laciniata'*, *Tilia tomentosa* Moench., *Elaeagnus argentea* Parsh., *Juglans mandshurica* Maxim., *Physocarpus opulifolius* (L.) Maxim., *Sorbus sibirica* Hedd., *Rhododendron dahuricum* L., *Swida alba* L., *S.a. 'Argenteo marginata'*, *Swida sanguinea* L., *Syringa vulgaris* L., *Syringa josikaea* Jacq.f., *Prunus divaricata* Ldb., *Spiraea vanhouttei* (Briot.) Zbl., *Spiraea japonica* L., *Spiraea media* Frieder. Schmidt., *Populus tremula* L., *Populus canadensis* Moench., *Padus maackii* (Rupr.)Kom., *Philadelphus coronarius* L., *Ph.c. 'Auraeus'*, *Philadelphus lemoinei* Lemoine., *Ostrya virginiana* (Mill.) C.Koch., *Ostrya carpiniflora* Scop. , *Malus domestica* L., *Malus niedzwetzkyana* Diek., *Fraxinus excelsior* L.

These two groups of plant species can be useful for creation of verdure and optimization of urban and industrial landscape in the areas air pollution by NO_x, NH_3.

Lastly, the third group are the gas-sensible species: *Pseudotsuga menziesii var. glauca* Schn., *Picea pungens* Engelm., *Picea abies* (L.) Karst., *Juniperus sabina* L., *Juniperus communis* L., *J.c. 'Hibernica'*, *Abies concolor* Lindl. et Gord., *Pinus mugo* Turra, *Pinus sylvestris* L., *Taxus baccata* L., *Laburnum anagiroides* Medic., *Catalpa speciosa* Warder., *Catalpa bignoioides* Walf., *Cladastris lutea* (Michx.f.) C.Koch., *Aristolochia macrophylla* Lam., *Clematis vitalba* L., *Mahonia aquifolia* Nutt., *Sorbus aucuparia* L., *Rhododendron catawbience* L., *Rhododendron luteum* Sweet.

Tree and shrub species effectively absorb nitrogen gaseous air pollutants (see Table 4).

TABLE 4. Nitrogen contents in leaf some species of tree and shrub in control (A) and in zone of air pollution by NO_x, NH_3 (B)

Species	Test	Nitrogen contents,% dry wt			
	area	protein N (C)	nonprotein N (D)	total N	C/D
Robinia pseudoacacia L.	A	2.072	0.120	2.192	17.267
	B	3.892	0.368	4.260	10.576
Sambucus racemosa L.	A	2.332	0.752	3.084	3.101
	B	3.521	1.250	4.771	2.817
Acer negundo L.	A	2.004	0.448	2.452	4.473
	B	3.515	0.500	4.015	7.030
Populus bolleana Lauche	A	1.269	0.240	1.509	5.287
	B	2.226	0.292	2.518	7.623
Partenocissus quinquifolia L.	A	1.548	0.380	1.928	4.074
	B	2.325	0.489	2.814	4.755
Acer saccarinum L.	A	0.697	0.204	0.901	3.417
	B	0.893	0.215	1.108	4.153
Elaeagnus angustifolia L.	A	1.601	0.253	1.854	6.328
	B	1.585	0.432	2.017	3.669
Hippophae rhamnoides L.	A	2.087	0.140	2.227	14.910
	B	4.400	0.660	5.060	6.607
Acer platanoides L.	A	1.514	0.092	1.606	16.456
	B	2.710	0.132	2.842	20.530
Acer pseudoplatanus L.	A	0.851	0.075	0.926	11.342
	B	1.423	0.100	1.523	14.230
Philadelphus coronarius L.	A	1.800	0.248	2.048	7.258
	B	2.359	0.284	2.643	8.306
Tilia cordata Mill.	A	3.822	0.370	4.192	10.330
	B	4.121	0.630	4.751	6.540
Caragana arborescens Lam.	A	1.332	0.334	1.676	3.880
	B	1.588	0.568	2.156	2.796
Salix alba L.	A	0.898	0.160	1.058	5.612
	B	0.861	0.360	1.221	2.392

5. Summary

There are a number of general points one can draw from the preceding information. Chronic air pollution by SO_2 leads to violation of physiological parameters of Scots pine (*Pinus sylvestris* L.) needles. This violation consists of considerable reduction of carotenoids, in acidiation of cellular contents and reduction of cytoplasm buffering, in activation of peroxidase tissue activity, in inhibition of primary photosynthetical reactions (Hill's reaction activity, cyclical and noncyclical photophosphorylation) in connection with sulfur accumulation in the needles (from 0.11 to 0.17%) until the appearance of visible symptoms of injury. The revealing changes of metabolism are the consequence of forest stability reduction in air polluted atmosphere in Belarus.

Sulfur accumulation in the needles *Pinus sylvestris* L. in the larger part of Belarus is not at the critical level. The author's study confirmed the usefulness of Scots pine *(Pinus sylvestris* L.) as a bioindicator (pollution accumulator) of environmental pollution with sulfur.

Various plant species respond to air pollution in different ways. For the environment optimization by means of gardening, the most promising are plant species which combine a high level of tolerance with clearly expressed air pollution accumulation capabilities. Biological accumulation and neutralization of poisonous gases should be considered as a means of smog control, in addition to conventional engineering methods, as a factor in atmosphere aftercleaning.

6. References

1. State report about environment condition in Republic of Belarus (1985), in V.F.Loginow (ed.), *Air pollution,* Ministry of Natural Resourses and Environmental Protection of Belarus Publishers, Minsk, pp. 29-31.

2. Nihlgard, B. (1985) The ammonium hypotesis - an additional explanation to the forest dieback in Europe, *Ambio* **14**, 2-8.

3. Sergeichik , S.A. (1994) *Woody Plants Resistance under Technological Air Pollution Pressure,* Nauka i Tekhnica Publishers, Minsk.

4. *Forest landscapes of Belarus* (1992), E.A.Sydorovich (ed.), Nauka i Technica Publishers, Minsk.

5. Kizino , F.E. and Lebedev, V.A. (1990) *Forest wealth of Belorussia,* Uradzaj Publishers, Minsk.

6. Huttunen, S., Laine, K. and Torvela, H. (1985) Seasonal sulphur contents of pine needles as indices of air pollution, *Ann. Bot. Fennici* **22**, 343-359.

7. Molski, B. , Bytnerowicz, A. and Dmuchowski, W. (1983) Mapping air pollution of forests and agricultural areas in Poland by sulphur accumulation in pine (*Pinus silvestris* L.) needles, *Aquilo Ser. Bot.* **19**, 326-331.

8. Molski, B. and Dmuchowski, W. (1990) Distribution and movement of selected elements in Poland using pine needle analysis, in W.Grodzinski, E.B.Cowling and A.Breymeyer (eds.), *Ecological Risks - Perspectives from Poland and the United States,* National Academy Press, Washington, D.C., pp. 215-231.

9. Dmuchowski, W. and Bytnerowicz, A. (1995) Monitoring environmental pollution in Poland by chemical analysis of Scots pine (*Pinus sylvestris* L.) needles, *Environmental Pollution* **87**, 87-104.

10. Gavrilenko, V.F., Ladigina, M.E. and Handobina, L.M. (1975) *A big practice on plant physiology*, Vischaja shkola Publishers, Moskow.

11. Sydorovich, E.A., Arabej, N.M., Busko, E.G., Alechno, A.I., and Piculik, N.I. (1991) *Pollution of the Forest Landscapes of Belarus with Sulfur- and Nitrogen- Compounds and Heavy Metals*, BelNIINTI Publishers, Minsk.

12. Linzon, S.N. (1983) Long -term effect of sulphur dioxide on forest growth, *Aquilo Ser. Bot.* **19**, 157-166.

13. Wallace, R.G. and Spedding, D.J. (1976) The biochemical basis of plant damage by atmospheric sulphur dioxide, *Clean Air* **10**, 61-64.

14. Nikolaevskij, V.S. (1979) *Biological Basis of Plant Gas-Resistance*, Nauka Publishers, Novosibirsk.

15. Lorenc-Plucinska, G. (1982) Influence of SO_2 on CO_2 assimilation and carbon metabolism in photosynthesis processes in Scots pine, *Arboretum Kornickie* **27**, 285-310.

16. Barachtenova, L.A., Nikolaevski, V.S. (1988) *Effect of Sulfur Dioxide on Plant Photosynthesis*, Nauka Publishers, Novosibirsk.

17. Bytnerowicz, A., Dmuchowski, W. and Molski, B. (1990) The air pollution accumulation capabilities of some tree species in the vicinity of the chemical plant in Torun, *Rocznik Dendrologiczny* **33**, 15-27.

18. Sergeichik, S.A. (1984) *Woody Plants and Optimization of Industrial Environment*, Nauka i Technika Publishers, Minsk.

19. Sidorovich, E.A., Kulagin, Y.Z. and Sergeichik, S.A. (1981/1982) Biological aspects of optimization of the environment, *Rocznik Dendrologiczny* **34**, 25-36.

20. Ilkun, G.M. (1978) *Air Pollution and Plants*, Naukowa dumka Publishers, Kiev.

21. Tarabrin, V.P., Kondratuk, E.G. and Bashkatow, V.G. (1986) *Phytotoxic Effects of Organic and Nonorganik Pollutants*, Naukowa dumka Publishers, Kiev.

Position Papers

Working Group 2

"Forest Growth, Expansion and Management"

FOREST ACCESSIBILITY AND SUSTAINABLE DEVELOPMENT

F. ATANASIU, Dipl. Eng.
The National Institute of Wood (INL)
7, Sos. Fabrica de Glucoza
Bucharest 2
Romania

Abstract

This paper analyzes the conditions of Romanian forests (extrapolated to Eastern European countries) in connection with forest accessibility (density of forest road network) and with possible extraction technologies, based on forest management regulations in force vis-a-vis the situation of industrialized countries. Deliberations during NATO -ARW WORKSHOP from July 1996 will provide new contributions for better understanding and solutions, leading to a suitable wood production coupled with a stability of the ecosystems.

The analysis which we propose is based on ascertaining the situation of forest from Romania compared with that from other countries of Europe and in the world, in order to draw some particular or even general conclusions.

1. Some Forest Data

The evolution of forest area in Romania is:

- 6.100 mil ha in 1944
- 6.450 mil ha in 1957
- 6.400 mil ha in 1996

representing about 28% of the country's area

In 1944 the structure of forest by species was:

- 25% coniferous wood;
- 75% broad leaved wood (beech representing 34% and oak 19-20%)

For the same year, the distribution of forest depending on altitude was:

- 58% at the altitude exceeding 700 m;
- 33% at the altitude between 150 m and 700 m;
- 9% plain, up to 150 m.

R. M. Gutkowski and T. Winnicki (eds.), Restoration of Forests, 179–194.
© 1997 *Kluwer Academic Publishers.*

At present, the growing stock is estimated at about 1.3 billion cu.m., with an annual increase of more than 32.0 mil cu.m.

Considering the structure by species, the Romanian forests are presently composed of:

- 31% coniferous (mostly spruce);
- 31% beech;
- 12% oak;
- 15% various hardwoods;
- 5% various softwoods;
- 6% others.

The first regulations concerning management and exploitation of forest appeared at the end of the XVIII-th century and then, with the formation of the first specialists, the Forestry Code, as a fundamental law, appeared in 1892, with successive revisions in 1962 and 1996.

We can say that the principles of the scientific forestry have a long time tradition in Romania. Our forestry school is well known in the world for its highly experienced scientists and engineers (e.g. Prof. Chiriþã, Stinghe, Drãcea) who have beneficial effect on forest management.

The strong reduction of forest lands from the XIX-th century and beginning of the XX-th century is due to unfavorable conjectural events and also to a wild harvesting in the absence of suitable forest road development.

The author will discuss in detail this problem which is critical for the optimization of the steps to be taken in this situation, as compared to other East European countries, both in respect to logging and forest reserves.

2. Forest Roads

At the level of 1994 the forest railway network amounted to about 3000 km. During 1944-1955 period more than 1200 km were constructed and 40 km permanent cableway systems added to the permanent transport facilities. In 1959 the forestry sector owned 4340 km railways, in majority narrow track gauge of 760 mm. The strong expansion of the railway network was largely due to the fact that after World War II, our country underwent an intensive reconstruction which required important wood quantities and suitable transport facilities in a short period.

The expansion of the forestry railway network encountered problems which appeared soon after. So, the railways could not penetrate in the side valleys where the great forest units, especially exploitable stands, existed. At the same time, the railway required high investments and the operation costs were also important.

In contrast to the railways, the forest roads for motor vehicles which can be pushed inside the felling areas and thus reducing at great extent the extraction distances, allow for the expansion of intensive silvicultural treatments, make possible harvesting of all the forest products, require small investment and the transport cost is lower. Thus, the road transport is efficient even in the condition of logging of relatively reduced wood amounts.

Beginning with 1956 - 1957 a special interest was placed on forest road development due, mainly, to the development of tractor and truck manufacturing industries.

Simultaneously, the activity for restraining and dismantling the wood consuming temporary devices (chutes, wooden pavements) and their replacement with tractor roads and cableways was implemented.

The forests were State owned, the development of forest road networks covered easy accessible territory, aiming at the opening of forest areas, so as to allow the continuous harvesting of both the main and secondary products.

Gradually, the fixed cableway systems have been put out of operation and at the same time the majority of forest railways became inadequate. Owing to the expansion of the national hydropower system, the wood transport by rivers has been restricted to a minimum. Today it is still used but only in a reduced proportion in secluded areas (under 1.5% of total transport volume). Alert endowment of forest with roads during 1960-1970 (more than 1600 km/year) raised important problems with regard to designing and execution of these roads which called for some theoretically clear solutions. The technical progress in the road building has been realized by mechanization of works, advanced methods of site organization, use of prefabs, etc.

On a theoretical level, the concept of a "forest road network" and also the concept of "optimum density of transport roads" emphasizing the importance of correlation, at the technical and economic level, between the extraction road network and the permanent road network have been substantiated.

The problem of provision of the whole forest covered area with roads constituted a permanent preoccupation of foresters in the last twenty years, but, due to the lack of financing sources and to some technical considerations (loss of productive land, disclosure of some stands sensible to wind etc.) the steady development of forest road networks has been hindered.

The forest road network includes about 30,000 km with an average density of 6.2 m/ha placing Romania much below the levels of other European countries such as Austria (36 m/ha), Switzerland (40 m/ha), France (26 m/ha) and Germany (45 m/ha). With the existing forest road network there are rendered accessible just 4.1 mil. ha (65%) for a maximum extraction distance of 2 km (average distance is 1.2 km), while the rest of 2.2 mil ha (35%) are remaining out of this limit. Unfortunately, after 1989 the road building was terminated and only resumed in 1994 with 60 km forest roads.

Research concluded by the Faculty of Forestry and Wood Exploitation - Brasov, with regard to the optimum density of the forest road network indicated 12-20 m/ha are necessary. However, the forestry organizations from our country considered only 9-10 m/ha necessary for this period of time.

As a consequence, this poor road network gives rise to :

- improper correlation of the yearly cut with the allowable cut for each cutting unit which leads to excessive cutting in the more accessible forest logging units;

- failing to undertake the forest management operations in due time, with adverse effect on stand growth and -phyto-sanitary conditions, in addition to important wood quantities being left in the forest;

- impossibility to extend the selective cutting system at the required level;

- high production costs due to greater extraction distances;

- imbalance in distribution of the growing stock by age category, with a reduced share of exploitable and pre-exploitable stands.

The debates during the works of various organizations, have pointed out these shortcomings and underlined the requirement for a further development of forest road networks so as to reach the balanced opening of all forest reserves, for average extraction distances of 200 - 700 m and creating in this way the necessary premises for a scientific management and ecological reconstruction of stands. These proceedings have also given a special attention to the fact that the forest roads issue is of vital importance for the rational management of the forests.

Regarding the further development of forest road networks, research carried out by the "National Institute of Wood" - I.N.L. revealed that for obtaining the optimum density and for reaching the above parameters, it is necessary to provide another 45000 km of forest roads. For the first stage, up to the year 2000, 8500 km should be built, i.e. about 1700km/year respectively. It will be necessary at first to solve the problems of financing of this road construction program, by various sources : self-financing, state budget, extra budgetary funds (establishing of tax duties for merchantable trees, foreign loans, etc.). It is also to be noted that the forest roads in Romania have multiple functions, not only for forestry, attesting their public investment character as well.

The reduction of the extraction distance simplifies the logging technology making thus possible the use of cableways for short distances (up to 500 m) with 1t maximum load for secondary products and 1.5 - 2t for main products, working together with articulated forest tractors with 33 - 47 kW engine power for main products and 18 - 21 kW for secondary products. Animals can be used efficiently for short distances (wood extraction up to cable-way or tractor sites).

It is stressed that under the present environmental conditions, the further development of the forest road networks should be made by maximum environment protection. It is known that the roads built inside the forest are directly and indirectly prejudicial for the environment: firstly, due to the works necessary for road building and secondly, for the road traffic and the alterations appeared due to the construction works.

The steps considered appropriate to be taken are:

- avoidance of exposure to dominant winds after cutting the strips required for road construction;

- diminishing, as much as possible, the road building areas by balancing of cut and fill volume so as to limit the necessary stores and borrow pits;

- strictly observation of the cutting strip width as it is specified in the projects;

- execution of embankments one year before the consolidation of the practicable lanes;

- planting, grassing for the improvement of the aesthetically aspect of the banks;

- vegetation masking of stone quarries at the end of their exploitation;

- protection shields, by tree planting, along the forest roads outside the forest;

- removal of salts from materials used during winter for skidding prevention so as to protect free water and radicular system of trees and shrub near the roads;

- improvement of the fuel and admission systems used by the motor vehicles.

For limiting the working time of the open sites in the forest, the road execution with torrent correction operations have to be correlated. Layout of forest roads must be made so that to provide slope stability.

Finally, it is emphasized, as a general truth, that in the absence of the optimum forest road network, a proper forest protection and a sustainable forestry development are not possible. Forest specialists, working in extremely hard conditions, must find the best methods for granting the role of forests as a permanent wood source. Longer extraction distances, at present time, lead to technologies which cannot be matched with forest protection. The modern logging technology used in developed countries including equipment such as tree harvesters and forwarders can be used only for suitable ground conditions (small slope), and requires 200-400m collecting distance. In the case of Romania, due to hard conditions, about 2.2 mil ha (35%) remained unharvested so far, so that there are still compact forests in various mountain areas.

3. Wood Extraction Technologies

Each logging technology includes some methods and means for wood extraction which have direct impact on forests.

Wood in various shapes and dimensions (trees and parts of trees, stems, logs, branches) determines the solutions, the power of vehicles, loading method, etc., which will affect the geometrical characteristics of roads. The sites for rough conversion and sorting require provision of stores, operating spaces etc.

The concept of forest provision with permanent and movable transport facilities in Romania evolved, in time, towards a continuous opening of forest, especially by valley roads. Knowing that the increase of the average extraction distance is directly proportionate with the density of the transport networks and their location, as compared to forest stands, the forest roads designed rationally have reduced the distances and the efforts of the wood extraction operations, increasing thus the efficiency.

To obtain a minimum extraction distance, the forest road network must be suitable to the land survey and peculiarities of the forested area to be harvested. Provision of the forest reserves with transport ways in order to obtain average extraction distance as close as possible is beneficial both for the uniformity of machinery system used and of the logging methods.

Possible extraction distances:

- by tractor, 1200 - 1800 m;
- by cableway system, 1500 - 2000 m;
- by animals, 300 - 500 m.

Logging technology and especially the operating method are in continuous evolution with suitable impact on the elements of strategy and structure of forest road network. Within close periods of time (1960 - 1975, 1975 - 1990, 1990 - 1995) when the density of networks was not changed sensibly, the methods used in Romania were:

- rough conversion at the stump for final assortments;
- trunks and masts logging;
- tree lengths and parts of trees.

The evolution of the logging technologies, although they have been influenced by each other, has been marked usually by silvicultural considerations and economic policies.

Fig. 1 - Logging technologies in practice in Romania during the years 1960 - 1990

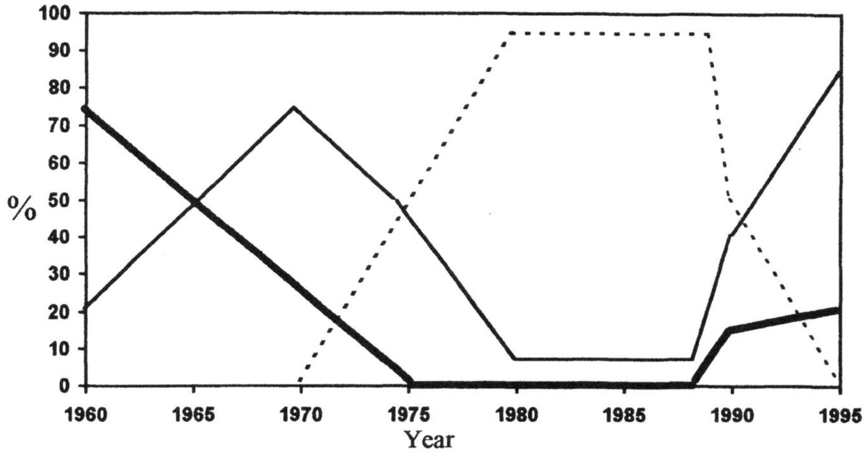

INL - P.Boghean & Al. Pavel

Legend: (thick line) Conversion and sorting to the stump place; (thin line) Trunks and masts; (dotted line) Full or partially cut tree

From the above diagram it results that trunks and masts logging is more and more used.

From the recent data, the share of the wood extraction from the total cost of logging operations is about 39%, between the alternative use of the tractors and cables, their exists an insignificant difference (3%).

Solutions for wood extraction take into consideration the physical-geographical diversity, and try to use at the best, the relief energy and are aimed to achieve technical and economic efficiency under given conditions. At the same time, the progress in the conception of installations and equipment and in the organization of extraction operations, keep the pace of the advances for optimum levels of forest road networks which evolve however, slowly. Provision with forest roads and the logging technology used influence largely the structure and the characteristics of the machinery system and especially the extraction method.

From the mechanical systems used for wood extraction, the most sensible to the density of forest road network, determining in fact average gathering distance, are the cable installations. The first cable systems used in Romanian forests (1950)

which were not properly provided with forest roads at that time, were the semi-permanent type cable systems installed on distances of more than 2 km. As far as the roads have gradually been built, the machinery system included also mobile cable systems FP-2 for distances up to 2000m and FPU 500 for distances up to 500m. At present, the extraction distance with the cable system FP-2 could be reduced to 1500m.

Extraction distance (length of system) is the main characteristic of the cableways. Considering the present density of the forest road networks, the cable systems FP-2 and FPV-500 are used at their maximum limits. The test of using a tower cable system for distances up to 400m. (FUMO 403) with greater mobility and short assembling time has not given the expected results. The use of such a system and of other similar ones depend on the extent of further construction of forest roads.

In the general context of an existing forest road network, the technical means attached to collecting lines are directly implied, with a determined share, by the rhythm of the process flows as well as by the economic efficiency up to the transport stage. Due to this fact, the establishing of the collecting lines is done depending on factors either individual or combined.

The natural and technical-economic factors from Romania are especially imposing the problem of interfering the tractor with the cable system. Configuration of the relief, specific to a large majority of the mountain forests (61%), brings these two types of equipment as optional when sometimes are difficult to be substantiated. For the same average extraction distances, the spreading out of tractor roads inside the felling area lead, due to road sinuosity to a covering area 2-3 times greater than in the case of a cable system. This aspect is strictly economic and the selection of one or another type of the proposed solutions requires a more complex analysis. It can be appreciated that the cable system has an increased capacity for being adapted in case of large extraction distances and it is well accepted due to the environment protection requirements.

In the case of the articulated forest tractor, the extraction distance considered efficient from technical and economic point of view is 600 - 800 m. Beyond this distance the efficiency is reduced sensibly. The performances of skidding tractors represents from the technological point of view, an adaptation to the specific characteristics of the felling sites in the respective country and to the present level of road construction.

Disadvantages:

- excessive felling in the more accessible production units;
- limiting of selective cuttings;
- preferential use of concentrated extraction;
- high production costs, manpower, materials and energy;

➢ improper utilization of forest products.

In time, the lack of a proper density of forest road networks imposed the adoption of uneconomic solutions either for forest management or wood extraction such as:

➢ damage of soil, seeds and standing trees, beyond the normal values;

➢ important wood resources insufficiently managed and chal with much delay in the economic circuit.

Among the most important technical and economic factors intervening in the development of a forest road network a mention is made for the methods for the collecting and transport of wood.

The insufficient transport network is regularly compensated by a higher level of mechanized transport. But it is restrictive due to energy shortage and their environmental negative influence.

4. Transport By Trucks

The wood transport is executed predominantly by forestry trucks (91%). The development in our country of the auto transport has been stimulated by various factors such as: relatively reduced costs by comparison with railways, the energy sources, increased accessibility and mobility of road means by comparison with other solutions.

At present, some factors have lost their value, since the cost per km of forest road has increased greatly as the fuel costs grew gradually. The necessity to reduce the specific fuel consumption determined the changing of forest road transport means structure in the direction of introducing those with bigger capacities. This increase influenced adversely the road wear, leading thus to the increase of costs for maintenance and repair.

The share of the main type of motor vehicles used for wood transport is covered by those with a loading deck of 14 t (32%), truck- and - trailers of 20 t (30%) and of 16 t (19%). For the future, it is envisaged the step up of the modernization of transport facilities which shall include the improvement of the superstructure and especially the provision of hydraulic cranes. The preoccupation of our specialists is directed towards the establishment of new transport facilities based on some types of tractors with increased power and loading capacity.

Regarding the forest road network and its importance on methods and techniques used for rational wood harvesting, the following conclusions can be drawn:

- forest road density, as to optimum necessities under the conditions of our country which is stipulated to 13.3 m/ha in the first stage, is at present not sufficient for an efficient utilization of forest products;
- logging technologies include greater average extraction distances with adverse effect on production costs;
- extraction facilities, and especially cableway systems, register a decrease in number due to long erection lengths and are replaced by other equipment with adverse ecological influence;
- tractors and cable systems, considered the basic equipment for wood extraction, operate on large distances marking a limit beginning of the areas of inefficiency;
- the average extraction distance of 1.2 - 1.4 km leads to a heavy machinery system with negative influence in the flow: cable systems compared with tractors and, in case of other means, animals compared with skidding, are better framed with requirements for forest protection;
- the transport operations also exhibit difficulties in both diversification of types and dimensions and in the provision with modern equipment for loading - unloading specific for modern technologies.

5. Sustainable Forest Development

Being an extremely vast subject let's focus on some more important subjects.

As it can be seen, the biggest problem in our forests is the acute lack of roads which between 1945 and 1989 led to the intensive exploitation of the easily accessible areas which resulted in a severe imbalance in the homogeneity of stands, according to silvicultural norms. The necessary step which was taken was the decrease in the annual cut to about 1/2 of the allowable cut (about 32 mil cu.m/year) and thus the rebalancing of the situation in the whole exploited forests. The great chance is, as mentioned above, that 1/3 of forest areas remained intact. Moreover, the standing trees in our country represent a relatively important volume of about 1.3 billion cub. m. i.e. 7% of the European volume of forests.

Forest management was an important factor even from the beginning of the XX-th century, with a renewal every 10 years. Presently, the Forestry Code, approved by the Parliament in 1996, became compulsory and must be applied entirely to comply with environment regulations worldwide.

Certainly, today it is imperative to treat the forest in conjunction with the economic and social aspects, observing the present regulations for wood utilization. This new concept of complete analysis of the problems at the regional and national level (Master plan) is more and more necessary.

After 1989, I.N.L. - Bucharest, on the special order of the logging and wood transport companies, undertook such works based on the inventories of the forest and of the assets and elaborated the operation programs to secure sustainable development and profits. Generally, the need for funds is quite extensive and it is only partly covered until now. We are still awaiting interested foreign investors as well as the completion of the mass privatization for the revitalization of the entire forest sector.

Paradoxically, the poor accessibility of roads resulted in important reserves of wood. This wood reserve is important for attracting foreign funds and several firms from USA and Asia are willing to invest in Romania and, in this case, some problems with the modernization of the operation system in our forests will be solved. Recently, a Canadian investor said that in his country the forest has been exhausted and he is willing to invest in Romania, a statement which appeared strange to the author, same as if he would have said that the Atlantic Ocean no longer has water.

Owing to the fact that in the industrial sector some capacities of 200,000-300,000 cub. m./year will be set up both for the production of particle board and fiberboard (smaller capacities are no more efficient), special problems will arise to ensure the wood supply. This will lead to a pressure for organizing an industrial type of forest activities with severe steps, both silvicultural and environmental.

Many of the solutions applied in our country for wood exploitation so as to answer the silvicultural problems are based on research conducted for more than 50 years which has been carried out by INL and ICAS (Silviculture Institute). The research is aimed at forestry and soil protection, torrent correction, machinery system, etc. some with remarkable success in forest management. By participation in the implementation of some woodworking complexes (logging - woodworking units) in tropical regions from Asia and Africa, our specialists were in direct contact with the problems of tropical forests, which from the point of view of exploitation and preservation rise barriers ten times larger than those of the temperate forests. For a period of more than 10 years the author encountered these problems in SRI-LANKA and NIGERIA, which were partially solved but which provided a reason of meditation vis-a-vis the forestry subjects in the world industrial epoch. Beside the national research programs, the international research programs (such as COST, NATO and EUREKA for better utilization of wood and energy from wood) shall be taken over and implemented.

Nowadays, the problems are more complex and with increased variables and restrictions but, it is possible to find answers especially for securing the stability of ecological systems in the forests, only by well developed mathematical models, optimum ratio of mechanization in the forests and road densification, to confine oneself only to these major ones.

The possibility to secure enough funds in the Eastern countries for research leading to practical solutions, including their transfer to real economy is a critical need. Another need, is to create multidisciplinary teams since each specialty follows its own way, especially for problems connected to biodiversification, but without solving the problems in a satisfactory manner.

Genetic engineering will also have a substantial contribution for the improvement of resistance at various wood species especially those with low resistance, such as oak, which in our country presents dying phenomenon in an alarming proportion.

There is also the problem of the Quality Management (ISO 9000) and of the Ecological Management (ISO 14000) as well as those of "Certification", Monitoring, etc. as determined from various conferences. About 11600 ha. as plots for forest supervision and control, based on proposed norms, have been set up in Romania.

It is also necessary to enhance the ethics of the statistical data which are correlated as well as the partisanship shown by various organizations which participate at the decision taking level, so as to secure the sustainable forest development concept.

6. Conclusions

1. The subjects discussed at this NATO Advanced Research Workshop are of vital reality for Central and Eastern European countries, which might influence the entire wood production and ecological system of forests in the world.

2. From this brief analysis it is concluded that in some countries the forest conditions and the operation system have a national specific so that it is necessary to provide a strategy in two or three steps in short, medium and long run so as to improve the extent of wood production, environment protection etc.

3. Compatibility of phenomena, in connection with the environment protection, is so large that it is necessary to find suitable funds for carrying out research on a large scale, by teams from more countries and geographical zones, to save the forests which are the lungs of the earth.

I am convinced that this meeting within the program of NATO - ARW will result in many solutions in connection with these general subjects.

7. Pertinent Observations From ARW

7.1 GENERAL

The four working groups developed the following topics:

W. Gr. 1⇒ pollution factors in forest depletion

W. Gr. 2⇒ forest growth, expansion and management

W. Gr. 3⇒ wood utilization; new products and structures

W. Gr. 4⇒ social and economic consideration on forest resources

There were presented papers of high scientific and practical importance analyzed either separately or jointly during the workshop. In the end, the drawn up syntheses established pertinent directions to be followed, especially in the scientific research.

7.2 MAIN ISSUES AND SOLUTIONS

When considering the core problem such as "forest sustainable development" the other topics are in an organic interaction. One can mention some important issues, as follows:

Forests:

> necessity for strict application of silvicultural methods.

> realization of an ecological logging.

> provision of a modern infrastructure: forest road network and machinery system.

> forest development at the required economic parameters.

> provision of forest multifunctionality by optimization of the basic values of forest.

> provision of management for all the forest activities, ensuring profitability similar to that of trading companies.

> genetic engineering shall have a greater contribution to the improvement of species less resistant to various pollutants.

> provision of development programmes taking into consideration both local and regional factors as well as the drawing up of a "Master Plan."

> creation of software programmes to control a large number of technical, economic and social factors for an effective management.

Forest depletion:

> air pollution and mass transportation.

> wind falls (slashes) because of climatic anomalies.

> attack of insects and fungi because of the climatic conditions.

> soil degradation due to erosion and mining.

> major concentration of pollutants in some areas from East and Central Europe.

> stress produced by human factors.

> low combustion of solid fuels.

> necessity of alternative energy source (solar, hydro and wind energy).

All these, ask for cool desulfurization, high combustion for biomass, etc.

Wood constructions:

> wood has in general, good ecological properties and low energy input.

> necessary frequent utilization of plain and complex wood structures for industrial constructions, housing, bridges.

> wood is as good as any other material and can be used under high-tech conditions for multi-storied buildings, and wide span constructions.

> testing methods shall secure the control of specific strength and therefore "stress grading" in the sawmills appears necessary.

> regulations at international level for wood constructions are requested.

Forest socio-economic considerations:

> forest sustainable development is strictly connected with economic feasibility.

> demand and supply concept can also be applied to wood products.

> the forest has a social function by its ecological values ensuring the welfare of a large part of the population.

> the traditional cultural barriers must be removed by intensive mass media action for the acceptance of new wood housing solutions with low energy inputs.

Scientific and practical conclusions:

> forest development, becomes a strong request not only for the Central and East European countries but also for the whole mankind owing to forest depletion in an alarming proportion.

> stopping of forest depletion by eliminating the major polluting factors through technology transfer and provision of funds at the international level for researches.

> coalition of all the research efforts at the international level for the carrying out of scientific researches to obtain answers to all the problems identified by scientists participating at this workshop.

> coordination of global research activities by multiple scientific programmes at international level, NATO, EUREKA, COST, UN-FAO, etc. so as to avoid some parallelism, in order to put into practice the research works in these emergency fields, efficiently.

Interesting study tours were organized by NATO-ARW, with the aid of Dr. Winfried Duffner of Thurn und Taxis, in the Bavarian forests, which can be regarded as natural laboratories owing to the utilization of modern silvicultural methods, logging and management. At the same time, the interesting visit of the up-to-date sawmill, near Saulburg, run by process computers, proved the possibility of obtaining sawn timber products qualitatively compatible with the new concept of wooden prefab houses.

We consider that due to the thorough preparation of this Workshop exceptional results were obtained for which we must thank the NATO organizers as well our Bavarian hosts for the excellent conditions offered.

8. References

Atanasiu, F. (1963) Case prefabricate (Wood prefabhouses) - Edit IDT

Chiru V. (1980) Utilaje pentru exploatări forestiere (Forest Exploitation Equipment) - Ed. didactică °i pedagogică

Copăceanu D., Bălănescu E., Ghica P., Rusu Gh. (1983) Tehnologia exploatării lemnului (Logging technology) Ed. Ceres

FAO - (1989) Conservation guide 13/5 - Road design and construction in sensitive watersheds - United Nations

Furnică H., Beldeanu E. (1985) Exploatarea pădurilor cu elemente de industrializare a lemnului (Forest Exploitation and Elements of Wood Industrialization) - Ed. Ceres

Râmbu Ion (1978) Tehnologia prelucrării lemnului (Woodworking Technology) - vol. I-II. Ed. Tehnică

FORESTS OF UKRAINE AND IMPROVEMENTS OF FOREST MANAGEMENT

M.V. CHERNYAVSKY
Ukrainian State University of Forestry and Wood Technology
Kobylianska str.1, 290005, Lviv-5, Ukraine
E-mail : tschern @forest. lviv. ua

Abstract

Presently, the Ukraine is a sparsely wooded country. Woodlands constitute 14.5%. The total forest area is 9.942 mln ha, 8.620 mln ha are covered by forests, and the total volume stock is 1.320 mln m^3. The Ukraine area covered by forests can create about 120 mln tons of organic substances annually, draw from the atmosphere more than 234 mln tons of carbon dioxide, and emit up to 182 mln tons of oxygen. It is necessary now to work out the complex measures concerning "The Ukraine's nature restoration." Forest management systems and forest growth systems exist in Ukraine.

The following conception is proposed concerning the role of the forest: Comprehensive preservation of existing forests and intensification of their ecological role is the basis of the biological diversity of the world, ecological balance of the planet, and stable development of every state. This conception can be common for all European countries.

1. Introduction - Ecological Situation and the State of Forests in Ukraine

The ecological situation and the state of the Ukraine's forests are stipulated by the great technologic emissions of harmful substances into the atmosphere, soil, and water pollution. They are also stipulated by the consequences of the radioactive pollution after the accident at Chernobyl nuclear power station. The total level of polluted area is calculated by the indexes of all kinds of pollution:

- for atmosphere-by main and specific polluters;
- for surface water-by organic-toxic properties, and sanitary routine;
- for soil by conditional quantity of pestycydes and their detoxication, as well as by the level of radioactive pollution by density of Cs-137 (Ci/km^2).

Taking into consideration the pollution influences to the human life activity (Baranovsky *et al.*, 1995) all these territories are distinguished into:

- conditionally clean (49,100 km^2);

R. M. Gutkowski and T. Winnicki (eds.), Restoration of Forests, 195–204.
© 1997 *Kluwer Academic Publishers.*

- little polluted (114,800 km^2);
- polluted (121,200 km^2);
- very polluted (116,700 km^2);
- extremely polluted (61,000 km^2);
- of ecological troubles and ecological disasters (7,400 km^2).

The soil resources of Ukraine occupy 60,370,000 ha. Primarily the highly-fertile soils and favorable climate conditions stimulated active agricultural activity and, as a result, very high economic development of Ukraine's territory - 92%. The extent of agricultural soil development is about 70%, and the level of plough-land is one of the highest in the world - 56% (Krysachenko, 1991; Lishchytovych, 1993).

Due to the transfrontier transfer of acid-forming technogenic compounds, acid rain is observed in formerly favorable regions at great distance from the emission sources (*UN conference..*, 1992). Presently, 36% of forest stands in the Ukraine have various kinds of damage due to air pollution. After the accident at Chernobyl nuclear power station in 1986, about 2.2 mln ha of forest stands in Ukraine were subjected to radioactive pollution (Lishchytovych, 1993). The most polluted forests (except the 30 km zone) are in Polissya and in the northern part of the Zhytomyr region (32,000 ha with a density of soil pollution over 15 Ci/km^2).

The radiologic situation in the damaged regions has been determined by J-131, Cs-134, Cs-137, Sr-90, Pu-239, and Pu-249. As numerous studies have shown, the highest quantity of isotopes got into forest cenosis, where they deposited. High concentrations of pollutants exist in all the components of forest ecosystems; forest litter and moss-lichen cover are characterized by maximum indices. During the last years most isotopes moved from the litter to the soil.

It is necessary now to work out the complex measures concerning "The Ukraine's nature restoration." Corresponding works should be based on the whole conception of ecological and cultural development of Ukraine. Although not yet achieved, progress is being made (Krysachenko, 1991).

2. Forest Traditions and Achievements

During the first thousand years A. D., forests covered almost the entire territory of modern Ukraine, with the exception of the steppe zone where they occupied insignificant areas. In the territory of the ancient Ukraine-Rus, forests occupied areas in the steppe 2-3 times, in the Left-bank of forest-steppe 3-5 times, and in Podillya - 5-7 times more (it is historical fact) than today (Beauplan, 1651; Ivanitskyj, 1936, 1939). As far back as eighteenth century the woodedness of forest-steppe reached 16.2%. Comparely with first millennium in nineteenth century it decreased in 4-4.5 times (Gensiruk, 1992; Gensiruk *et al.*, 1995).

During the last millennium the woodeness of the Zaporizhia and Mykolaiv regions decreased 10%, Dnipropetrovsk, Donetsk, Kherson and Odesa regions -

11-20%, Poltava, Kharkiv, Kirovograd regions and Crimea - 21-50%, Luhansk, Sumy, Chernigiv, Kyiv, Cherkasy, Zhytomyr, Rivne, Ternopil, Khmelnytsky, Ivano-Frankivsk and Zakarpattia regions - 51-70%, Lviv, Volyn, Chernivtsi and Vinnytsya regions - more than 70%.

Due to historical, geographical and geopolitical factor, Ukraine is a forest country, which has its own forest traditions and achievements.

Experience and technologies of forestry in steppe (from head of kozaky S.A. Lehkostup, who in 1804-1818 created more than hundred acres of pine plantations on Kharkiv region, expeditions by Prof. V.V. Dokuchayev in 1892-1907 in steppe where 444 ha of forest was planted) do not have analogies in the world forest practice. In 1950-1960 in the Kherson region the pine plantations were planted at the sands. The original ways and methods of forestry agromelioration, plans of mixed species of trees, regimes of planting different cultures in forest-steppe and steppe, ways of having forest recreation on the radioactive territories are worked out and realized in Ukraine. They also do not have analogies in the world within forest science and practice (Chernyavsky *et al.*, 1992).

3. The Forest Resources and Forests Use

Presently, the Ukraine is a sparsely wooded. Woodlands constitute 14.5%; the total forest area is 9,942 thousand ha; 8,620 thousand ha are covered by forests; the total volume stock is 1,320 million cubic meters, including 122.8 million cubic meters of mature stands; an average stock of mature stands on 1 ha is 252 m^3. It is a forest deficient state which cannot provide itself with its own forest resources or products of wood processing and must be oriented towards their import.

During the last half-century, the area covered by forest in the Ukraine increased by almost 30% - from 4.4 million ha in 1940 to 6.2 million ha in 1993. The age structure of the forest resources is very uneven: currently young forest areas are 44% and the deficit of mature (7%) and seasoned (11%) forests is chronic.

The total wood increment is 24.4 million m^3, forest use is 14.4 million m^3, and incremental use is about 60%. In some regions (the Carpathians and Podillya) a high degree of incremental use (over 75%) merely indicates a tendency towards the exhaustion of mature forest stocks and an excess of allowable intermediate produce intensity.

During the last half-century, Ukraine's forest species composition hardly changed: the areas of pine and oak plantations increased somewhat, especially those of young and middle-age stands, as well as the areas under beech stands. The Ukraine's forest age structure has good prospects but there are few mature stands. This testifies that there are no reserves for an increase of forest use in the future.

An average annual area of major produce coupes from 1960-1970 was 55,000 ha, in 1971-1980 - 29,400 ha, in 1981-1989 - 27,700 ha and from 1990 - 1993 about 24,000 ha, and the sowing and planting of forests was carried out in the area: in 1960-1970 - 94,800 ha, in 1971 - 1980 - 42,400 ha, in 1981-1989 - 37,700 ha, and from 1990 - 1993 - over 39,000 ha.

This means that the effectiveness of artificial regenerations was and remains low, forest culture experience of the previous decades is being lost, methods and technologies of cutting aimed at natural forest regeneration have been unjustly ousted. All of this leads to unjustified expenses on forest regeneration, on the one hand, and to the state forest gene pool impoverishment, on the other.

According to the parameters which characterize forest provision (the stock per 1 ha of wooded lands is 171 m^3; the stock per capita is 20 m^3; increment per 1 ha of forest area is 4.0 m^3; increment per capita is 0.47 m^3; annual use volume per capita is 0.28 m^3; annual increment use is 60%, and forest use intensity is 1.40%), the Ukraine is one order lower than leading countries - exporters of wood (Popkov *et al.*, 1992). Stock and increment correlated to 1 ha of area covered by forest, first of all, depend on the age, species, and evaluation structure of the forest resources. According to these parameters, Ukraine is among the leaders in Europe. This testifies to the high productivity of Ukrainian forests and good conditions for organization and intensive forest management.

More than 200 species of trees and bushes naturally grow in the forests of Ukraine (Chernyavskij *et al.*, 1994). Pine forests occupy the largest area (36.4%), followed by oak forest (27.6%), spruce stands (9.3%), and beech stands (9.1%). Most of the plantations have high stand density (near 42% of all territory) and bonitet (80.3% of all area). As a result of intensive attendance for clear-cutting Ukraine has a paradoxical situation: stocks of ripening stands are higher than aging stands. This difference is perceptible in forests from first group where the medium stock ripening stands are 11% higher than in aging stands. It means a great disproportion in forest management system.

Presently the forest resources of Ukraine are divided into two groups: the first one occupies 3,453 mln ha and the other 3,722 mln ha. Forests which, on the whole, fulfill nature protective functions belong to the first group and are subdivided into the following categories of protection: water protection; protection; sanitation and health-improvement and the natural forest reserve territories. Forests, which, along with ecological, are of exploitation significance and in order to preserve protective functions, continuity and inexhaustability of which the regime of limited use is established, belong to the second group (*Forest code of Ukraine*, 1994, article 36). Forests, which perform mainly the water protective functions constitute 9% of the forest of the first group: protective - 39%, sanitation and health-improvement - 51%, nature reserve - 6%. The singling out of groups and categories of protectiveness has been carried out according to "the functions performed and to the purpose of forests", and if there was a necessity, according to the regime of management depending on the purpose of the forests.

4. Ecological Role of Forests

The Ukraine's area covered by forests can create about 120 mln tons of organic substances annually, draw from the atmosphere more than 234 mln tons of carbon dioxide, and emit up to 182 mln tons of oxygen. This capacity of Ukraine's forests is now within 46% of the necessary utilization of carbon compounds and air oxidation.

Forests are one of the key components of continental ecosystems in the formation of favorable climates and in the provision of stable development conditions. All types of forest ecological systems play an important role in maintaining of economic processes (power engineering, industrial raw materials, product consumptions), and positively influencing the development of agriculture, especially dry crop-growing.

These properties are most effectively seen when optimum woodedness is reached, which according to different data, is now for Ukraine from 19%-20% to 30% (Dyachenko, 1991; Gensiruk, 1992; Chernyavskij *et al.*, 1994); and a balanced correlation between arable lands, forests, and meadows. The deformed crop-growing, typical for the Ukraine, differs substantially from many countries of Europe.

Natural territories and areas - biosphere reserves, natural reserves, national natural parks, regional landscape parks, game reserves, natural monuments; artificially created areas - botanical gardens, arboretums, zoological parks; and parks - park-monuments, garden art; all belong to the Ukraine natural reserve fund (*UN conference*, 1992). The current network of the natural reserve fund includes 15 biosphere and state reserves, 3 natural parks, 17 regional landscape park, 1799 game reserves, 2725 natural monuments, 482 forest genetic reserves, and 1728 reserve objects of other categories with different regimes of protection - from complete non-interference of man to certain limits of economic activity. The total area of the natural reserve fund is 1.664,900 ha.

5. Forest Management Systems

A forest growth system involves complex measures, which are carried out in time of clear-cutting rotation on every plot not going out to the limits of exact kind of forest, and which is directed on the planting and maintaining the plantations in state. Also, the results of forest management are the most consistent with the rational forest management. There are technological variants inside the limits of forest growth systems, which differ by the concrete combination between all kinds of forest measures and technical means of their realization.

Forest management systems are based on a zonal-typological basis, taking into consideration the division of forests into groups and categories of their protection, distinguished management parts, determined forest ripeness age, ways of clear-cutting and recreation of forest plantations, standards of using forest resources. These systems are also divided into groups and categories with observing other organizational and technical measures. According to the types of forests, the Ukraine plans, realizes and constantly improves it, taking into consideration the experience of forest management and science achievements.

Some features and the character of forest management are negative. Some of all these are:

- insufficient differentiation of forest management systems according to the desired forest function, (special purpose forests) categories of protection and types of forests;
- absence of the long-term prediction on using forests for categories with different levels of protection;
- low level of mechanization during the clearing and thinning;
- no well-founded routine of forest exploitation in forests of different aiming function;
- inoperative technologies for the regenerative clear-cutting;
- very small volumes of forest recreation by natural seed way and by the improved genetic material;
- absence of economic substantiation for forest management systems which take into consideration the multi-functional role of forests;
- realization of the monitoring of the forests all over the territory;
- absence of finances for forest reproduction and tending of woods in the Chernobyl zone.

The 4-level structure of management (ministry - regional administration - state forest management - forest division) does not favor better forest management organization. Obviously some of these levels have many specialists, which are unnecessary.

That situation has its own objective principles. Today it is very important to study new approaches for the forest management as quickly as possible. And these approaches will be based on the extensive and valuable experience of practical forest division, and on the recent science achievements.

6. Concept of Forest Policy

On Jan. 21, 1994 a new "Forest Code of the Ukraine" was adopted, and forests in the Ukraine are the property of the state (*Forest code..*, 1994, article 6). According to the new Law, the forests of the Ukraine are its national wealth, and according to their purpose and location perform chiefly ecological (water protection, protection, sanitary, health-improvement, recreation), aesthetic, educational, and other functions; have exploitation purposes and are subjected to state registration and protection (article 3).

The main requirements concerning forest management organization are also envisaged (article 34). That is: forests preservation, their protection from fires, protection against harmful insects and diseases, improvement of water protection, protection, climate regulation, sanitation, health-improvement, and other useful properties of forests for the purpose of health protection of the people, and environmental improvement; continuous and rational use of forests in order to systematically meet the requirements of production and population in timber and other forest products, extended reproduction, improvement of species composition and forest quality; improvement of productivity, and improvement of forest production efficiency on the basis of unified technical policies, and scientific and technological achievements.

As a matter of fact, this is the strategy of forest growing and the role of forests in Ukraine. It should be noted that the principle of continuous and rational use of forests with their dominant water protection, water regulation, climate regulation and environment creating role on the basis of the complex management, started to be realized in 1959.

Despite some extent of declarativity (principles of continuity, steadiness and progressiveness as a strategy of forest management often were infringed to satisfy different modern trends and campaigns such as "spruce mania," "poplar mania" and now "fir mania," inexhaustibility of forest use is also a conditional, relative thing; continuous, rational use of the forest is also possible only as an ideal) this approach was conductive to the affirmation of the state forest policy as a realization of the global ecological role of forests.

Under conditions of constant timber shortage (annually about 20 mln m^3 ; Dyachenko, 1991), people and the state realize the great significance of the forest as an important ecological factor which stabilizes the environmental situation. Only due to the forest significance, long ago realized by thousands of foresters and millions of people, did it become possible to fix by law on the state level the role of forests in the Ukraine and on the planet.

The Ukraine state forest policy is the total combination of legal, economic, administrative, and forest management measures directed at forest preservation and regeneration, rational use, and enhancement of the polyfunctional role of forest as a global ecological factor in the interests of the nation. This policy also determines the strategy for forest preservation, forest use, and the role of forests on a planetary scale.

Thus, taking into consideration the previous material, the author proposes the following conception concerning the role of the forest: "*Comprehensive preservation of existing forests and intensification of their ecological role is the basis of the biological diversity of the world, ecological balance of the planet, and stable development of every state*". It can be improved after our common discussion on this conception.

7. National Scientific-Technical Program "Lis" ("Forest")

A conception, structure and forms of realization of National science-technical program "Lis," which determines the perspective directions of rational use, preservation and reproduction of Ukrainian forest resources in the context of requirements adopted in special resolution of world conference on environmental problems and development "Rio-92" (Tunytsya *et al.*, 1995) is formed in Ukraine. National scientific-technical program "Lis" envisages stage by stage (in accordance with economic possibilities) the enlarged regeneration, preservation and rational use of forest resources on the basis of modern equipment, advanced technologies, and the best organizational-economic forms of management. The program consists of four closely interrelated blocks including:

- problems of forestry as a specific, independent branch of the agrarian sector of economy;
- problems of forest industry as a specific,independent branch of extractive industry;
- problems of woodworking, forest-chemical, pulp and paper, and other branches of processing industry;
- problems of improvement of organization and management in forestry, forest and woodworking industries.

The unique nature of the program "Lis" is, in the author's opinion, evident in the fact that it includes:

- a mechanism of unification of relationship of forestry, forest and woodworking industries, branches of wood processing;
- a mechanism of a worker's interest in resources economy, and their preservation on the basis of steady development;
- a mechanism of considerable self-financing for fulfilling of the scientific tasks.

The project of the national scientific-technical program "Lis" was considered, adopted and supported by the State Committe of Ukraine on the problems of science and technologies (1993), Collegium of Ministry of Forestry of Ukraine (1995), Presidium of the National Academy of Sciences of Ukraine (1995) and by many other institutions, departments and ministries.

8. General Concept

Forests of each European country have their own specific planetary importance. That is why, their preservation and recreation are not only one country's, but our general business. Because of that, it is suggested to adopt the conception as the working document of the workshop. Later, it would be possible to address with it to the public, governments, majors of different countries and other interested transnational formal or informal organizations. The only guarantee to make our environment and health better is to come to a mutual understanding, to cooperate and to realize the role of the forests.

This is the conception of: "Comprehensive preservation of existing forests and intensification of their ecological role is the basis of the biological diversity of the world, ecological balance of the planet, and stable development of every state." It would become the coordinated strategy of forest management all over the world. General planetary role and meaning of forests demand from us an adequate general planetary approach for their preservation, keeping, rational use and recreation.

The real working out on this conception taking into account the specifics of forestry management in every country would be the plan of forest management systems based on the following main parameters:

- optimal woodedness;
- forest distribution according to their aiming functions;
- species composition and forest structure;
- ways of clear-cutting (the main care) and regeneration according to the aiming functions;
- economic expedience and choice of forest management according to the natural and economic conditions on the base of constant development of the country;
- ecological and climate-regulative role of the forests;
- preservation of protected plantations and virgin forests (primary), as a base of biodiversity;
- possibility to substitute the forests and other forest resourses;
- traditions and achievements in forestry of each country;
- improvement of the structure organization of forest management within every country;
- the results of ecomonitoring of the forest state.

The separate components of the forest management systems are elaborated in every country quite well. Today, having used the experience of countries-leaders, we have to generalize these achievements, but because of the forestry which is an art and creative branch of activity, not to standardize them. According to the national science programs in this context it is possible to put the European problem of forest management systems in different zones, depending on side types. It would be something like a European program of forest ecomonitoring.

Majors of our countries like to reiterate that Europe is our common house. If we, as foresters, would accept this or the similar (the author is far from the thought that just this conception will be approved by the majority) conception and program of preservation and rational utilization of European forests, it could be a base or a part of our common conception and scientific program. It would be a good deed and good example of colaboration for noble purpose. At the same time we will take into consideration the experience and achievements of forest management of every country.

Ukraine could be an organizer in elaboration of the forest management system in oak-tree forests.

9. Acknowledgements

At the end of my report I thank Professor Tomasz Winnicki (Poland), Dr. Yurij Koltun, Professor Grygorij Krynytsky and colleague Taras Kuleshnyk (Ukraine) for the great assistance in preparation of this report.

10. References

Baranovskyj,V., Pyrozhenko, K. & Shevchenko,V. (1995) Medic-ecological maps of Ukraine, *Greenpeace* 1, Kyjv, 32.

Beauplan de, G. (1651)*Description of Ukraine (5 maps).*

Chernyavsky, M. V. & Gensiruk S.A. (1992) Development of the silviculture science in the Ukraine, *Forests of Ukraine*, Naukova dumka, Kyjv, 26-60.

Chernyavskij, M. & Shvadchak, I. (1994) Reservation of the Ukrainian Forests and ecological Balance (Trends and Prospects), *Conservation of Forests in Central Europe*, Zvolen, 123-130.

Dyachenko, Ya.Ya. (1991) Efficiency of intensive reproduction and rational use of forests raw materials, *Lisove gospodarstvo*, 2, 2-4.

Forest Code of Ukraine, 21.01.1994, *Voice of Ukraine, 13.04.1994*, N68 (818), 8-14.

Gensiruk, S. A. (1992) *Forests of Ukraine*, Naukova dumka, Kyjv, 408.

Gensiruk, S.A. & Nyzhnyk, M.S. (1995) *Geography of Ukrainian forest resources*, Svit, Lviv, 122 .

Ivanitskyj, B. (1936) *Forests and Forestry in Ukraine*. Vol. 2, Works of Ukrainian Scientific Institute, Warszawa, 45(7), 123.

Ivanitskyj, B. (1939) *Forests and Forestry in Ukraine*. Vol. 1, Works of Ukrainian Scientific Institute, Warzsawa, 45(11), 204.

Krysachenko, V. S. (1991) Anthropogenic pressure on Ukrainian nature and its protection, *Ukrainskij Botanicheskij Zhurnal*, 48(2), 5-14.

Lishchytovych, L. (1993) Ecological-economic problems of Chernobyl accident consequences liquidation in the Ukraine agricultural complex, *Ojkumena*, 2, 56-61.

Popkov, M. Yu. & Savushchyk, M. P. (1992) Timber provision, forest fund and forest use in Ukraine. Kharkov, 111.

Tunytsya Yu.Yu. & Krynytsky G.T. (1995) On the conception of development of the national scientific-technical programme "Lis," *Forest science and education on the boundary of XXI century: contemporary state, problems and perspectives*. Lviv, 75.

UN conference on environment and development Brasil-92, *Ukraine national report (1992)*. Chas, Kyjv, 42.

THE ROLE OF FOREST TREE BREEDING FOR THE 21ST CENTURY

L. PAULE
Faculty of Forestry
Technical University
SK-96053 Zvolen
Slovakia

Abstract

Increasing demand on wood as a raw material will also demand corresponding methods to achieve this goal. One of the cheapest alternatives to achieve increasing demands on wood, both in quantity and quality, is to improve tree breeding. In the next century significant changes in quality of the environment are expected (*e.g.* consequences of environmental pollution, global warming etc.). Since the breeding goals (*e.g.* maximum yield) could be achieved only in defined ecological conditions, for undefined ecological conditions a new breeding goal – maximum adaptability – is proposed.

1. Introduction

The twentieth century is characterized by several changes in forest management practices which significantly affected the production and resistance potential of central European forests. Central Europe was under the influence of the forestry in German speaking countries and due to this influence large areas of mixed forests of Germany, Bohemia and western Poland were transformed into the Norway spruce and Scots pine monocultures. The consequences of this forest management practice, mostly on the ecologically inappropriate sites, are clear enough and they are accompanied by ecological disasters (windbreaks, snowbreakes, insect calamities, higher air pollution damages etc.).

In the middle of this century the broader application of tree breeding started. From the beginning, only the utilization of seed from approved plus stands was the main breeding activity in central European countries. Later the establishment of seed orchards was applied in forestry practice. In several cases it led to application of genetically narrowed reproduction material which consequently destroyed the resistance potential of these stands.

The last two decades of this century were, in contrast, under the influence of the biodiversity issues and nature close forestry, trying to solve the dilemma whether we need forests for wood production only or also for fulfilling the tasks of multiple-use forestry.

1.1. WHY FOREST TREE BREEDING

One of the principal aims of forest tree breeding has been to increase the produc-

R. M. Gutkowski and T. Winnicki (eds.), Restoration of Forests, 205–213.
© 1997 *Kluwer Academic Publishers.*

tion potential of forest ecosystems. In many cases, this aim was fulfilled. In poplar breeding and in some other European countries also with other tree species (*e.g.*, maritime pine in France) significant genetic gains were reached regarding the yield.

In many central European countries (including Slovakia and the Czech Republic), the application of forest tree breeding in forestry has been based mostly on the application of simpler breeding procedures including plus trees and stands selection and consequent establishment of seed orchards and seed stands. Application of both breeding methods led to narrowing of the genetic basis of the breeding populations, although in seed orchards the minimum number of clones were doubled in the last decades. In none of conifers, has the degree of domestication been as it is in the case of poplars.

In the last decades forest tree breeding has frequently faced criticism in connection with resistance breeding. Forest tree breeding had been expected to produce resistant genotypes to survive in the polluted areas but this stage has never been reached. The intensity of air pollution increased much faster than the possibility to get satisfactory breeding results.

1.2. GENE CONSERVATION

The issues of tree breeding have been frequently changed by gene conservation. In many cases the same methods are used but the aims of each differ. The principal aim of gene conservation is to maintain the genetic basis of forest tree populations as broad as possible to create the basis for future utilization of genetic resources in breeding programs and forestry practice.

1.2.1. *In situ conservation*

Forest trees are long-living organisms and due to this fact gene conservation *in situ* can be considered as the principal method of **dynamic gene conservation**. Forest reserves, gene reserves, approved stands linked with natural regeneration are the best tools for gene conservation *in situ* which can guarantee the maintenance of the broadest genetic basis.

1.2.2. *Ex situ conservation*

In cases where the *in situ* methods of gene conservation do not work safely, the *ex situ gene conservation* is the appropriate tool for gene conservation. In this category belong gene conservation in forest units planted outside the natural sites (sometimes even outside natural range), colonial archives, seed, pollen and tissue banks, etc. Some of the *ex situ* conservation methods are better than *in situ* ones for they do not require as much space as the long-living trees. Application of both methods in gene conservation of forest trees can be best compared with the application in agricultural crops (mainly annual plants).

2. Current Aims of Tree Breeding

2.1. BREEDING PROGRAMS

Numerous breeding programs are currently active in Europe. Most of them are short-term ones, but in several countries long-term breeding programs are also ongoing. Short term breeding programs include *e.g.*, the application of the seed orchards and subsequent testing of the selected reproduction material. In contrast, long-term breeding programs include hybridization, selection and subsequent testing of the best units.

In ongoing breeding programs various growth (yield) and adaptive traits have been used for culling the best units, *e.g.*, quantitative and qualitative wood production, resin and latex yield, late flushing, air pollution resistance etc. Usually, these traits are combined and the breeding programs are usually aimed at numerous traits simultaneously, rather than at a single characteristic.

One of the most common breeding aims in Norway spruce is the selection of the fast growing genotypes which at the same time perform well in stress conditions. The direct selection of the resistant genotypes failed, but it is still possible to select relatively tolerant genotypes which perform well in stress conditions.

2.2. BREEDING FOR UNKNOWN ECOLOGICAL CONDITIONS

In contrast to most of the agricultural crops, forest tree populations usually are subjected to the following major environmental challenges:

- Tree populations are exposed to a great heterogeneity of environmental conditions in time and space: trees are extremely long-lived and grow in natural environments which reveal complex and highly variable site conditions. Generation cycles of trees and sites deviate significantly from those of agricultural crops.
- The complexity of environmental stress conditions cannot be controlled efficiently: Prophylactic pest control in agriculture is not possible, fertilization and other measures to improve site conditions are strictly limited in their efficiency, and impacts in frost cannot be avoided.
- Future environmental stress can deviate substantially from the present conditions because of long term complexity of environmental stress, of interactions among stress components, and because of its non-predictable dynamics. Global warming will interfere with abiotic and biotic stress conditions and possibly further increase environmental variability and dynamics.

  The populations are characterized by how they respond to environmental changes such as:
- **adaptedness** is physiological constitution of the individual, or the genetic (allelic or genotype) structure of population, which allows to survive in the given ecological conditions. Adaptedness is a **state.**
- **adaptation,** (evolutionary adaptation) which means a **process**, which leads to the adaptedness by the changes of the genetic structure of populations,
- **adaptability**, which means genetic potential allowing the population to adapt

to changing environmental conditions.

Tree populations usually reveal extensive genetic variation within populations. The summarized results refer to studies which include different gene loci, number of loci, sample sizes, materials and ontogenetic stages. Furthermore, the mode of geographical variation and mating systems vary considerably among species.

2.2.1. *Air pollution*

The impact of air pollution on forest tree populations has been investigated for a long period in many countries involved (*e.g.*, Germany, Czech Republic, U.S.A etc.). The causes of the air pollution damage are rather complex, but they have a significant genetic background. It is rather improbable that we will be able to breed for high tolerance and resistance potential of individuals, but the possibilities to create breeding population with higher proportion of tolerant individuals based on broader genetic variation seems to be probable.

In general, the impact of pollution on genetic structure of tree populations is rather difficult to predict, although several attempts to simulate and to model these processes have already been made (Degen 1995).

2.2.2. *Climatic changes*

Recently, a new dimension has been introduced into the breeding program. Specifically, breeding is linked directly with the expectation of the occurrence of the global climate changes. In this respect not only the genetic structures of populations growing in contrasting climates (cold *vs.* warm) are compared, but also the processes following the seed transfer towards warmer, or colder climate are investigated as in the older provenance trials as well as in newly established ones.

The traits being used for breeding for stress conditions due to air pollution as well as future climatic changes are growth or adaptive traits (flushing) as well as traits linked with reproduction processes.

3. Forest Management Practices Leading to the Improvement of Resistance Potential

3.1. NATURAL VERSUS ARTIFICIAL REGENERATION

From the genetic point of view, the establishment of forest stands plays an extraordinarily significant role. There is no doubt that natural regeneration secures the best transfer of genetic information between the parent populations and the progeny. In numerous studies, however, the differences between the genetic structure of the parent population and its progeny were found, which could be attributed to: (i) the changes in the mating and reproduction processes, and (ii) the natural differences in the genetic structure of these two ontogenetic stages. Significant differences in the genetic structure of the parent populations and the progeny can be expected in years with poor seed crop, or in the case of small population size (small number of flowering trees).

Long regeneration periods could produce such a genotype structure, which is not a product of one reproduction (flowering, or seed year), but a product of some subsequent seed years. Genetic diversity of the next generation could thus be larger than that of the parent one.

In artificial regeneration, even without the application of the improved material, several factors can lower the genetic diversity of the artificially established forest stand, sometimes even negatively. The number of trees from which the seed was collected, the way of raising the reproduction material, seed and reproduction material transfer, and consequently even the way of the establishment of forest stand can contribute to the narrowing of the genetic structure (sometimes in rather unpredictable ways even in the case of seed collection from plus and approved stands). In the case of the application of bred material the narrowed genetic diversity is not a drawback, for we used tested material for given ecological conditions.

When comparing the application of the natural regeneration of indigenous populations and of the bred material, the bred material has a narrowed genetic diversity and adaptation potential. It has, however, higher yield (or higher values of other characteristics), than any natural population but its adaptation potential is narrowed and corresponds to the narrow array of ecological conditions.

In case of unpredictable change of ecological conditions in the near and far future the results of the classical tree breeding (tested only in a limited number of ecological conditions) could not be used. We have to focus *a priori* on such a genetical material which is characterized by good growth and adaptation potential in a relative broad spectrum of ecological conditions.

If we have to answer the question if the natural regeneration will be a sufficient tool to react to a quick change of ecological conditions (*e.g.* climatic, air pollution) the reply will be negative. Natural regeneration can only produce the reproduction of the population optimally adapted for given ecological conditions, but adaptation to the new conditions can only be done by the evolution process (selection) and not as the change within a single generation. Furthermore, if we presume such a change of ecological conditions which will require changes of species composition, then this change cannot be reached by natural regeneration

3.2. CHANGE OF SPECIES COMPOSITION *VS.* PROVENANCE CHOICE

If the changes of local climate leading to the dying-off of present (local) populations adapted to present climatic conditions are presumed, then we have presume *a priori* the necessity of the replacement of present populations for those with higher ecological tolerance to new (at present unknown) ecological conditions. If the changes are quick enough, we cannot come out from original populations and their natural regeneration with subsequent selection of non-adapted genotypes, but we have to replace these populations with the new ones originating from regions with similar climate or ecological conditions.

Possible reaction of different provenances to new conditions is given by provenance tests, which compare different provenances on the same site or a set of provenances on different sites with the aim: (i) to investigate the geographic variation of growth and adaptive traits and the adaptation potential, and (ii) to investigate the limits and direction of possible seed transfer. The second topic is especially important for prediction of measures against the possible climatic and ecological changes.

3.3 PROVENANCE TRIALS

The first internationally based provenance trials were orientated towards the research of growth traits and adaptation potential of the provenance set in different conditions. Their reactions on the seed transfer towards the warmer or colder climate was investigated from the point of view of growth and adaptive traits.

Table 1. Review of more important international provenance trial with economically important tree species established on contrast sites

Tree species	Year of establishment	Number of provenances	Number of sites
Picea abies	1938–39	36	15
Pinus sylvestris	1938	53	20
Larix diecidua	1942	56	15
Larix decidua	1959–61	25	20
Picea abies	1964–68	1100	22
Picea abies	1972	21	29
Fagus sylvatica	1982	188	15
Fagus sylvatica	1995	161	22

Since the 1930s an array of international provenance trials with economically important tree species (mainly conifers) was established. The most recent provenance trials with *Pinus sylvestris* and *Fagus sylvatica* covered the sites with warmer climate, *i.e.*, those which could be applied in northern latitudes after warming up of climate. The provenance trial with *Pinus sylvestris* covered mainly lowland provenances, *i.e.*, those which were in the past under similar selection character.

Two provenance trials with *Fagus sylvatica* possess a similar aim and they are important from two points of view: (i) importance of the European beech as the replacement for non-indigenous coniferous stands in some parts of Europe (*e.g.*, Germany, western Poland, Czech Republic, etc.), and (ii) reaction estimates to expected climate changes.

3.4. GENOTYPE × ENVIRONMENT INTERACTIONS

Evaluation of provenance trials with a higher number of provenances replicated in more sites offers estimates of the genotype × environment interactions. Estimates of the genotype × environment interactions are important for higher effectiveness of the breeding programs as well as for better selection of provenances for an uncertain future. Not only top ranked provenances are promising (according to their growth and adaptive traits), but also those with broadest ecological amplitude, which are usually in contradiction with genotypes best adapted for given ecological conditions, are promising

4. The Tasks for the Next Century

In the second half of the 20th century, the forests of central Europe have faced significant changes of environment quality, mainly due to air pollution. Although several activities leading to the decreasing of the air pollution have been started on the international level and positive results have also been reached, the present state of the central and eastern European forests is not satisfactory.

The prognosis for the improvement of ecological conditions (intoxicated sites, acid precipitation, acid soils, etc.) is not promising. The health status of the forests due to direct pollution and subsequent influence of other biotic (insects) and abiotic factors (*e.g.*, drought) lead to some extent to catastrophic disasters and to mass dying-off of the forests.

Naturally, improper forest management practices often were one of the sources of the unsatisfactory health state of forest stands and even of larger forest complexes.

4.1. FORESTRY PRACTICE

The reasonable application of rules on seed procurement and seed transfer (on the national and international level) are one of the prerequisites of the improvement of forest stands. Uncontrolled seed transfer should be prohibited and only scientifically proved limits and directions of seed transfer should be applied in forestry practice. Evidence of seed origin (transfer) can serve as the proper decision tool, both in case of satisfactory as well as of unsatisfactory results of the seed transfer.

New breeding aims should be incorporated into the breeding programs. They should consider not only the maximum genetic gain in growth and yield traits by narrowing of the genetic basis and genetic variation, but they should also be aimed at the broadening of genetic diversity as the breeding aim for uncertain future.

4.2. FORESTRY RESEARCH

Within forestry research the new applications of the population genetics (genetic monitoring, evaluation of genetic resources) should be applied on routine basis as the decision tool. Within the breeding programs new trends in provenance research (tests for new ecological conditions, tests in simulated ecological conditions, G×E interactions) have to be applied. Within the breeding programs new breeding aims (adaptation potential) should be developed and formulated.

4.3. INTERNATIONAL COOPERATION

In the field of the international cooperation, more than ever before, new tasks should be implemented. Procurement and testing of the breeding material should be based on the international basis to reach a variety of testing sites as well as to get a variety of tested objects available. Internationally conducted breeding programs are another aim in the field of the international cooperation. Exchange of experimental material and finally exchange of know-how are further tasks.

Following the Ministerial Conferences on the Conservation of Forests in Strasbourg and Helsinki the EUFORGEN program for the realization of the S–2 resolution on genetic resources has been implemented since 1994. At present it is aimed at four model tree species, namely, black poplar, Norway spruce, noble hardwoods and cork oak. In the near future it will also be applied to other conifers of economical importance (Scots pine) and broadleaves (oaks and beech).

5. Is There a Chance to Improve Forestry?

It would be a too optimistic to believe that tree breeding would be the only option to rescue the forest of the 21st and later centuries. We are really dealing with the forest trees long-living in the variable and permanently changing ecological conditions which are difficult to predict.

That is why the best option seems to be the combination of the tree breeding with the classical silvicultural practices aimed at nature-close forestry. In this field the excellent activity is shown from the results of Pro Silva movement. Establishment of plantations with the improved material especially on abandoned agricultural land would be the best option to produce the raw material and thus to save the natural forests and forest with protection functions.

Following the biodiversity issues (ecological, species and genetic diversities) seems to be the best tool for forest management for uncertain future.

6. Conclusions

Uncertainty is the principal factor influencing the regeneration of forest stands for the future, in which the more pronounced ecological and climatic changes are expected. It will be the case, until sufficiently valid models of climatic changes are developed.

Independently from the present conditions (or the conditions in the period when forest stands will be regenerated) the forest stand will develop up to its maturity in ecological conditions which can significantly deviate from present ones. Any choice of provenances, or of other reproduction material requires the presumption of the stability of ecological conditions.

The best strategy for the reproduction of forest stands for uncertain future will be maintenance of the high genetic diversity, selection and breeding of "generalists" (*i.e.* provenances with broad ecological amplitude), as well as to be prepared for a possible transfer of provenances , or of their mixtures, from the regions with warmer climate.

7. References

Davis, M. B. Zabinski, C. (1992) Changes in geographical range resulting from greenhouse warming: Effects of biodiversity in forests, in R. E. Peters and T. E. Lovejoy (eds.) *Global Warming and Biological Diversity,* Yale University Press, pp. 297–308.

Degen, B. (1995) Ökologische Genetik. Vorläufiger Ergebnisbericht des Fachgebietes. Bundesumweltamt, Berlin.

Geburek, T. (1994) Genetische Strategien für das forstwirtschaftliche Handeln angesichts klimatischer Änderungen, in T. Geburek, F. Müller & U. Schulze (eds.) *Klimaänderungen in Österrecih. Herausforderung an Forstgenetik und Waldbau.* FBVA Berichte **81**:19–36.

Gömöry, D. (1992) Effect of stand origin on the genetic diversity of Norway spruce (*Picea abies* Karst.) populations. *Forest Ecology and Management* **54**:215–223.

Ledig, T. (1986) Heterozygosity, heterosis, and fitness in outcrossing plants, in M. E. Soule (ed.) *Conservation biology: the science of scarcity and diversity,* Sunderland, Sinauer Associates, 5:77–104.

Paule, L. (1995) The biodiversity of forest in the Western Carpathians., in J. Paulenka and L. Paule (eds.), *Conservation of Forests in Central Europe,* Arbora Publishers, Zvolen, pp. 33–55.

Paule, L. (1996) Genetic conservation of European beech (*Fagus sylvatica* L.). *Forest Genetics* **2**(3):161–170.

Paule, L. and Gömöry, D. (1993) Die Bedeutung der Urwaldreservate für die Untersuchung der Evolution und Genetik der Waldbaumpopulationen, in Saniga, M. and Š. Korpeľ (eds.), *Symposium über die Urwälder,* Technical University, Zvolen, pp. 61–66.

MANAGEMENT OF FOREST ECOSYSTEMS IN THE KRKONOŠE NATIONAL PARK, BLACK TRIANGLE REGION, CZECH REPUBLIC

O. SCHWARZ
Krkonoše National Park Administration
543 01 Vrchlabí
Czech Republic

Abstract

Management in the Krkonoše Mountains is affected by the destruction of forest ecosystems caused by emissions, and simultaneously by the necessity of protection and support of the natural values of the area, as well as other non-wood-producing functions. The situation is complicated by the ill-calculated conversion of the original forest ecosystems to the even-aged spruce stand, which was carried out in past. The soil acidification caused by the atmospheric deposition is further increased by the decomposition of the spruce litter. The priority objective of the management is a reconstruction of the forest ecosystems to a form close to that of original condition, preferring natural regeneration and processes, preservation of the gene pool of individual autochthonous woody species, preservation and support of biodiversity of ecosystems and a reclamation of the soil threatened by erosion. The region of Krkonoše has been categorized according to natural values, effects of emissions, site conditions and according to the preferences of non-wood producing functions. The composition of the tree species, as well as the technologies used are modified with respect to these categories. From an economics point of view, the management is loss-making. Not even the most urgent forestry activities can be covered using the profit from a sale of timber. A part of forestation and research focused on a success and economical as well as ecological efficiency of the realized measures is covered by the Dutch foundation Face, that "buys" the effect of the CO_2 binding in the biomass of trees. Another financial support is provided by the Ministry of Environment of the Czech Republic. However, some of the important activities (preservation of the gene pool of the Krkonoše woody species, conversion of the spruce plantations which were established improperly in the course of the last 20 years) have not been funded as yet.

R. M. Gutkowski and T. Winnicki (eds.), Restoration of Forests, 215–226.
© 1997 *Kluwer Academic Publishers.*

1. Introduction

The Krkonoše Mountains are the border mountain range between the Czech Republic and Poland. They are located directly in the "Black Triangle" Region. Natural values of the area on both the Polish and the Czech side are protected, by legislation, as a national park and its buffer zone. In 1992, the entire area of the national parks and the buffer zones was proclaimed a Bilateral Biosphere Reserve of the UNESCO. The unique biodiversity of the Krkonoše Mountains is related mostly to their exceptional geographical location, elevation and geomorphology. Therefore, the Krkonoše mountain range had formed an important junction linking the northern tundra, pushed back repeatedly by continental glaciers to the south, and the alpine and subalpine ecosystems which have expanded northwards from the Alps.

The land area of the Czech part of the biosphere reserve is 54,787 ha (36,358 ha the national park, 18,430 ha the buffer zone). The Forestry Department of the Administration of the Krkonoše National Park ("the Administration") manages even the most valuable ecosystems (alpine tundra, subarctic peat bogs nowadays classified among the wetlands of the Ramsar Convention, glacier cirques, dwarf pine stands, original mountain spruce stands and mixed forests, and a substantial part of the secondary florid mountain meadows valuable from the biodiversity point of view). An economic effect cannot be the aim of the management. The main objectives of the forest ecosystems treatment, defined by the above mentioned features, are the protection of biodiversity, a landscape enhancement, and educative and recreation functions (8-10 millions of "beds days" registered annually). Other preferred functions are protection of soil and climate and management of water supplies.

In 1984, the Krkonoše National Park was classified by the (IUCN) among the most endangered national parks of the world, because of the destruction of ecosystems by emissions, a high attendance rate and the impact of the ill-calculated management in the past.

2. Impact of Emissions and the Zones of Endangering

It is very well known, that the decline of the Krkonoše forest ecosystems was caused by the complex of interplaying emission and ecological impacts (phytotoxic compounds in the atmosphere and soil, deterioration of the soil properties, failures in the nutrition of plants, pathogenic organisms, climatic impacts, etc.). The effects of these impacts were combined with the effects of the improper management in the past, particularly the conversion of the original composition and structure of forest to the even-aged spruce stand. The conversion was carried out in an area larger than 75% of today's forest ecosystems of the Biosphere Reserve (the Czech part). About 50% of the forest ecosystem area was totally destructed.

The apparent damage to ecosystems by emissions began in 1977. Concentrations of sulphur compounds in the atmosphere have been measured since 1980. After 1991, the concentrations of airborne SO_2 significantly decreased (below 10 μg/m3, i.e. below the values critical for the spruce stands at extreme sites). The concentrations of SO_4^{2-} in aerosol decreased below the values measured in 1980 [1]. Sulphates are the dominating factor of the emission impact in Krkonoše [2], and the decrease of their concentration was probably caused by the reduction of emissions from the lignite-burned power plants in the Black Triangle Region. Thus, the emission situation has been altered significantly. We anticipate, though, a long-term persistence of the forest ecosystems decline, caused especially by soil processes affected by the atmospheric deposition.

Up to now, the Administration did not have information sufficient for the assessment of the decline rate of the individual stands, resulting from the emission load. Therefore, it was difficult to plan forest regeneration (more than 93% of timber harvest in the period 1982-1992 was a random logging). It was impossible to perform the natural reproduction of the forest ecosystems, and since the plants for an artificial regeneration use to be prepared as long as 4 years in the Czech conditions, it was often impossible to use a genetically more convenient plant stock for the forestation.

In order to enable a planned cultivation of the forest ecosystems, the area of the Krkonoše Mountains has been differentiated into 4 zones of endangering, according to an assessed longevity of mature spruce individuals. The classification is based on the emission rate, site conditions and a current state of the forest stands at the locality. In the A zone of endangering, the longevity of mature spruce individuals is estimated to no more than 20 years since the beginning of the effects of a given emission level. In the B zone it is 21-40 years, in the C zone 41-60 years, and in the D zone 61-80 years. No locality at which the mature spruce longevity would exceed 80 years has been found in Krkonoše. The zoning is illustrated in Fig.1 (tax allowances are defined by law for the A and B zones).

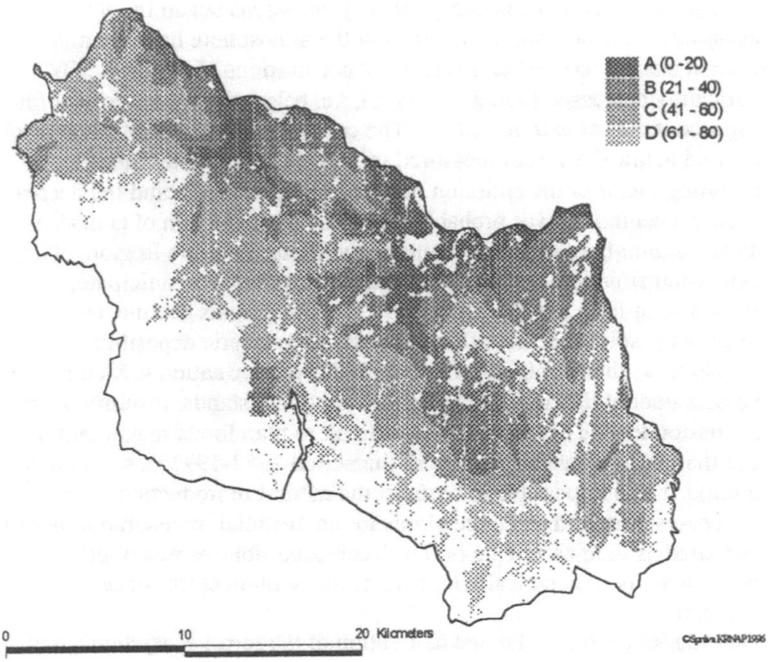

Figure 1. Zones of endangering by the emission impact in the Biosphere Reserve Krkonoše (the Czech part)

However, even the categorization does not help the Administration to plan the forest ecosystem treatment with a sufficient accuracy and with regard to the long four year period of the planting stock preparation. Therefore, a detailed research of emission impact has been carried out, and an extensive research of the soil environment affected by the emissions is currently underway. Models of a forest ecosystem development at different scenarios of the development of emission in the Black Triangle Region are being prepared.

3. Management

The Administration wants to moderate the negative impact of the emission and ecological effects of the forest ecosystems treatment, based on practical experience and results of research. Management is based on site conditions, information about the emission load and knowledge of the response of ecosystem to an emission load. The priority objective is the maximum preservation of the ecosystem, species and intraspecific biodiversity. To achieve this goal, a natural regeneration and processes are used as frequently as possible, in parallel with the protection of rare plant and animal species or selected problematic ecosystems.

3. 1. SITE CONDITIONS AND A TREE SPECIES COMPOSITION

The basic planing unit is the set of forest types. It comprises complex information on a soil, elevation and aspect and it is characterized using soil pits and phytocenological survey. The code of the set of forest types consists of the forest altitudinal vegetation zone (in the Krkonoše Mountains 5. to 9.), determined by the elevation and by the expozition, and of the ecological series determined by soil properties. The area and distribution of forest vegetation zones in Krkonoše is illustrated by the Fig.2.

Figure 2. Distribution of forest vegetation zones in the Biosphere Rezerve Krkonoše (the Czech part)

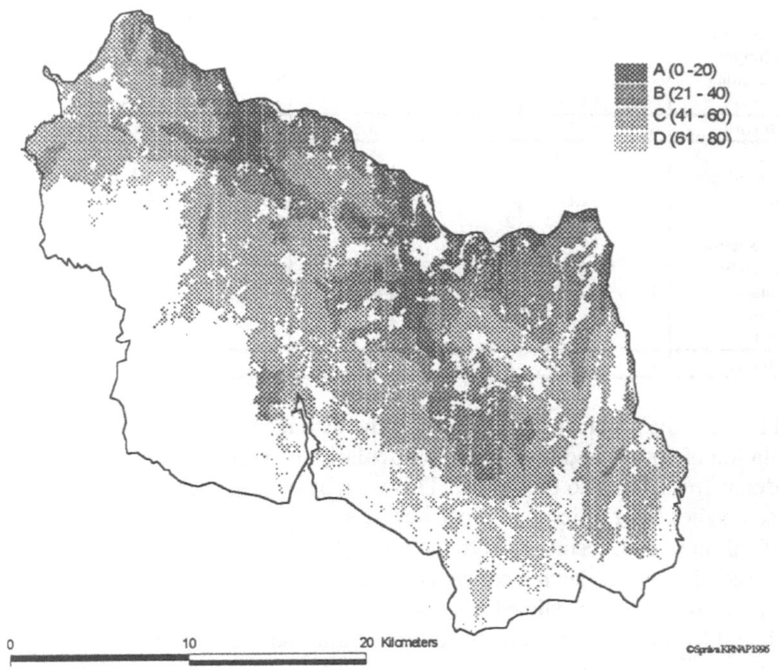

The natural tree species composition (climax and seral stages converging to it), is known for each of the forest types. The natural tree species composition is taken into consideration if possible (especially with respect to the emission conditions), during the reforestation, and reproduction of today's mature stands. It is also considered in the course of rescue and regeneration of the biodiversity of those stands which were recultivated improperly and converted to the spruce monocultures in the period of escalating ecological catastrophe in the past 20 years. Further treatment of the forest ecosystems is also carried out with respect to the natural tree species composition and stand structures. A comparison of the current tree species composition to a natural one is [3] listed in the following table:

Woody species	**Composition of Woody Species in %**						
	Natural						**Current**
	Forest vegetation zone					**Total**	
	5	**6**	**7**	**8**	**9**		
Picea excelsa	2,7	35,6	69,6	93,5	20,4	53,7	86,7
Pinus mugo	0	0	0	0	79,6	8,2	6,9
Larix decidua	0	0	0	0	0	0	0,9
Abies alba	37,8	21,4	10,5	0	0	12,4	0,1
Coniferous	**40,5**	**57**	**80,1**	**93,5**	**100**	**74,3**	**94,6**
Fagus sylvatica	54	40,4	18,9	+	0	22,6	2,6
Acer pseudoplat.	2,7	+	1	0	0	0,3	0,6
Betula sp.	+	2,6	0	0	0	1,1	0,9
Alnus sp.	2,7	+	0	0	0	0,1	0,4
Sorbus aucuparia	0	0	0	6,5	+	1,6	0,6
Ulmus glabra	+	0	0	0	0	+	+
Fraxinus excelsior	+	0	0	0	0	+	0,2
Salix sp.	0	0	0	0	+	+	0,1
Decideous	**59,5**	**43**	**19,9**	**6,5**	**+**	**25,7**	**5,4**

The table makes clear that the goal of the management is to reduce the population of spruce and to enhance particularly the population of deciduous trees, but also that of fir. This way, by the reduction of the spruce needles litter, microbial processes in soil may be enhanced and the acidification of soil, resulting from the atmospheric deposition, might be decreased. Moreover, such a conversion of the tree species composition is a rational response to a predicted warming of the climate. Introduction of beech and fir into the spruce monoculture or into the clear-cut areas which have arisen due to the emission-caused forest decline, is cost demanding.

3. 2. APPLICATION OF A GENETICALLY CONVENIENT PLANTING STOCK

An autochthonous material is employed preferentially in regeneration, since at the same sites and emission conditions, the autochthonous stands are more viable than alochthonous ones [4]. In the autochthonous and provenancely convenient stands, the natural reproduction is preferred, while in provenancely inconvenient ones, the reproduction is carried out artificially, using a genetically convenient material. At a shortage of a convenient generative stock, the Administration employs propagation by cuttings. Establishment of appropriate parent trees requires further expenses.

3. 3. THE NATIONAL PARK ZONATION AND NON-WOOD PRODUCING FUNCTIONS

The area of the Krkonoše Mountains has been divided into 4 zones according to the natural values: I. (4,152 ha), II. (3,046 ha), III. zone of the national park, IV. is the buffer zone. The zoning is shown in Fig. 3. All the measures in the zone I are directed to a maximum reduction of the civilization impact on the ecosystems. An exploitation of timber is ruled out and therefore, the measures lead to a financial loss. With respect to recommendations of the scientific board of the National Park, measures will be taken in some cases even in the I.zone (removal of the Pinus mugo stands), in order to create conditions for effects, which in other cases need not exist (frost soil forms).

The goal in the zone II is the management using well-considered actions (especially measures against the spread of bark beetle). If possible, the stands affected only slightly by emissions should be preserved. The stands which have been considerably damaged by the emissions are regenerated exclusively using an autochthonous plant stock in order to achieve the condition close to that of the natural composition. It is permitted to take a part of timber from this area, but the wood substance is usually deteriorated and transportation in the mountain conditions is costly. Thus, the management in this zone is loss-making as well.

In zone III and in the buffer zone of the National Park, the management is economically efficient to some extent, however, the effect is reduced by the limitations imposed on the applied technologies, because of the preference of the non-wood-producing functions. A part of the forest stands has been identified as the important forests from the biodiversity point of view. Some of the rare plant species and selected problem phytocenoses are supported in these localities by the regulation of the management at the expense of the economy , if it is required by the character of the herb layer. Another restrictions of technologies are necessary in the stands with a preferential water-management role of the forest, as well as the protection of springs of mineral waters, or in those with a landscape, hydrological, climatic, recreation or educative function. The limitations imposed on technologies reduce the opportunity of yielding a maximal economic effect.

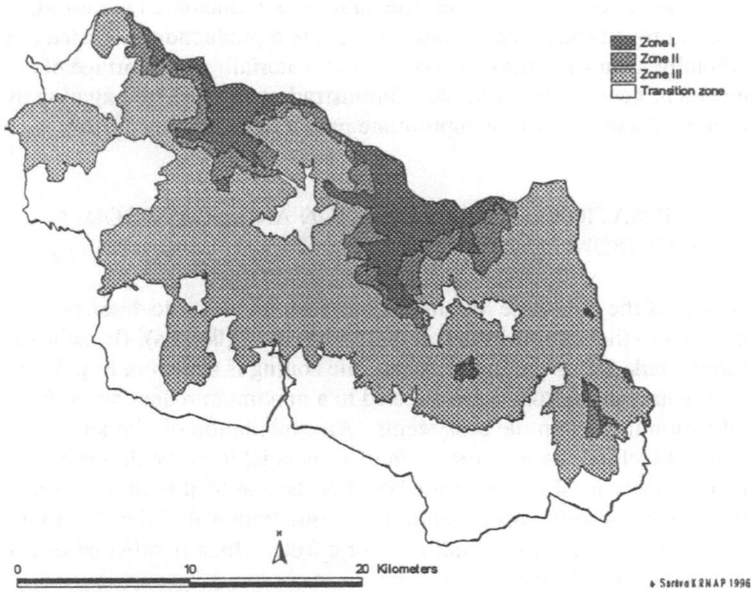

Figure 3. Zones according to the natural values in the Biosphere Reserve Krkonoše (the Czech part)

Among other very important objectives of forest management is the protection of soil in the localities endangered by intraskeletal erosion, using specific biotechnological and natural technological measures, and preventing the formation of new avalanche trajectories threatening settlements. A sufficient amount and efficient application of research information is a prerequisite of a successful recovery and reconstruction of the forest ecosystems and their non-wood-producing functions. Geographical information systems are being used by the Administration of the National Park for the evaluation of a part of research activities and for planning and record-keeping of the realized measures [5].

4. Research

The Administration co-ordinates a part of research activities, sets and covers some of the tasks. Thus, the research activities are directed towards a practical use of the results in the planning and realization of the measures and management.

The general objective of the research projects and monitoring is a support of the Krkonoše forest reconstruction by a sufficient amount of research data, to formulate recommendations for planning and performing of forestation and for the protection of plantations. The results of research are applied in the change of the forest working plan, which is being processed currently now. These changes are focused particularly on the regeneration of biodiversity of the forest ecosystems, based on tending of genetic variability (including the intraspecific variability) and a directing of the forest cultivation to the ways close to the nature. In research projects, a maximum application of the natural regeneration has been emphasized as well as the employment of a genetically convenient planting stock and the economy of forestation. The Research Institute of Forest Management and Game, Research Station Opočno, the University of Amsterdam, the Administration of the National Park and many other Czech and foreign institutions participated in the research projects.

The basic aims and results of the realized research are:

1) Monitoring of the emission impact and development in the area, monitoring of the condition of ecosystems and the dynamics of their development. The map of distribution of the total atmospheric deposition [6] is shown in Fig. 4.
2) To determine the development of the health of the spruce stands in future using the retrospective development of their healthy status (satellite photographs and ground investigations) and regarding the development of emissions and ecological factors. To enable a planned regeneration of stands affected by emissions, to prevent the elimination of those stands which, despite the apparently critical health status, have a chance to survive and to enable the planting of a sufficient amount of the seed stock in an appropriate genetic quality.
3) To check the presumptions of the natural regeneration in the relation to the soil environment, ground layer and trees and to design the scope of methods for the natural regeneration.
4) To design a technology for the conversion of the stand composition, particularly for the introduction of the target woody species (with respect to the site conditions) into the spruce cultures and the employment of pioneer species in the conversion of the composition of stands.
5) To map the localities and to design the technologies for the regeneration on the soils endangered by intraskeletal erosion and new avalanche trajectories (especially underplanting and employment of natural ameliorating materials) and to test the technology of spruce layering.
6) To create and test a technology for the rootage of cuttings from the mature trees and establishment of a parent tree (including the clone archive) for the further propagation by cutting. The technology that has been designed and tested allows for rootage of approximately 50% of cuttings.
7) To gain clone mixtures of the individuals which are relatively resistant to the effects of the current emission and ecological factors.
8) To identify the dependence of the humidity conditions development and the herb coverage for the different site conditions (complexes of forest

types) and to design the forestation technologies including an optimal timing of the realization of the forestation.

Figure 4. The map of distribution of the total atmospheric deposition in the Biosphere Reserve Krkonoše / Karkonosze (the Czech and the Polisch part)

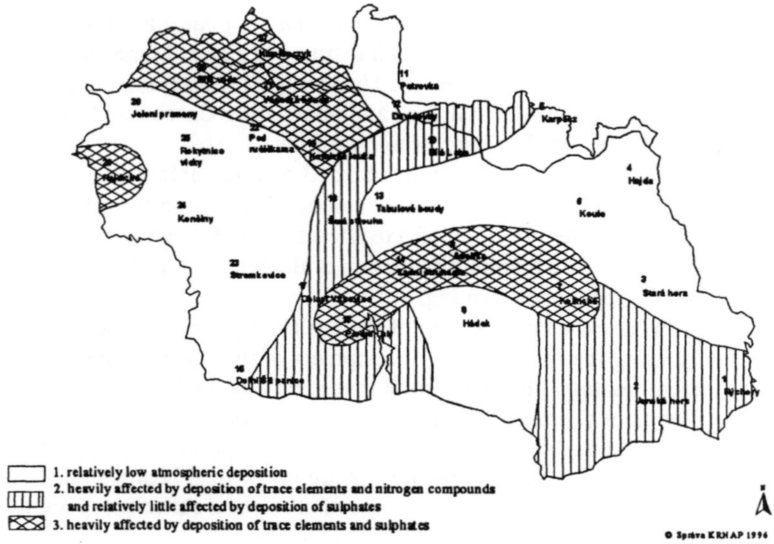

The forest management approach also uses also a number of results of monitoring in other fields of interest, e.g. botanical inventory investigation, records of significant geomorphologic phenomena, monitoring of the damage of the woody species outplants on the emission clear-cuts by Microtus agrestis, etc. The research activities and monitoring have been financially supported in particular by the Dutch foundation Face, as a support of forestation projects, and by a grant of the Ministry of Environment of the Czech Republic.

5. Financial Support of the Forest Ecosystems Management

The low quality of the timber output, high costs of the harvest, skidding, forestation and treatment of cultures and also the consideration of the non-wood-producing functions makes forest management an economic loss.

Since 1992, a substantial part of the forest ecosystem reconstruction in Krkonoše has been funded by the Dutch foundation Face (Forest Absorbing Carbondioxide Emission), which uses this way of "buying" the effect of the CO_2 binding in the biomass of trees. Without this support, the forest ecosystems reconstruction in Krkonoše would be possible only in a limited scale. The Face foundation was established by association of the Dutch power engineering companies Sep (The Electricity Generating Board) and it is financed by the energy consumers (0.00018 USD per used up kilowatt hour).

Up to 1994, the forest management in the Krkonoše National Park was supervised by the Ministry of Agriculture, that used to cover the financial deficit using the profits from the sale of timber in other regions of the Czech Republic, which are profitable from the point of view of the forest management. In 1994, the right of management in the Krkonoše forests was transferred to the Ministry of Environment and the financing of the deficit became troublesome. With the exception of tax allowances, the state is not in charge to support the management either in the forests damaged by emissions, or in those with priority non-wood-producing functions. The shortage of funding resulted in the Administration not being able to carry out some important activities in the cases where the delay may result in further distress, e.g. the rescue of the genetic pool of the autochthonous populations of the Krkonoše spruce. The complex of the emission and ecological impacts led to a genetic and often even a physical elimination of the mountain spruce populations (7). In this case, the Administration applied unsuccessfully for a financial support to the World Bank in the scope of the Global Environment Facility (Forest Biodiversity Protection Project GEF 05/21685 POL as well as the GEF Program - Czech Republic Biodiversity Project).

6. Recommendations

1. Immediate

1.1. To categorize the forest stands according to the requirements of non-wood-producing functions and according to the impact of emissions

1.2. To provide a state financial support according to the individual categories and with respect to the current and desired condition of the ecosystem. The funds for such support should be obtained either from the state budget or rather similarly to the Face foundation, by an increase of the costs of power.

1.3. To provide a state financial support for using timber as a building and construction material, which will be one of the ways of the efficient absorption of CO_2 emitted to the atmosphere by the burning of fossil fuels.

1.4. Forest management must be based on the results of the practically oriented research and monitoring.

1.5. To determine how much of the wood biomass may be taken away from the forest ecosystem without damage to its functions.

2. Prospective
2.1. To seek less power-consuming technologies
2.2. To seek more "clean" sources of energy
2.3. To seek new ways of application of timber
2.4. To decrease the prices of timber for building and construction.

7. References

1. Schwarz, O., (1996), Results of measurments of sulphid compounds concentrations in Krkonoše Mts. *Monitoring and Management of Ecosystems on the area of Krkonoše National Park*, Proc. Conf. Opočno, April 15-17 1996. In print

2. Hošek, J., Kaufman, R., Schwarz, O., Hofmeister, J. (1996) Total Atmospheric Deposition in Krkonoše and Jizera Mts. (1994 - 1995). *Atmospheric Deposition and Forest Management*, Field excursion guide for the NATO-Advanced Research Workshop, Špindlerův Mlýn, Krkonoše (Giant) Mountains, Czech Republic, April 23-26, 1996

3. Lokvenc, T., end all., (1992), Reforestation of the Krkonoše Mts., Krkonoše National Park Administration and Forest Research institut Opocno, 111 p., 1992

4. Schwarz, O.,Vacek, S., Andil, P., Hošek, J., Kaufman, R. (1996) Ecological relations of spruce damage in Krkonoše. In S. Vacek (ed.), *Monitoring and Management of Ecosystems on the area of Krkonoše National Park*, Proc. Conf. Opočno, April 15-17 1996. In print.

5. Schwarz, O. (1995) Application of geographic information system on analysis of natural environment and forest stand as a precondition for regeneration of forests in the National Park Krkonoše. In K. Matijka (ed.), *Investigation of the Forest Ecosystems and of Forest Damage. Processes in Forest Ecosystems and their External Functions*. Proc. Workshop Opočno, April 25-27, 1995, pp. 68-77.

6. Hošek, J., Kaufman, R. (1995) Total Atmospheric Deposition of Environmentally Significant Substances on the Area of Krkonoše National Park and Landscape Protected Area Jizerské hory Mts.- Vegetation Period 1994. Research report, (in Czech), Project *Rehabilitation of the forest ekosystems in the National Park Krkonoše Mts.*, Krkonoše National Park Administration, Schwarz, 1996

MANAGEMENT AND UTILIZATION OF DAMAGED FORESTS IN CENTRAL AND EASTERN EUROPE

G.F. ZAK
Atlantic Environmental Services, Inc.
188 Norwich Avenue
P.O. Box 297
Colchester, Connecticut 06415, USA

Abstract

Management of the degraded forests in central and eastern Europe requires the development of an action agenda. At least one forest research project and management approach should be initiated and/or continued. The effort should be results-oriented and consistent with ecological balance. Conclusions should guide the application of remedial measures to convert forests to different species, or stabilize the health of existing forests. These measures will contribute to the overall goals of expansion of the European forest, both as a carbon sink and to invigorate the forest products industry in countries of the former United Soviet Socialist Republic (USSR). In particular, neutralization and fertilization of forest soils should be more fully investigated and probably more widely applied. Orderly conversion of softwood monocultures to climax stands should be encouraged. The development of pollution-resistant genotypes should be accelerated. This paper provides the rationale for these forest growth, expansion, and management recommendations.

1. Introduction

1.1. BACKGROUND AND RELEVANCE

Defoliation of trees inconsistent with the typical effects of pathogenic agents or insect predation has been observed and studied in central and eastern European forests since the late 1970s and early 1980s. Research aimed at identifying the cause has provided evidence so convincing that certain European governments, such as those in Poland and the Czech Republic, consider air pollution as the major cause of defoliation and forest decline. Other European governments group air pollution with other factors predisposing forests to decline [1], probably because the effects are not so pronounced.

A number of sources, including automobile exhaust, are thought to play a part. However, industrial emissions of sulphur dioxide to the atmosphere are believed to be the main catalyst of forest decline, via the wet and dry deposition of acidic pollutants. "Acid

R. M. Gutkowski and T. Winnicki (eds.), Restoration of Forests, 227–235.
© 1997 *Kluwer Academic Publishers.*

rain," while percolating through the soil, mobilizes otherwise less mobile inorganic compounds, such as magnesium, and lowers the pH of the soil itself. Soil organisms, mychorrhiza, and nutrient availability may be affected. This process can result in:

- tree malnutrition and possible death, because vital inorganic nutrients are depleted (through leaching);
- decrease or cessation of activities by soil organisms that are beneficial to tree growth; and
- degradation of potential drinking water sources (groundwater).

The major source of atmospheric sulphur dioxide is the combustion of high- sulphur coal [2]. International conventions and political realignment have led, in some cases, either to the shut-down of coal burning plants or installation of pollution reduction technologies [3]. These efforts have reduced levels of atmospheric pollutants in central and eastern Europe, but forest decline is expected to continue as a result of ongoing deposition of pollutants. The 1990 estimates of defoliation, compiled by the United Nations, are presented in Figure 1.

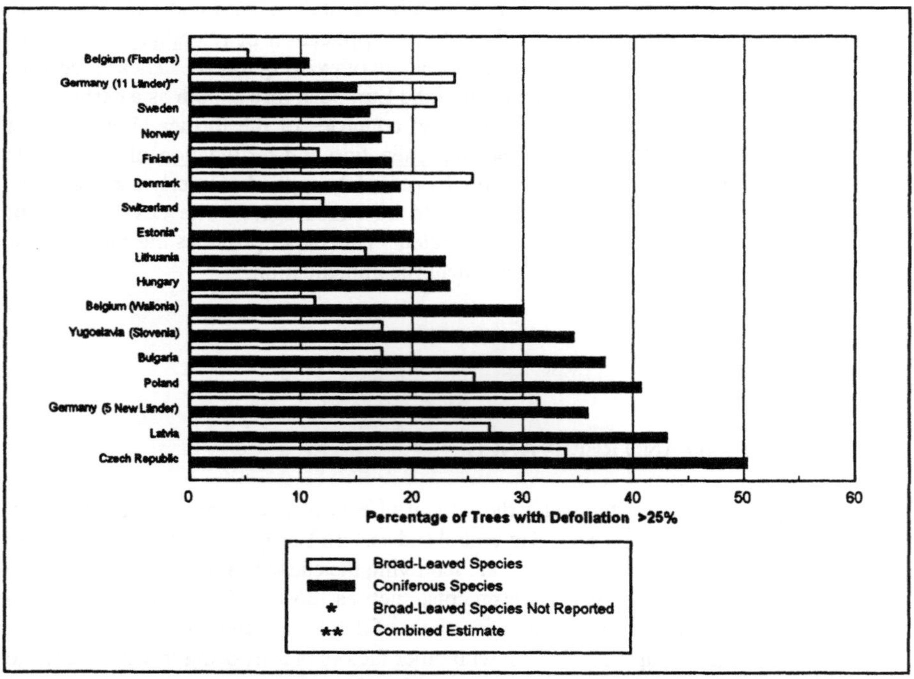

Figure 1. Relative defoliation (1990) for broad-leaved and coniferous species in selected European countries (Derived from United Nations, 1992).

Two generalities can be inferred from Figure 1: defoliation is most severe in coniferous species and in the most eastern of the eastern European countries. The defoliation may also be caused by severe insect predation or other factors such as tree age. However, successful insect predation is often a function of poor health, which may be induced by pollution. Defoliation may also occur in younger trees stressed by pollution. In addition, the normal decline exhibited by trees of increasing age may also be accelerated by the effects of pollution on the micro-site.

The relevance of these conditions is their economic and ecological effects, which are local and global. The risks of forest decline include degraded or lost value of forest products, potential loss of forest flora and fauna (biodiversity), and decreased sequestration of carbon. Social health and climatic stability, as well as genetic diversity, are threatened. Since the complete halt of fossil-fuel use is not considered a viable alternative in the short-run, forest management methods to reduce and/or reverse widespread defoliation and associated ecological and environmental losses are required.

1.2. THE CURRENT FOREST SITUATION

A significant reason for forest decline in central and eastern Europe is species-related. Following World War II, when European forests were exploited to support the war effort [4], afforestation projects relied heavily on the planting of conifers. Coniferous plantings have remained a method-of-choice in afforestation efforts, although genetic stock has not always been of local origin. The wisdom of these deliberate monocultures has always been the basis for argument [5].

Theoretically, coniferous monocultures lend themselves to efficient management. Often, they are more productive than broad-leaved stands. Planning is less complex, and economies of scale can result in greater profits. But the inherent lack of diversity leaves them more susceptible to destructive agents. A monoculture weakened by pollution may be open to catastrophic insect damage.

This has occurred in many forests in central and eastern Europe, especially in the area where the borders of Germany, Poland, and the Czech Republic come together (known as the "black triangle"). As a result, forest managers have had to apply unplanned sanitation and clear-felling to salvage trees before death ruined their value [6]. Broad-leaved stands have also been affected; however, the apparent impact is not so devastating.

1.3. INTERRELATIONSHIPS

Physical and chemical characteristics such as forest elevation and soil buffering capacity can apparently influence the degree of defoliation and reduced growth. At the same time, increased levels of carbon dioxide and nitrogen support soil and physiologic processes that may result in increased tree growth [7]. However, soil acidification (resulting from acid rain) generally accompanies pollutant deposition, and the acidification may render other necessary growth nutrients or compounds unavailable or make toxic compounds more available. As part of the ongoing NITREX program, Gundersen and Rasmussen (1995) [8], as well as others, have provided details of the interactions and behaviors in soils and trees at several forested catchments in Europe.

2. Remedial Action Agenda

The most severe defoliation has occurred in central and eastern European forests composed of a considerable percentage of coniferous species. Many of these stands are of cultured generation; of the cultured stands, many are overmature [9]. This overmaturity, combined with the reduced health associated with pollution, suggests that clear-cut harvesting and conversion to pollution-resistant climax species is one approach that will help the European Community reach the goals stated in the abstract of this paper. For those coniferous stands that are immature (on a financial maturity basis), a different approach is recommended. The general decision path is illustrated in Figure 2.

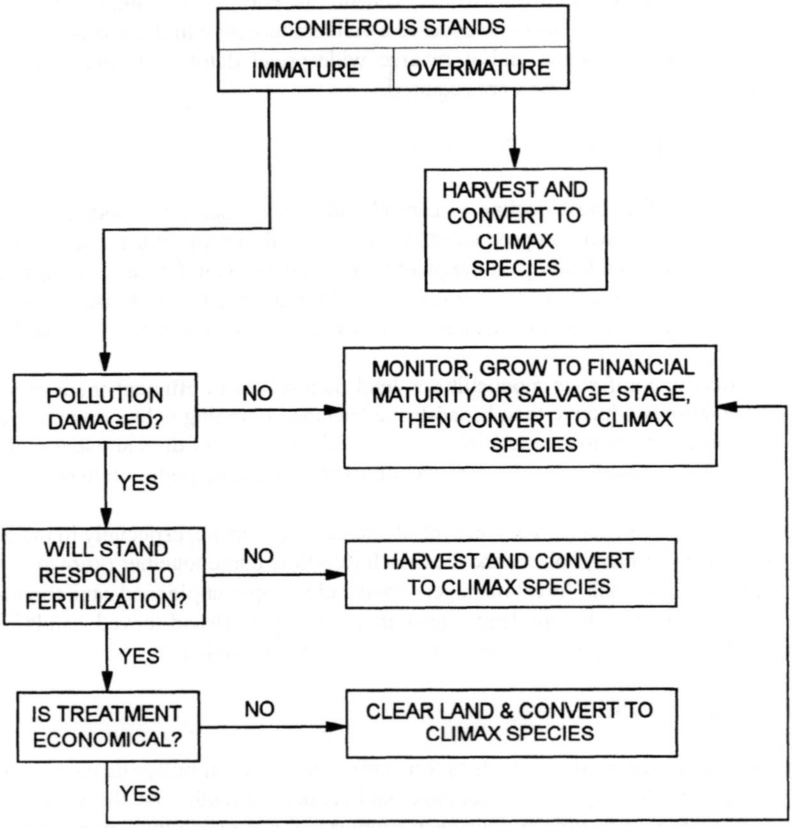

Figure 2. Flow hart for management of pollution damaged coniferous stands

2.1. NEUTRALIZATION AND FERTILIZATION

Research conducted by Glatzel, et al. (1992) [10], Schneider and Zech (1990) [11], and others demonstrates that the health and growth of declining Norway spruce (*Picea abies*) stands can be improved with the application of neutralizing and/or fertilizing agents. Specifically, manipulation (increase) of soil pH and provision of nutrients that have otherwise been transformed or leached, seem to contribute to the improvements. Others, such as Haug (1990) [12], have observed mixed results.

To counter the decline and potential death of forests that have not reached maturity, remedial neutralization and fertilization of soils must occur (where possible and appropriate). Unfortunately, the mixed results of fertilization experiments indicate that the state of the art in forest soil management is far from the "cookbook" stage. As such, the action and interactions following dispersion of neutralizers (such as dolomite limestone) and fertilizers (such as magnesium) should be carefully studied. This study should be conducted in a manner consistent with potential large scale, widespread application. One method follows.

- Delineate those forests worth growing, but at the greatest risk of decline. This delineation can be based on economic, social, or ecological factors, or a combination of factors.
- Define the predominant nutrient and pH imbalance(s) in those forests chosen for treatment and/or those stands *within* the chosen forests. The definition would be based on analysis of soil and tree samples.
- Determine the best mix of neutralizers and fertilizers for the defined forests and stands.
- Make aerial applications with a pelletized composite that disintegrates over time, releasing neutralizers and fertilizers. Helicopter application can be used in areas with exaggerated topography.
- Monitor the results with an agreed upon, systematic, region-wide scale for determining success or failure.
- If the forest is a coniferous monoculture, convert to climax species subsequent to harvest of the conifers. If the existing forest is composed of climax species, artificially regenerate following harvest with local, improved genotypes.

This method should initially be applied on a "pilot-scale" basis, with careful monitoring of the results. If the *overall* results are favorable (i.e., forest survival and acceptable growth), the method should be more widely applied. An increase in groundwater nitrates and/or other undesirable responses may be likely. It may also be possible to manage these responses.

The ideal goal is development of an efficient system to apply remedial treatments, prescribed on the basis of soil type, slope, aspect, elevation, species, and symptoms of decline. Developing an effective method for classifying soils, conditions, and treatments will be the key to success. This is the main area where additional research and technology will be useful. Ultimately, the homogenous traits which enabled the decline in monocultures could also support their deliverance.

2.2. CONVERSION TO CLIMAX SPECIES

Kuusela (1994) reported that ancient climax forests (those with a relatively stable species mix, which maintain themselves indefinitely) in many parts of central Europe were approximately 80% to 90% oaks and beeches, and that current forests are 60% to 70% coniferous [13]. This conversion from a climax to temporary mix is typical where humans have settled; the notion is that the different (but indefinitely maintained) species mix is more orderly and efficient to manage. Yet, as described in previous subsections, indefinitely maintained monocultures pose risks, and some of these risks have had recent effects. Therefore, Kuusela recommended conversion to local, natural climax species, where appropriate.

There are distinct or potential advantages in converting to improved climax species, as follows.

- Climax species are naturally better adapted to the site and climate.
- When the climax is primarily composed of deciduous species, less deposition of pollutants will occur (because decreased surface area results in decreased interception of pollutants).
- If the climax is primarily deciduous, there may be greater carbon storage potential (hardwoods are denser, and may replace coal or oil as a home fuel source more frequently than softwoods, thereby reducing *net* carbon release).

Of course there are disadvantages. Conversion to and initial maintenance of climax broad-leaved species is likely to be labor intensive. In addition, downwind countries could receive increased deposition of pollutants because the new broad-leaved stands will intercept less. But anecdotal information suggests the presence of an able labor pool, and cited literature indicates ongoing reduction of pollution emissions. Another difficulty will be the necessity to shift to different silvicultural methods. This shift is less dramatic, on a relative scale, than the conversion to climax stands. It also presents the opportunity to employ more recently developed ecosystem management techniques. This perspective is appropriate, considering the ecological rationale for conversion to climax species.

2.3. DEVELOPMENT OF SUPERIOR GENOTYPES

Improved seed of local provenance may not be available for a number of years, because cross-breeding of superior traits requires the careful selection of parent trees. That selection is followed by breeding and cross-breeding. But the time required to develop improved seed should not delay the establishment of new climax stands. The seed for these stands should be collected from apparently healthy, local seed trees. While non-climax monocultures are being converted, the task of improving seed can be ongoing, for use in future conversions.

Initially, the effort should be concentrated on the collection of seed from native, apparently unaffected trees in stands that are both pollution and insect-damaged, as these agents appear to be the problems most likely to persist. The search for improved seed should also include parent material with other desired traits, such as rapid growth and drought resistance. Obviously, the seed must be collected before affected stands are clear-felled and regenerated.

3. Pertinent Observations from the Advanced Research Workshop (ARW)

My own approach to the forest management issues considered at the ARW was a problem/solution approach. As such, I recommended potential solutions, as described above. They may be useful at the right place and time. However, during discussions with my colleagues, it became clear that my recommendations were developed in a vacuum.

3.1. ARW RESULTS

Political and economic forces represent the greatest hope and the greatest threat to any forest. In wealthy, developed countries these forces are balanced to a certain extent. But as the countries in transition struggle to develop, they may look to the forest for hasty generation of currency or products. The deleterious effects of these practices, especially in tropical forests, are well documented.

Otakar Schwarz (Director, Forest Authorities, Czech Republic) described the current government policy to restore the dead climax forest in their portion of the "Black Triangle." The original climax forest took hundreds of years and specific ecological processes to develop. This is not easily reproduced by humans, and there is no guarantee that the attempt will work. The economic resource (which is scarce to begin with) might be better used for something else, because reforestation and forest succession will eventually take place on its own in most locations. This is especially true if the policy of the Polish government (which owns forestland at the boundary of the Czech forest) is not coordinated with the Czech government.

Ladislav Paule (Forest Geneticist, Technical University, Slovakia) described how the development of genetically superior tree species has a strong aspect of risk, because 50 years from now (when the "improved" genetic stock may be ready for harvest) the superior traits may not be relevant to the newest markets or environmental conditions.

Kazimierz Rykowski (Forest Ecologist, The Forest Research Institute, Poland) reminded us that the forest, however we treat it, will always serve as a place for animals and trees to live and grow. He reminded us, as well, that as a result of our activities, what lives and grows will be that which can adapt to the site. It may not be what we want. More importantly, ***it may not be what we need.***

Julius Natterer (Wood Architect, Ecole Polytechnique Federale de Lausanne) demonstrated how construction and occupation of wooden buildings is less costly, less ecologically harmful, and more supportive of human mental and physical health, than buildings constructed with concrete and steel. Yet it is discriminated against, worldwide, by zoning laws which presume that wooden architecture is unsafe.

I pointed out that human population plays an important role because most pollution problems were created by humans. The extent is correlated with the number of people and their lifestyles. Based on my direct involvement with investigations at many polluted sites, clean-up is technologically impossible at this time. As such, some land and groundwater is destroyed and may never be safe again.

The point of these anecdotes is that focussed, solution-oriented recommendations are not likely to improve overall management and utilization of the forest resource if activities between countries are not coordinated. Coordination will not be effective if the public and their leaders do not understand ecology and the implications of various management practices.

3.2. PREDOMINANT ARW CONCLUSIONS

The ARW participants came together with the goal of reducing the problems in the forests of central and eastern Europe, with the use of science and technology. However, when considering the apparent challenges most participants agreed that appropriate policy and maintenance of necessary ecological functions must come first. Application on a *regional basis* is crucial, because forests and their products (clean water, clean air, etc.) do not pay attention to political boundaries. Science and technology are best regarded as tools that may add productivity and/or efficiency to existing processes that we know work.

3.3. GENERAL RECOMMENDATIONS

In order to successfully pursue sustainable management of central and eastern European forests, political, economic, religious, and educational barriers must be addressed, as follows.

- Development of a regional task force to examine and resolve conflicting environmental, forest management, and wood utilization problems in each of the participating countries.

- Development of an acceptable model or decision matrix that can generate the relative economic, environmental, and ecological effects that different forest management regimes will have on a particular forest.

- Development of educational centers which teach the fundamentals of ecology and the environment, and highlight the role of nature as the source of all humankind's water, food, air, and building materials.

- Consideration of the effect of human population on natural resources. Organizations which recognize the impact of human population on the stability of ecosystems and natural resources must say so. The minimum goal should be to balance the non-population control decrees of other organizations with significant influence.

4. References

1. The United Nations, Economic Commission for Europe (1992) *Air Pollution Studies 8, Impacts of Long-Range Transboundary Pollution*, The United Nations, New York.

2. Worldmark Press, Ltd.(1988) *Worldmark Encyclopedia of the Nations, Europe*, John Wiley and Sons, Inc. New York.

3. International Union for Conservation of Nature and Natural Resources (1991) *The Environment in Eastern Europe: 1990*, Information Press, Oxford.

4. Worldmark Press, Ltd.(1988) *Worldmark Encyclopedia of the Nations, Europe*, John Wiley and Sons, Inc., New York.

5. Kuusela, K. (1994) *Forest Resources in Europe 1950-1990*, Cambridge University Press, Cambridge U.K.

6. Czech Geological Survey (1995) *Forest Die-back in Affected Regions of Czech Republic*, Acidic Deposition in the Black Triangle Region, Proceedings, 5th International Conference on Acidic Deposition Science and Policy, Czech Geologic Survey, Prague, Czech Republic.

7. Kuusela, K. (1994) *Forest Resources in Europe 1950-1990*, Cambridge University Press, Cambridge U.K.

8. Gundersen, P. and Rasmussen L. (1995) *Nitrogen Mobility in a Nitrogen Limited Forest at Klosterhede, Denmark, Examined by NH_4NO_3 Addition*, Forest Ecology and Management 71, 72-87.

9. Kuusela, K. (1994) *Forest Resources in Europe 1950-1990*, Cambridge University Press, Cambridge U.K.

10. Glatzel, G., Kazda, M., and Sterba, H. (1992) *Effects of Air Pollution on Mineral Nutrition of Norway Spruce and Revitalization of Declining Stands in Austria*, Water, Air, Soil Pollution 61, 309-323.

11. Schneider, B.U., and Zech, W. (1990) *The Influence of Mg Fertilization on Growth and Mineral Contents of Fine Roots in Norway Spruce (Picea abies L.) Stands at Different Stages of Decline in Northeast Bavaria.* Water, Air, Soil Pollution 54, 469-477.

12. Haug, I. (1990) *Effects of Fertilization with MgSO4 and $(NH_4)2SO_4$ on Soil Solution Chemistry, Mycorrhiza and Nutrient Content of Fine Roots in a Norway Spruce (Picea abies) Stand*, Water, Air, Soil Pollution, 54. 453-468.

13. Kuusela, K. (1994) *Forest Resources in Europe 1950-1990*, Cambridge University Press, Cambridge U.K.

Position Papers

Working Group 3

"Forest Utilization for Raw Materials and Value Added Products"

WOOD PRODUCTS IN STRUCTURES AND SOCIETY - A TASK FOR AN INTERDISCIPLINARY SCIENTIFIC APPROACH

Problems and Prospective of Better Utilization of Forests for Structural Raw Materials and Value Added Products in Experience of a Country in Economic Transition

M.F. DRDÁCKÝ
Academy of Sciences of the Czech Republic
Assoc. Research Center for Historic Structures and Sites (ARCHISS Telč)
ÚTAM - CLEM
Vyšehradská 49
128 49 Praha 2
Czech Republic

Abstract

The contribution - a position paper - is concentrated on an interdisciplinary approach to a solution of the problem of better utilization of Central European forests for structural raw material and value added products. An experience from the Czech Republic is utilized to stress the complexity of problems, especially their dependency on technological, scientific, political, social and economical characteristics as well as historical backgrounds and structural design philosophy.

1. Central Europe Situation

For a sake of better understanding of the author's position and experience, there are summarized some basic data at the beginning.

1.1. GENERAL FEATURES

The Czech Republic continues in a long historical tradition of industrial utilization of forests and the production of wood per capita is one of the highest in Europe, (1.2 CUM of round wood - 1990). The following figures present Central European situation in production of round wood in comparison with world development tendencies. From the Figure 1 it follows that former socialist countries, especially Poland and Czechoslovakia, substantially decreased their production of wood in the period of economy transition. On the other hand, Austria, another Centre Europe country, made a substantial progress in the same period. In the early eighties the former Czechoslovakia produced about 19 millions of m^3 of the round wood, three

R. M. Gutkowski and T. Winnicki (eds.), Restoration of Forests, 239–250.
© 1997 *Kluwer Academic Publishers.*

fourth of this being coniferous wood and about 50% represented sawlogs or veneerlogs. The country exported about one fourth of the round wood production and this amount continuously decreased about six times during the following ten years, (see Figure 2, data in 1000 CUM). The import, on the contrary, figures a very low percentage (about 0.5%) and in Czechoslovakia it is slightly increasing, in the Poland decreasing. For example, in Austria import represents about 40% and in Hungary about 20% [1], (Fig.3, data in 1000 CUM). The world tendencies in round wood production are more dynamic in all parts of the world except of Europe, where the stagnation is substantially influenced just by changes in former socialist countries, Fig.10.

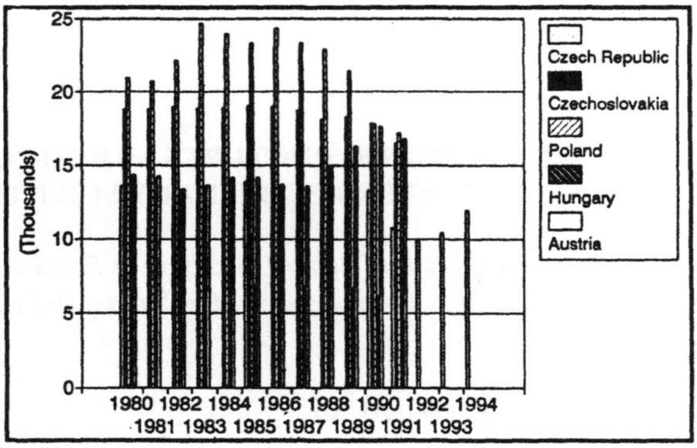

Figure 1. Comparison of round wood production in the Central Europe (in millions of CUM), FAO [1].

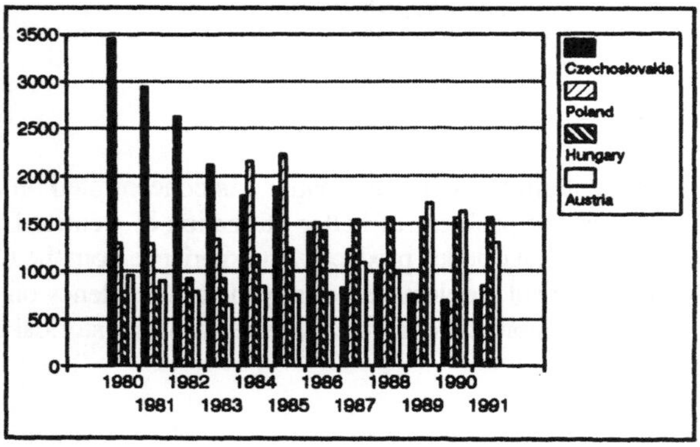

Figure 2. Round wood export in the Central Europe.

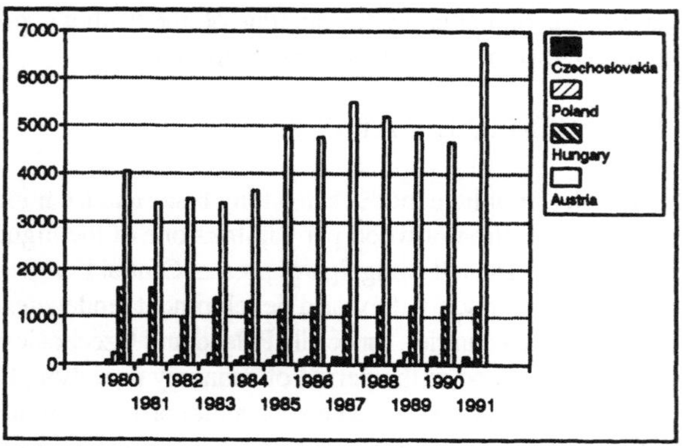

Figure 3. Round wood import in the Central Europe.

1.2. WOOD PRODUCTS DEVELOPMENT

1.2.1. Sawn Wood

Sawn wood production decreased in a similar manner to the round wood production in the years of economy transition, (see Figure 4). The products are green, rough and only visual graded lumber.

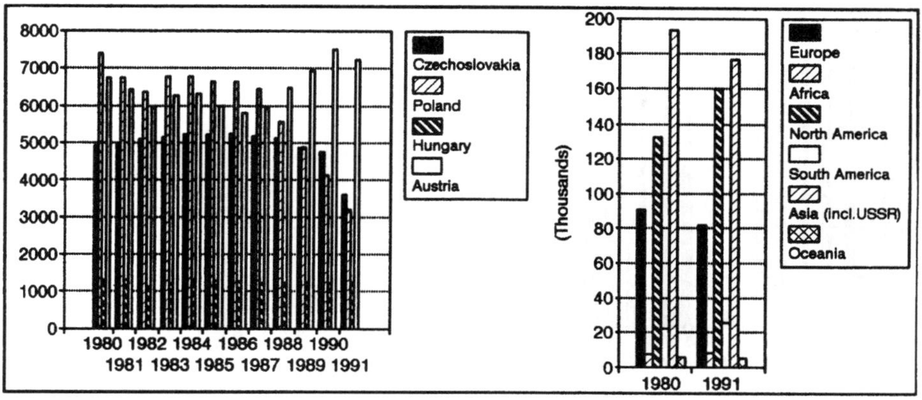

Figure 4. Sawn wood production in Europe, (left - in 1000 CUM), and in the world (in millions CUM).

1.2.2. Wood Based Panels

Veneer. In the production of veneer, the former Czechoslovakia represented one of the most efficient countries and it still keeps its leading position in the region.

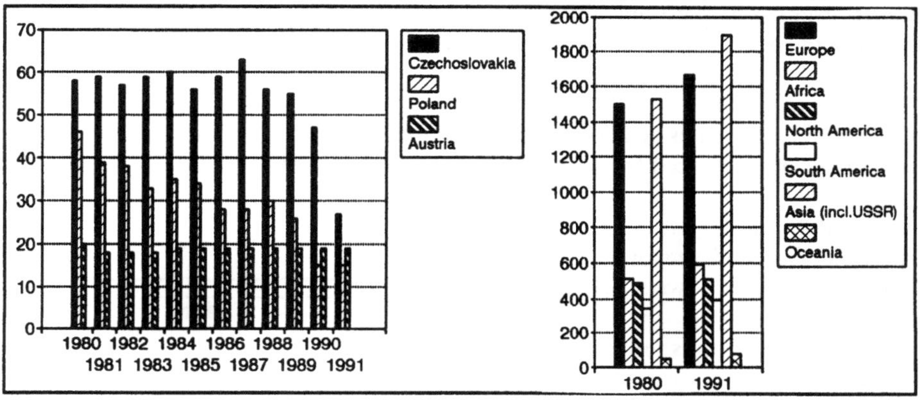

Figure 5. Veneer production in 1000 CUM in the Central Europe, (left), and in the world, (right).

Plywood. The same statement is valid for the plywood production, (Figure 6), and here is worth to see the rapid increase in Austria, (about ten times higher volume).

Fibre and Particle Board. The production of fibre and particle boards exhibited the same tendencies as the other products above, i.e. stagnation and slight (Czechoslovakia) or severe (Poland) decrease in comparison to slight continuous increase in Austria.

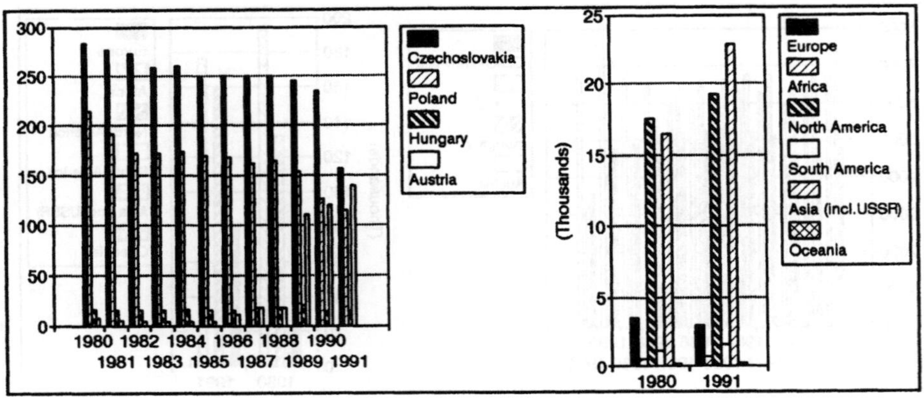

Figure 6. Plywood production in the Central Europe, (left - in 1000 CUM), and in the world (right - in millions of CUM).

1.2.3. Pulp and Paper

Figure 7 further illustrates the Centre European situation in the last decade in wood industry comparing here only the former Czechoslovakian and Austrian production of pulp and paper which also includes paperboard.

1.2.4. Jointed Wood

Wooden windows, gates and doors are widely used in the Czech building industry and they are mostly produced by small private companies or indi-viduals, (see Figure 8).

1.2.5. Poles & Sleepers

Poles, sleepers and similar products have been systematically replaced by concrete, prestressed concrete or steel elements in the Czech countries.

1.2.6. Engineered Timber

The utilization of timber in building industry was severely restricted by a governmental regulation policy in the former Czechoslovakia. Therefore, civil engineering systems

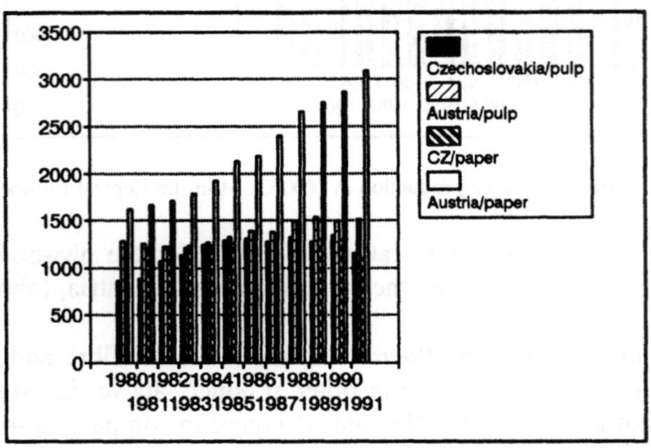

Figure 7. Pulp and paper production in 1000 MT.

have not been developed widely except for some utilization for temporary houses, mostly accommodating construction site utilities. For those buildings, wood wall and floor panels and roof trusses were locally produced. The restrictive policy prevented any major progress in engineered timber or wood based panels and only few modern technologies were introduced into the building industry.

Metal Plate Truss. There was adopted and transferred into the Czech countries a truss technology utilizing metal plates for connecting sawn solid lumber elements (Gang Nail System).

Glue-Laminated Wood. This technology has been used for large span industrial or sport buildings for many years and the design has been supported with a national specification. The structural system proved to be very successful even though the production is very sensitive to negligence or maintaining technology requirements.

Panelized Low Rise Buildings. This system has been mostly based on particle board type panels derived or transferred from German technologies. Problems with fenol-formaldehyde pollution retarded their development in the eighties, nevertheless in the last years some new systems occurred, which are based on other technologies, (i.e. "log cabin" systems, framed panels, composite boards with differential types of adhesives). According to a Czech database of building industry, 28 units producing prefabricated timber systems exist in the Czech Republic, (see Figure 8, which further shows the composition of other facilities [2]).

1.2.7. Wood Furniture

Among other wood products furniture plays the most important role and, in typical forest regions, it represents one of the leading industries, covering about 10% of employment opportunities.

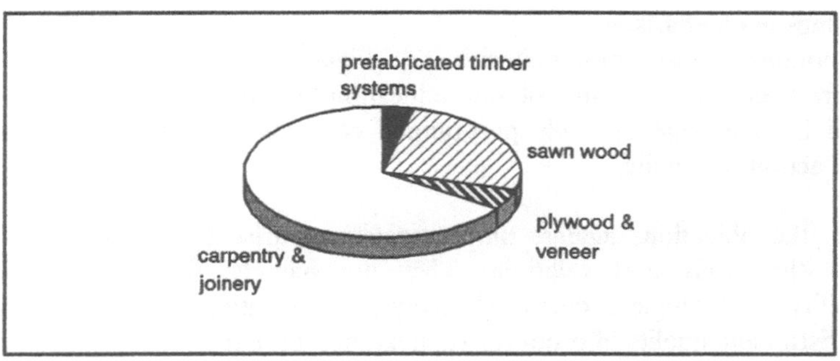

Figure 8. Main wood building industries in the Czech Republic [2].

2. Wood in Society - Needs for a Complex Approach

2.1. NEEDS FOR A CHANGE IN PUBLIC OPINION

The previous chapter illustrates a very high influence of political and economy issues on the development of wood utilization in society. Unsuitable former Czechoslovak state policy not only restricted applications of timber in building industry but also supported negative prejudices about wood and helped to create a public opinion which prevented a sound market development in the country when more liberal conditions occurred. Moreover, new ownership relations in forests and unequal price relations between the Czech Republic and Western Europe on one side changed the price of raw materials in a way which does not promote investments into innovative wood industries. On the other side, it made it more difficult to plan optimum use of forests, which needs new approaches and techniques. Here only a change of governmental policy can invert this non-favourable development.

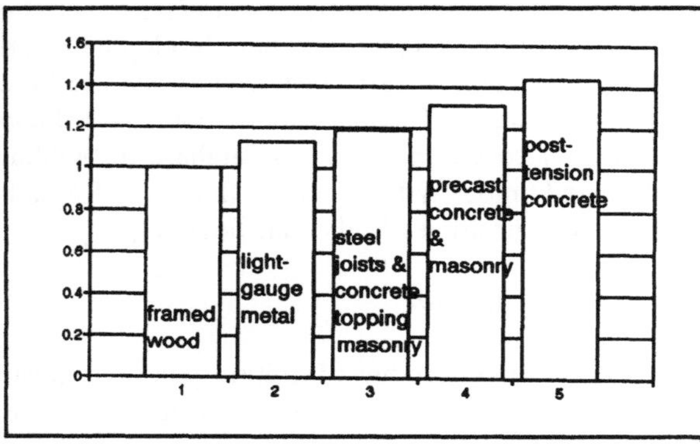

Figure 9. Cost comparison for a four-story building built of common construction materials.

Nevertheless, timber in buildings has a long historic tradition in the country for roofing structures and nearly no builder tries to utilize other materials for sloped roofs. Timber is also widely accepted for floors. In contrast, no layman and only few professionals believe it possible to build multistory wood-framed buildings, even though, economic case studies showed cost advantages of wood frames in comparison to other common construction materials, [3], (Figure 9). Such a lack of knowledge is apparently a defective feature of our education systems on all levels. It can be improved by a change of study programme contents and by means of targeted remedial action on public.

However, the objections against timber structures arise from a broader field of problems than only cost relations. They are namely: fire hazard, life span, vulnerability to biological decay, shrinkage, deformations, non-homogeneity of characteristics and quality of products, (appearance, etc). In recent years, there is one additional reason - a forest saving movement which is frequently based on emotions, mysteries and narrow thinking than serious scientific analyses taking into account wider interdisciplinary views.

2.2. WOOD AND ENVIRONMENT

It has long been known that, from the energy consumption point of view, wood is by far the best material for structural members loaded in compression, nevertheless, few investigations enhancing this knowledge have been performed. In 1991 a spacious timber research laboratory was constructed in Canada [4], covering an area of 64 m by 111 m. A study on environmental aspects has been carried out and presented in [4] with the following results:

> *The net energy requirements for the materials used for the basic structure of this new research building were about 600 MJ/m^3 as opposed to an estimated 1700 MJ/m^3 if the building had been constructed in steel, which represents 2.83 times more energy.*

It is worth to mention that the unit cost of a typical model steel structure is on average 1.5 times higher than of wood, which is a quite significant disproportion and shows that the ecologically non-friendly and damaging materials are still being indirectly promoted. Here the base of cost of structural materials is again a subject to be dealt with.

> *Further, it has been estimated that the production of carbon dioxide in the manufacture of wood for this building has been only about one third of that produced in the production of steel for the same type of building. In addition the carbon locked up in the cellulose, hemicellulose and lignin of wood will not be released into the atmosphere in the form of carbon dioxide as long as the wood does not decay or burn.*

Moreover, replanting of previously harvested forests increases the removal of free carbon dioxide from the atmosphere because the more vigorous growth of young trees results in a greater retention of carbon dioxide.

From those data it follows that a reasonable use of wood as building material causes less damage to environment and exhibits savings in both energy and cost. In other words, better and intensive utilization of wood from forests is a very effective way to save forests. Appropriate measures should be taken in the countries in economy transition and in the developing countries, where an increase of building activities is expected, especially for housing and infrastructure - the fields where wood could be successfully used.

3. Wood Products in Building Technology and Science

3.1. ENGINEERED WOOD PRODUCTS

Interdisciplinary approaches have proven to be successful in development of engineered wood products - new materials or elements with enhanced design characteristics namely structural composite lumber and prefabricated wood I-joists as well as composite or combined structural members, as e.g. wood - metal or wood - concrete composite structures, reinforced laminated wood, etc. Czech attempts in this respect are represented by investigations of composite wood joists - concrete slab assemblies for reconstruction and strengthening of old timber floors or by research into deep composite beams made of metal sheet corrugated webs and wooden flanges [5].

3.1.1. Structural Composite Lumber

Structural composite lumber (SCL) has been attracting more and more popularity because it gives to designers and consumers a product of high quality, better performance, versatility and economic efficiency. It is represented by engineered wood products denoted as laminated strand lumber (LSL), laminated veneer lumber (LVL) or parallel strand lumber (PSL) as well as by oriented strand boards (OSB) in wood based panels. The SCL products as well as prefabricated wood I-joists (PWIJ) optimize wood fibers to their full structural advantage. The production of SCL is still quite limited and a wider application outside USA is highly desirable. Naturally, the parameters of SCL are dependent on raw material species and full utilization of special characteristics requires appropriate design as well as codes and standards supported, in some countries, by investigations of local materials.

3.1.2. Reinforced Glue-Laminated Wood

The successful technology of glulam beams has been substantially improved by an introduction of fiber-reinforced plastic or high strength fiber-reinforced plastic material integrated in tension or high shear stress zones into structural elements. This reinforcement not only increases the load carrying capacity and the long term stiffness of structural members but also enables one to use less volume and low grade species for production of glulam beams. One of the most modern techniques utilizes prestressed fibre-reinforced plastic bands made of aramide, carbon or glass fibre in a plastic matrix. The success of this product is due to easy manufacturing, excellent performance and cost savings about 25%-30% compared to non-reinforced glulam [6]. The new system has a potential to replace steel and concrete in many applications.

3.2. WOOD CHEMICALS IN NEW COMPOSITE BUILDING MATERIALS

In the former Czechoslovakia considerable research attention had been paid to composite materials with macro-molecular matrix utilizing fyto-mass. One approach investigated a possibility to remove resin glue from production of wood base boards and utilize for the bonding a chemical destruction of surface layers of agglomerated wood particles. During such a process, the surface of wood chips is activated by an acid agent, which, at elevated temperatures, causes decomposition of polymer ligno-cellulose substance into components, (e.g. lignin), and conversion of other substances into monomer and oligomer ones, which enable a creation of a solid polymer structure in the course of synergic process [7]. The laboratory and pilot production results have shown a real possibility to apply this method in production of particle boards.

Another family of research tasks was concentrated on production and utilization of fural resin or its derivatives [8]. In the Academy of Sciences, Institute of Theor. and Appl. Mechanics (ÚTAM), two technical materials based on furol-fural glue were developed and verified in the building industry. The first one, called *Berol*, is a particular composite of a polymer-concrete or a polymer-mortar type. Its characteristics are dependent on the filling aggregates and they typically feature: tension strength of 8 - 10 MPa, compression strength of 40 - 60 MPa and modulus of elasticity of 15 - 18 GPa. The material is nonpermeable, with a very high chemical resistance, low thermal expansion coefficient, (about $1.5x10^{-5}$), and other characteristics well suitable for water engineering, sewerage systems, etc. A second particular composite, made of rubber granulate, is called *Gural* and for its production the crushed rubber wastes, as e.g. old tires, are utilized. A special processing ensures a very high strength of bonds between aggregates. Gural can be produced with prescribed and very special mechanical characteristics: Young modulus of from 30 to 300 (or even 3000) MPa, compression or tension limit deformation of between 70% and 3%, or 25% and 2%, respectively. Gural has a very low wear, excellent damping characteristics and a very high resistance against chemicals.

4. Enhancement of the Role of Wood in Society

4.1. TIMBER IN TRADITION

Timber is one of the oldest engineering materials and historic experience over centuries give excellent examples of its mode of use. The study of wooden structures, their joints as well as variety of forms and concept possibilities is one of the main sources for new understanding and promotion of wood in modern structures. Moreover, many historic wood and timber structures possess distinguished artistic and aesthetic values supported by the special material characteristics, which should be seriously discovered and disseminated. In some parts of the World, the population still masters skill, knowledge and crafts with natural approaches to the best utilization of wood, which might be fully understood only by cooperation of historians, etnography researchers and engineers.

4.2. LESSONS FROM WOOD

Wood and trees contain in themselves an enormous potential to educate humans and especially engineers. A sophisticated structure of wood on the microscopical level, a clever utilization of different materials in the macroscopic composite, chemistry and energy processes in creation of plants, growth of trees in accordance with loading and sustainability conditions - that all has to be not only learned by professionals but also widely disseminated for a broad audience. Wood and its source - trees and forests - thus might be utilized in mediating ecological understanding, thinking and behaviour.

4.3 OPTIMUM UTILIZATION OF NATURAL RESOURCES

In the case of Nature structures we admire the harmony of their shapes and their reaction under loads in agreement with the mechanical principle of minimum energy. Artificial systems require that an analogous principle of minimum effort necessary for the design, production, erection, function, maintenance and destruction should be formulated serving as an objective function for optimal structural efficiency. Nevertheless, optimization of an exploitation of natural forces and materials is a quite difficult task, because the criteria are not clearly defined. The author understands the optimization as an achievement of maximum degree of utility, (in a general sense), which at the same time fulfills the condition of a minimum energy input necessary for the realization of the objective, and also preserves the environmental and ecological balance [9]. The minimum energy input can be measured by the value of the objective cost function and therefore its minimization seems to be the most suitable criterion. However, the creation of a relevant objective function is a difficult problem and until now more attention has been paid to the methods of finding an extreme of such an objective function than to its construction. This question is one of most important challenges in a search for better utilization of natural resources and it should be solved again in a collaboration of scientists and engineers from different fields.

5. References

1. (1992) *FAO Yearbook - Forest products 1980-1991*, FAO Forestry Series No.26 & FAO Statistics Series No.110.
2. (1995) *Czech and Slovak Building Industry from A till Z* (in Czech), Building

Information Agency (SIA) Prague.
3. Sarkkinen, D.L. (1992) Multistory Wood-Frame Buildings: Cost and Structural Design Aspects, *Wood Design Focus* **3**, 8-10.
4. Stieda, Ch. (1993) A Timber Research Laboratory, Canada, *Structural Engineering International* **3**, 77-79.
5. Bartoš, F., Bouška, F., Drdácký, M. (1988) Composite thin-walled girders with corrugated webs, in Proceedings *5th Conference on Metal Structures*, Vol.2, CNIT Timisoara, Timisoara, 13-18.
6. Tingley, D.A., Leichti, R.J. (1994) Application of High-Strength-Fiber-Reinforced-Plastic in Building Components of Low-Rise Wood Structures, OSU Report
7. Bareš, R.A., Vašíček, A. (1986) Glue-less composite material from plant particles - UNIFYTO, (in Czech), Research Report ÚTAM, 51 p.
8. Bareš, R.A. *et al.* (1990) Production of fural by means of a hydrolysis of plant raw materials, (in Czech), Coming Praha, 75 p.
9. Drdácký, M. (1979) Trends in optimization of shapes and structures in civil engineering, in Proceedings *Shapes and Structures in Engineering Defined by Physical and Environmental Aspects*, Praha, 44-51.

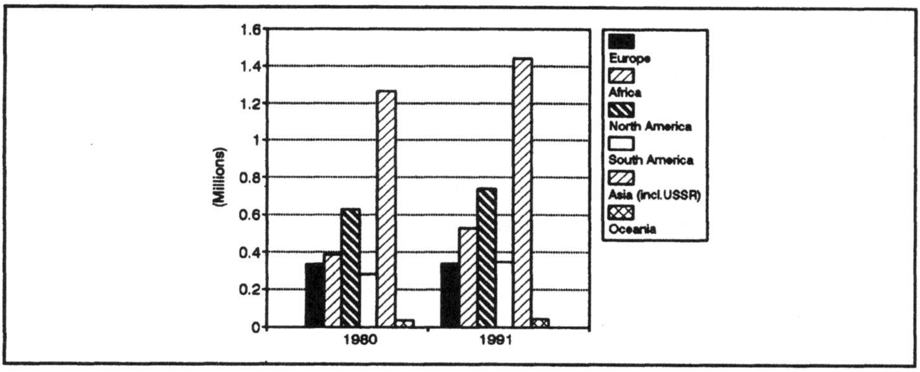

Figure 10. Round wood production tendency in the world (in 10^9 CUM).

Pertinent Observations from the ARW

The ARW meetings and discussions demonstrated a consistent demand for a higher use of timber in building industry as a necessary condition for conserving forests resources through better utilization of wood.

From the strategic point of view three items seem to be very important. They are:

- a need for development of appropriate modelling of economic and political (ecological) issues enabling an objective evaluation of timber structures compared to other materials;
- a need of establishing of regional interdisciplinary long term (five years) projects on all issues of wood engineering and forestry;
- a need for increased public awareness and professional knowledge on natural systems, especially utilization of renewable or low energy input materials, which could be done by means of educational programmes at all age levels.

As far as tactics are concerned, the contemporary socio-economic changes and needs in CE and EE countries could be utilized for a coordinated development of ecologically friendly technologies and their penetration into governmental policy and industrial practice by all means. The solution of housing problems is an ideal opportunity for this effort. However, involvement of timber into housing construction requires an interdisciplinary approach taking into account:

- local history and traditions in architecture and lifestyle;
- socio-economic failures experienced with so called "low-cost" houses programme in the USA and other countries;
- technologically successful examples of multistory solid timber houses as well as framed houses.

The process could be facilitated and accelerated only by society/government supporting programmes aimed to public opinion improvements as well as economy incentives.

Research should be oriented toward:

- utilization of local resources;
- development of progressive wooden structures fulfilling complex requirements for buildings or frameworks;
- development of wood-based materials for building industry;
- development and application of effective assessment methods and techniques for existing structures;
- reconstruction, repair and maintenance of historic timber structures.

HARMONIZATION OF FOREST RESOURCE UTILIZATION AND HOUSING DEVELOPMENT IN POLAND

M.A. GIZEJOWSKI
Institute of Building Structures
Warsaw University of Technology
Armii Ludowej 16, 00-637 Warsaw
POLAND

P. PASCHALIS
Department of Forest Utilization
Warsaw Agricultural University
Rakowiecka 26/30, 02-528 Warsaw
POLAND

Abstract

The authors' view on certain aspects of socio-economic considerations of the forest resource management and utilization, in regard to conditions in Poland, is summarized herein. An investigation into a joint treatment of the key aspects of forest resource prevention and sustainability, and the use of timber in the building industry sector of Polish economy has been overviewed. The paper constitutes therefore an attempt to position where Poland is and what is happening in forest resource management and timber industry development in the country, and whether the Polish economy would fit to the picture drawn for the Central and Eastern Europe by the natural environment global conservation strategy program. A number of different forest management and timber use related problems have been addressed which would help to direct the decision making process in order to avoid misconceptions and apprehensions in the utilization of forest resources in Poland.

1. Introduction

Since the creation of industrial society in 19th century, a concern about the aftermath, impact, residue and consequences of the industrialization process continuing over the 20th century is now being very much understood. A growing search for new material resources has done dramatic damage to the human natural environment affecting the global atmosphere, climate and life conditions. The global warming effect, the depletion of the ozone layer and the increase of earth's atmospheric carbon dioxide are the major problems awaiting for the solution within the framework of world's ecological program policy and by the cooperation of all the nations. These problems, well understood in technologically advanced countries, have to be spread out all over the world, especially delivered to developing countries and countries seeking economic reorientation towards the free market, product-effective industrial technologies and modern financial support systems. Mistakes of early world's industrialization policy may therefore not be repeated in technologically less advanced countries and may lead to the creation of their resource-friendly economies, harmonized with those of developed nations. Poland can be classified to as such a country. Poland, with all the other nations, has therefore to be involved in activities related to a global economy reorientation exercise towards

R. M. Gutkowski and T. Winnicki (eds.), Restoration of Forests, 251–269.
© 1997 *Kluwer Academic Publishers.*

conservation and better utilization of earth's natural resources. It would give a time margin to counter-balance the waste, nonsustainable and nonrecyclable resources in order to ensure the realization of global prevention strategy goals before it is too late. In this process, forest resource stabilization and better utilization become the most important challenges for multi-disciplinary experts in order to be successful in the development of resource-friendly world economy.

Poland is a country which, historically, has a rather good forest management policy balancing socio-economic demands and considerations. At least in this area, a rigidly conducted policy resulting from the fact that the overwhelming forest resources were state owned, helped to avoid an excessive deforestation due to the uncontrolled consumption. Such aspects as recreation, prevention of sustainable water and climatic conditions, conservation of ecological (green) zones in order to barrier and balance the rate of industrial pollution, conditions to ensure the wildlife renewable population, and finally the controlled industrial growth both in furniture and building sectors of national economy were paid proper attention.

The transformation of Polish society into the capitalistic mode activated new trends. The private sector is nowadays interested more in consumption than in long term renewable resource policy. The reason is simple - only better financial conditions may enable the future realization of long term investments and policies in the private sector. On the other hand, considering aspects of social wealthfare, the Polish national average income has not changed much, if not degraded. As a result, especially in rural areas and areas of high rate of unemployment, an illegal deforestation becomes the problem.

Although the majority of forest resources is still controlled by the government (Forest Service), the trends mentioned above may cause an unimprovable damage to the quality of environmental conditions. Other factors, such as growing social demands for mass-produced low cost housing, unavoidable restructurization of national economy (technological reorientation, expanding the network of highways and the bridge infrastructure) and pro export incentives are a threat for the realization of rational forest resource utilization policy.

A recent survey of forest resource conditions proves that the afforestation process has to be intensified in Poland to such an extent that it meets the Western European standards. The composition of species used in the afforestation process has to meet both the sustainability, renewability and biodiversity protection of forest resource. Aggressive promotional activities have to include the use of tree species of fast mass growth and application of new technologies for engineered wood products to be introduced for affordable housing. The virgin area concerned with the application of modern timber structures in building industry in Poland is still to be developed. The relevant educational activities, informative courses and technological transfer are of a great importance in order to be successful in achieving our socio-economic objectives of better forest recourse utilization.

In the following sections the issues discussed above are related to the problems of forest resource management and utilization in Poland and their implications on the development of construction market and wood based housing technologies. The figures presented in the paper are based on the results of professional monitoring activities carried out by the State Main Statistical Office (GUS), Forest Research Institute (IBL), Forest Management Offices (BULiGL) and the Housing Research Institute (IGM).

2. Forest Resource Management Record

2.1. FOREST AREA AND OWNERSHIP STRUCTURE

The total area of forests in Poland is 8,678.5 thousand ha. They are often small-area complexes. The state owned Forest Divisions administer 28 thousand forest complexes, 6 thousand of which are below 5 ha of area. The privately owned forest, occupying 15.8% of the total forest area, belongs to 1.4 mln owners.

On January 1, 1992, the total area of State Forests was 6,791,253 ha. Out of this:

Forested area	- 671,997,5 ha, that is 99.0%,
Non-forested area	- 71,278 ha, that is 1.0%,
The conifers occupy	- 5,374,999 ha that is, 79.2%,
The broadleaves occupy	- 1,416,254 ha that is, 20.8%.

The areal structure of age classes is as follows:

Non-forested forest area: 0.07 mln ha	- 1.0%,
Age classes I and II (1-40 yr old stands): 2.60 mln ha	- 38.3%,
Age classes III and IV (41-80 yr old stands): 2.64 mln ha	- 38.9%,
Stands more than 80 yr old: 1.48 mln ha	- 21.8%,

which gives altogether 6.79 mln ha being 100.0%.

From the survey data, it was evident that age class II prevails (35.5.%) while there is a disproportionally small fraction of forest stands of age classes I and V.

2.2. SITE STRUCTURE AND STAND SPECIES COMPOSITION

The dominant forest site type is the moderately humid coniferous forest, representing 30.2% of the total forest area. All the coniferous forest site types cover 63.3.% of forest area, the broadleaved forest site types - 33.9% and the humid alder site type - 2.8%, respectively. The proportion of lowland site types is 89.1%, the upland site types represent 3.6% and the mountain sites - 7.3%. In the State owned Forests, scots pine and larch occupy 70.9 % of forest area, spruce - 6,2%, silver fir - 2.1%, European beech - 4.4%, oaks, maple species and elm species, jointly - 6.3%, hornbeam - 0.3%, birch - 5.2%, alder - 4.2%, aspen, linden and willow - 0.2%.

2.3. VOLUME STRUCTURE OF FOREST RESOURCES

A continuous increase in the standing volume may be observed, the forest resources of the country amount to 1431.9 mln m^3, out of this 1293.7 mln m^3 are the State Forests resources. The gross volume of softwood large timber is 1,017,999.2 thousand m^3 (78.5%) and that of hardwoods - 275,275.2 thousand m^3 (21.5%).

Standing volume of stands in particular age classes is the following:

Non-forested forest: 0.5 mln m^3	- 0.0%,
Old-growth trees: 2.0 mln m^3	- 0.2%,
1-40 yr old stands: 188.8 mln m^3	- 14.6%,
41-80 yr old stands: 665.5 mln m^3	- 51.4%,
81 yr old and older stands: 436.5 mln m^3	- 33.5%

which gives the total of 1293.3 mln m^3 being 100.0%.

2.4. AVERAGE AGE AND MEAN GROWING STOCK VOLUME

The average age of all the stands in the State Forests is 55 years. It is highest in the case of beech (80 years) and fir (79 years) and lowest in aspen (26 years). The average age of all the stands is decidedly influenced by the average age of scots pine (52 years). The mean gross standing volume of timber per ha of all the stands in the State Forests amounts to 192 m^3/ha. Out of the more important commercial wood species is spruce which has the highest standing volume per ha - 260 m^3/ha while the appropriate figure for the main forest forming species in Poland, scots pine, is merely 183 m^3/ha.

3. Forest Resource Utilization Record

During the last few years the size of annual wood harvest has been close to the planned values and, consequently, a continuous decrease in the harvested wood raw material has taken place.

At present, the size of wood raw material harvest is determined in the State Forests by:

a) the prescribed cut (capability) of final felling and intermediate felling as calculated in the plan of forest enterprise management;

b) the possibility of disposal of wood.

The yearly net amount of principal yield is about 17.4 mln m^3 of large timber, including 13.4 mln m^3 of softwood large timber and 3.9 mln m^3 of hardwood large timber. About 9.2 mln m^3 of timber is harvested during final cutting and 8.0 mln m^3 in intermediate cutting. Out of the latter, 0.2 mln m^3 come from tending of forest cultures, 6.3 mln m^3 - from thinning in premature stands and about 1.5 mln m^3 - from forced sanital thinning.

At the country-wide scale, the principal yield in the State Forests was 111.9% of the prescribed cut for years 1968-1995, varying from 128.4% in 1983 down to 89.8% in 1991.

In particular categories of forest utilization the figures were:

a) the final cut size was on average 93.2% of the capability and it varied from 67.7% in 1990 up to 106.6% in 1974; in 1991 it was 78.8.%;

b) the intermediate cut size was 143.7% on average and it varied from 102.1% in 1990 up to 213.6% in 1983; in 1991 it was equal 103.1% of the prescribed cut.

Some changes occurred in 1992; a total of 20.34 mln m^3 of wood was harvested, including 18.96 mln m^3 of large timber, and the capability was exceeded of 10.2%. In the period 1968- 1993 clear disproportions existed between the amount of harvested softwood (117.7% of capability) and hardwood (92.2% of capability). This disproportion has been vanishing in the last three years. The 1971-1995 data show, besides, that the disproportion between the size of harvested softwood and hardwood was especially evident in intermediate cutting; it was lesser in final cut. This has been to a large degree due to the necessity of accidental felling following the 1980-1985 frequent outbreaks of nun moth and hurricanes - these disasters occurred mainly in the coniferous stands.

The assortment structure of the harvested wood raw material results from the prescribed cut size and cutting categories. In 1990, reduced demand for wood raw material contributed to the limitation of wood harvest size. Taking the year 1994 as a reference point, a significant boom in the domestic wood market should be noted. The activation of the free market mechanisms and execution of the rights of private ownership have caused, in turn, that the size of wood harvest in non-state forests to get out of control. The amount of wood harvested in 1993 in private forests is estimated at about 2-3 mln m^3. So, one can assume that altogether some 19.5 mln m^3 of round wood were harvested in Poland in 1995. It is equivalent to 0.5 m^3 of annually harvested wood per inhabitant. Considering the wood harvested in the State Forests only, the appropriate number will be significantly lower - 0.44 m^3 wood per person.

The volume of large timber from the final cut constitutes 47% of total volume of harvested timber (28% in clearcuts). The most commonly used method of wood harvest is the trunk method (56% of the total volume of harvested timber), then the method of entire bole (21%), the sortment method (18%), the method of entire tree (3%) and the method of loose wood (2%).

4. Housing Development and its Implication on Timber Demand

4.1. BACKGROUND AND PRIORITIES

To anyone who has even a passing acquaintance of Poland and East Europe, it will come as no surprise to learn that the primary building materials here are bricks, other small-size wall and floor ceramic or ash by-product based elements, and concrete, especially heavy duty large precast panels used for multi-family residential buildings. Poles have been handed down a long waiting list for living space, best expressed in the one and a half million families who do not have their own home to live in. There are only less than 300 residential places for each thousand people, whereas in some other European states the number is 400-500.

The plans and goals of housing policy in Poland in the 1980's failed to be implemented due to mistakes committed as early as the concept development stage. While in the developed countries lightweight construction technologies based on timber and other wood based materials were extensively used and improved, in Poland ways to solve the problems of housing construction were sought in the area of reinforced prefabricated concrete construction systems. The support was given to cost and energy-intensive, solid construction methods which necessitated the use of heavy transport and erection equipment. The consequence of this policy was housing estates composed of multi-storey apartment buildings made of prefabricated reinforced concrete panels. The resident is neither comfortable nor healthy in such an estate and the buildings themselves start to pose a number of technical problems cause by frost penetration, poor insulation and durability of prefabricated elements at their joints.

Anyone even with elementary technical knowledge can foresee the magnitude of repair and reconstruction problems facing the engineering staff and increasing in the course of deterioration processes within the large-panel housing stock. Funds appropriated for maintenance, repairs and modernization of facilities may exceed original investments. Land lost because of large panel development schemes will be difficult to recover in favor of more cost-efficient forms of development. Demolition and attempts at recycling of large panel materials may prove to be too expensive or even economically unjustified. What is necessary in Poland now, more than ever, is to change the stereotype of thinking represented by the engineer and the future user; to promote and popularize good and proven solutions adopted in the developed countries.

Years of depreciatory treatment of timber construction in the country have resulted in a huge psychological barrier - not only on the part of the public which shares the view that a wooden building represents questionable value at a relatively high price, but also among the general engineering staff. The existing systems of higher education in engineering do not promise any improvement of this situation. These factors impede the development of the timber construction sector to a much greater extent than a lack of industrial back-up facilities for the implementation of such construction methods. Due to these circumstances the most urgent issue is to start the process of "recovery" from the area of education and promotional activities aimed to popularize modern concepts of lightweight timber construction. It is only under conditions of properly shaped awareness of people within the engineering profession and suitable economic conditions, which have already been provided by market economy, that one can hope for the success of investment activities in the area of timber construction market in Poland.

4.2. CONSTRUCTION MARKET AND OBSERVED TRENDS

Although the Polish construction industry has changed beyond recognition in the past five years, the home accessibility crisis has not been overcome and every year less and less home units are being built. Figure 1 shows the number of home units finished and directed for occupancy, and newly started in the years 1990-1995. The huge, ineffective state-run enterprises which monopolized the market by involvement in the construction of multi-storey precast reinforced concrete residential estates have been broken up into many small and now well managed private companies and joint ventures. This changed the construction methods, the standard of home units and the architecture of dwelling developments.

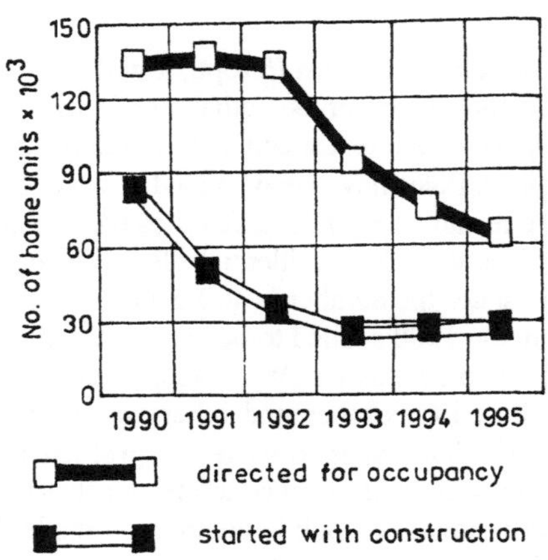

Figure 1. Trends observed in housing development

Presently, hardly any tall buildings are being constructed from prefabricated reinforced panels. The diminishing production of prefabricated heavy constructional wall and floor elements is being used in low-rise dwelling settlements of better architecture. Small and unfriendly arranged flats in such buildings are being allowed for larger living space with conjunction of individual interior arrangements offered. Economical reasons force however the precast reinforced concrete home factories to die. A mass production of such substandard buildings is being replaced by high-quality residential low-rise apartment blocks of increasingly pleasant architecture. A bulk quantity of multi-family residential estates are now being made out of conventional small-size structural components, thermally enhanced for exterior wall applications. The modern wood frame technology, most of all, gains the popularity. This is a tangible evidence that investors are now controlling the way in which their money is being spent and enforcing a worldwide accepted vision of effective construction systems. Figure 2 draws attention to the change in construction systems used for residential purposes. The tendency observed in construction systems is that in multi-family dwellings the share of heavy panel technology decreases every year giving room for refined traditional and other technologies, the latter in one-family houses to be dominated by wood frame bearing constructions. This positive break through resulted in the change in the building material market. The market reaction was that the consumption of proven customer friendly building materials has been a growing feature over the past years.

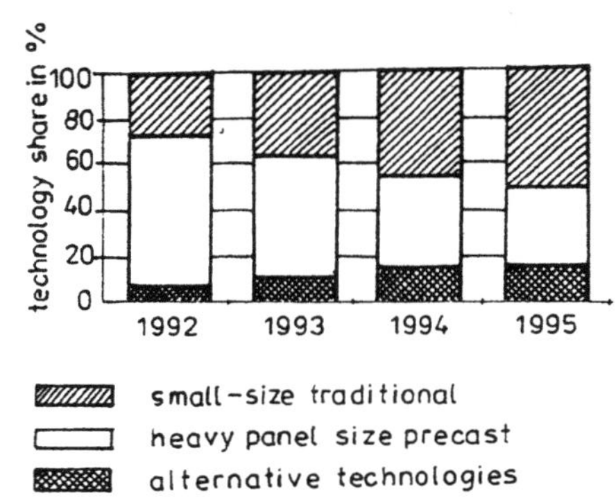

Figure 2. Tendency observed in applied construction technologies

Since 1992 the average production growth of building industry in Poland has been slowly increasing and the forecast for years to come is even better. One can expect that this tendency has to be dynamically improved in order to allow financing of the housing industry according to the demands of the free market economy. The home-building industry is however incapable of fulfilling the requirements of the market, mainly because the legal and financial systems are still crippled and inefficient.

In the near future, one can not expect a far healthier situation and the state will probably not be able to aid such forms of housing as private low-rise apartment buildings and detached houses. Polish banks will most likely continue to offer rather expensive credit inaccessible to many potential customers. Recently, a system of mortgages has been introduced which however will not play a significant role because of high interest rate on such credits. There is a paradox in the housing market in Poland that the potential customer is forced to pay the total uncredited amount. This amount is quite high, e.g. ranging from USD 350 to 700 per m sq. in multi-family apartments blocks for the Warsaw area. Considering that the lower bound corresponds to the monthly average salary in the sector of national economy, the home crisis in Poland seems unsolvable without affordable bank credit. This credit, however, would become accessible once inflation, currently at 1.5% a month, is reduced to a level accepted in developed countries. The distance between the number of home units directed to customers and those started with their construction is diminishing (see Fig. 1). Costly bank credits restrict the mass housing production on one hand. On the other hand they force customers to build stage-by-stage, depending upon their current financial ability; sometimes this routine takes over several years. An average construction investment cycle in the private housing becomes close to a year while doubled for building societies.

4.3. DIRECTION AND DYNAMICS OF DEVELOPMENT

The reduction in dwellings completed by housing cooperatives, enterprises and communal institutions has been associated with a stable situation in the private construction sector. Figure 3 illustrates the distribution of private housing share index for different states of Poland estimated for data collected in 1995. A growing tendency is being observed towards the expansion of regions characterized by the high private housing share index.

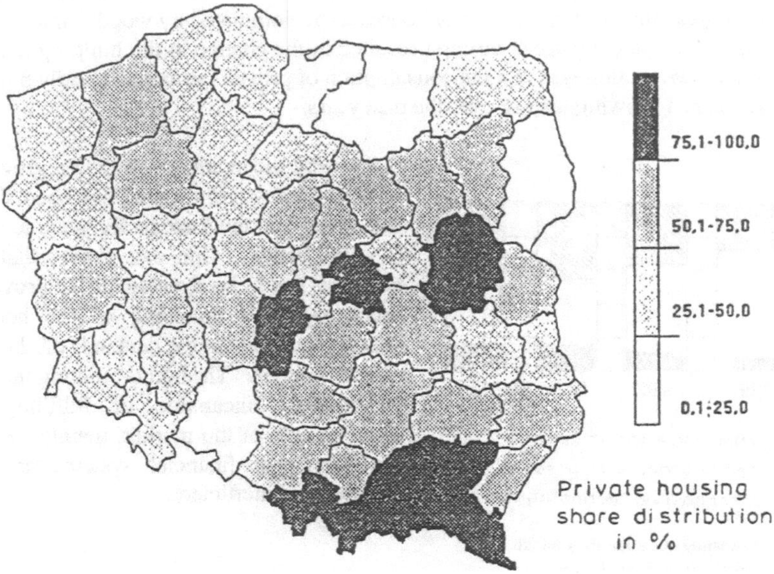

Figure 3. Distribution of private housing share index

Figure 4 shows the dynamics of private housing sector development in Poland in 1992-1995. The observed trend in home ownership status reversed the unhealthy situation in the Polish home market - since 1994 the dominant position has been allocated to private construction. The cost and energy effectiveness of construction systems are becoming the main factors stimulating the individual housing development. Especially in outlying city districts, large groups of terrace houses and semi-detached houses, the most popular of all in Poland, are increasingly appearing in number. This groups are constructed by small newly established companies, run by people with a foreign experience gained in developed countries, or those which have been emerging in the wake of the disintegration of large state enterprises, hence competing sharply within the framework of the new organizational and ownership formula. Although small building societies have made a considerable share in total dwelling constructions, their refined traditional housing developments are still much heavier in terms of weight when compared with the standards of developed countries, require more energy in production as well as during the service, and finally need the use of heavy building machinery. Because of that the building technology in Poland is being reoriented towards wood based building construction methods, especially in individual and small building society developments.

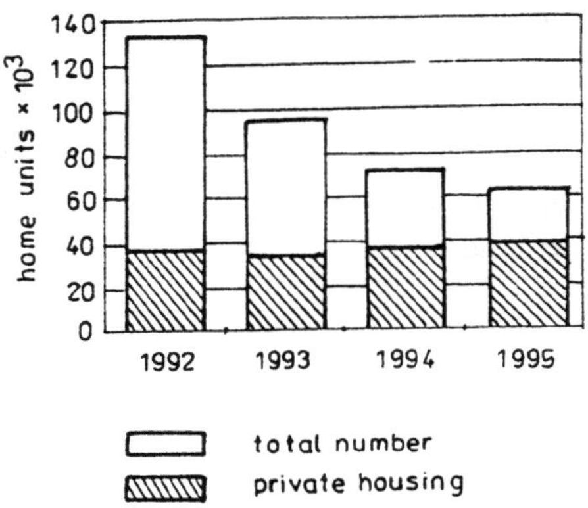

Figure 4. Dynamics of private housing development

The modern wood frame technology was wholly unknown in Poland until recent years. This changed when the American-Polish Home Builders Institute was established. The first show home built in the American wood frame technology was assembled by MURATOR as recently as in 1987. The Institute and MURATOR magazine have taken an active role in the promotional activities related to modern wood frame technology development in Poland. A number of practical wood frame courses have been conducted at the Warsaw University of Technology in the conjunction with the Foreign Agriculture Service, Washington D.C. and MURATOR magazine.

The Polish Association of Home Builders was created in order to enhance the promotion of energy-efficient and lightweight construction technologies. In 1995, for the first time, the International Convention of Home Builders, accompanied by the Build Better exhibition took place in Gdansk and this idea is continuing on an annual basis. The Polish Association of Home Builders has been created in order to enhance the promotion of energy-efficient and lightweight construction technologies. As a result of these aggressive promotional activities wood frame houses become a fashion among the Polish establishment, clients working for foreign companies or being in business for themselves. A number of carefully designed wood frame houses were already constructed in Poland for these model clients. The houses are intended as the buyer's high-quality residence places having the possibility of developing them in the future, therefore not a temporary investment. Such houses combine the concept of American wood frame building methods, the Polish tradition of useable roof space construction and modern tendencies in architecture manifesting in the creation of large and unobstructed interior living spaces. Being a permanent investment, they are however quite expensive and only a few of would-be clients from those who are on the waiting list for a living space can afford them.

The middle class in Poland, which is to grow increasingly strong and well organized, recognizes that home ownership is also, for them, an important element of their living standard. The number of potential customers would therefore significantly increase, if Poland had an adequately developed bank loan system. For this group of Polish society, increasing number of wood frame residential settlements are being started with and planned for further development. This new class of housing developments, which may be called affordable wood frame units, will require support from the industrial side and foreign investments in order to reconceptualize and systemize a new approach towards mass-constructed timber buildings in which a sustainability of the forest resource and development of modern wood based material industry are of a primary concern.

4.4. PROGNOSIS OF FOREST RESOURCE UTILIZATION

During the 1968-1995 period, the total forest area has increased by 335 thousand ha. Consequently, the volume structure, age structure, standing volume per ha and productivity of our forest have all changed too. The total gross volume of large timber is growing at a rate of 1.4% annually, which makes 15.2 mln m^3 or, nearly 64% of the mean annual increment. Also, the mean standing volume per ha has increased from 146 m^3/ha up to 192 m^3/ha, and similarly the average stand age - from 47 years up to 54 years. Forest utilization entered regions that were previously treated as impenetrable. Consequently, the area of inaccessible stands has decreased significantly. All these factors contribute to the expected increase in forest resources by year 2025. On the other hand, however, many limiting factors will also occur, among others:

a) a large area of industrial pollution;
b) the health status of our forests;
c) improper utilization of potential for production of mezoregions.

Considering all the factors listed above, one may envision the following prognosis of dynamics of forest wood resources: increment, prescribed cut size and harvest size in the State Forests. According to the prognosis, the mean stand age and standing volume (m^3/ha) will increase and, also, more advantageous will be the structure of harvested wood sortments (a higher proportion of the large timber).

Till the year 2025 the total volume of large timber in our forests will be on the increase, from about 1.28 billion m^3 (1.28×10^9) in 1985-1995 up to about 1.38 billion m^3 (1.38×10^9) in 1995-2005, about 1.46 billion m^3 (1.46×10^9) in 2005-2015 and, finally, up to 1.53 billion m^3 (1.53×10^9) in years 2015-2025.

The increment of large timber (gross increment in m^3) will maintain in years 1985-2005 at the level of 32 mln m^3 annually, while in years 2005-2015 it will drop down to 31.1 mln m^3 annually, and in years 2015-2025 a further decrease will take place - 30 mln m^3 annually.

A slow rise in prescribed cut (net value in m^3) is to be expected from about 17 mln m^3 in period 1985-1995, up to about 19 mln m^3 in years 1995-2025. The increase of the accepted prescribed cut will be the result of the clearly higher proportion of mature stands - from 9.5 mln m^3 in period 1985-1995 up to 10.4 mln m^3 in 1995-2005, 11.3 mln m^3 in 2005-2015 and, finally, 12.0 mln m^3 in period 2015-2025. On the other hand, a significant drop in the size of intermediate capability is expected, from the 9.2 mln m^3 in period 1985-1995, down the 6.8 mln m^3 in 1995-2025. The above prognosis is predicted on the basis that the logging process will be carried out on a rational cycle, under a sustained-yield management policy and introduction of wood based engineered products. Such a policy can provide a renewable resource for affordable and energy efficient mass housing construction sector in Poland, stability of which would be a factor enabling to resolve the country's housing problem inherited from the policies of centrally planned economy. In this context, Poland is considered to be potentially one of the largest and most important wood construction markets in Europe.

5. Better Forest Resource Utilization in Housing Industry

5.1. CONSTRAINTS AND DEMANDED CREATIVITY

There are mainly three groups of aspects related to a wider development of timber technology in the Polish building industry, namely human, socio-economic and technological. Issues related to all these three groups of aspects are briefly discussed in the following subsections.

5.1.1. *Human Aspects*

The younger generation, being more exposed to the West culture and more educated in terms of natural environment conservation, is more ready to accept timber utilization. Sociological surveys carried out in Poland among the young generation entering the productive age prove that they are willing to accept the timber technology provided that it would facilitate the realization of its model concepts of service function and esthetics. It may not therefore be a mass housing production similar to that of the West culture. Successive generations would probably not appreciate the life style of the previous generation, thus also its functional and aesthetic values. The construction market must also be flexible and open for an impact resulting from the creation of informative societies subjected to a very fast civilized development.

The traditional belief that the concept of multi-generation house facilitates a long term investment strategy, contributing mainly to the future generation family wealth, is slowly being replaced by practical consumption oriented values. The land value will soon become in Poland the factor mostly contributing to the cost of overall housing investment. Lightweight timber technologies are the most relevant because they facilitate the concept of a single-generation house. The house itself, if made out of timber products, may be easily disassembled, all the used timber directed for recycling and a new house, materializing the aspiration of new generation, quickly erected on the regained land. Such a cycle would fulfil the long term effectiveness of housing socio-demands.

The present decision-making generation is however confused by some misconceptions towards structural timber for building purposes. The prejudice is not restricted to the public at large but it also includes professionals. The building law is an old-fashioned in this substance, preferring massive and cost-ineffective solutions. To fight successfully in order to win over the reluctance to promote timber requires a better working knowledge about timber as an environmentally and user-friendly structural material, and the creation of human awareness for technological development conditions, the awareness accommodating the Polish inherited cultural tradition.

Human related educational activities are very much desired to be intensified in Poland. As concerns the scope of academic courses at technical engineering schools of higher education, timber construction has been pushed to a meaningless position, e.g. at the Warsaw University of Technology, the number of hours (obligatory and optional) of the courses in Concrete Structures (CS), Metal Structures (MS) and Timber Structures (TS) offered to students majoring in building structures was until recently as summarized in Table 1.

TABLE 1. The comparison of total number of hours for university structural courses

Subject	Basic course	Major course	Optional course	Total
CS	8	7	11	26
MS	8	5	19	32
TS	4	-	-	4

As a result of recent activities undertaken with the connection of courses on the American wood frame technology, one additional optional course on timber frame construction methods has been introduced to the university curriculum. Even this, it becomes obvious from an analysis of the above summary, prepared for the country's largest engineering university, that the main reason why timber construction has been underestimated is a lack of adequate engineering knowledge. As for investment, in our opinion it makes no sense to try and start from scratch; instead systems and technologies should be introduced to the domestic market which are proven by long-standing practice, and on that basis our own systems and technologies should be developed. For this purpose a broader research base will be necessary and new technical literature on this subject must be published. Viewed against the international background, the Polish base of timber construction is outdated and it needs profound restructuring. Market principles of the economy create a favorable climate for investment in this area also in Poland. In the construction sector emphasis must be put on the engineering training methods, timber construction systems and technologies as well as on the establishment of a modern scientific/experimental base which is a prerequisite for technological development of this branch of the construction industry.

5.1.2. *Socio-Economic Aspects*

Timber frame housing has to be price competitive if compared with other modes traditionally used in Poland as structural bearing systems. If the timber industry is serious about capturing and holding on a major portion of building market, it must remove the large fluctuations from its pricing. For instance, steel product suppliers are ensuring a stable-price policy which yields steel to become relatively cheaper if compared with the price increase of other building materials. In contrast, timber product suppliers are flat out when the building market is prosperous and only try to survive when it is not, showing their interest only in exporting or finding other alternative markets. In Poland, the recent interest shown by clients and contractors in accelerating the timber housing industry development resulted in an unexpectedly high increase in timber price, unjustified by the inflation index and rate of economic growth. Timber industry development may therefore be threatened also in Poland by decisions made on the basis of short-term profit-making policy. One can wish a stable situation when the major price controlling factors stimulating the development of timber housing construction market would reflect the long-term predictable balance between the resource sustainability and the volume of recycled and engineered products.

Another important issue is that the claim that the timber solution is less expensive is generally overused and often false. In fact, this term is ambiguous and may be read as a cheap price for a cheap looking product. Going further in this line, the mentioned way of marketing would, in the long-term, damage the reputation of timber as a structural material, value of which is being restored in Poland by the promotional activities of modern structural applications. It is also unwise to create an image that the application of timber building structure would result in a significant cost savings, since the bearing structure itself does not contribute substantially to the overall cost of the housing investment. Timber should be subjected to marketing as a competitive solution in building industry that can win because it best serves the needs of the client and has the potential to be environmentally friendly and sustainable. These important factors have to be emphasized. For educated societies, the environmental sustainability becomes the much more convincing measure of socio-economic benefits or damages, than other conventionally risen arguments. It recognizes that human activities over the generations and the responsibility for the environment are interdependent, the later having necessary social and economical determinants. In this context, timber compares favorable to other materials, being accessible in many parts of Poland, renewable, can be processed without environmental degradation impact and finally can be assembled without heavy equipment in order to encourage the efficiency of use.

5.1.3. *Technological Aspects*
An adequate level of technological advancement and quality of timber products and timberwork are prerequisites for modern wood technology, while a high level of proficiency is essential for sustainability. There is a relationship between the potential economic, social and environmental sustainability of timber and its application as a structural material since the building industry growth. There is still a room in Poland for growing and better utilization of timber products in building industry. In developed countries up to two-thirds of all sawn timber and probably a similar portion of other timber engineered products are directed to the construction market. Timber has to be used there intelligently and efficiently. In Poland, the waste of forest resource when timber is used for reinforced concrete formworks is enormous. Used formworks can not be recycled or reused. A better utilization of forest recourse is being required to enable the full recycling procedure applied to all the timber used in building industry. Conventional timber formworks have to be replaced by reusable ones, made of a material more suitable for this purpose. If timber is used wisely by the architects, structural engineers and contractors as a whole in long life, highly valued and intelligently designed engineering applications then the potential for sustainable technologies to develop is likely to be greater than if the material is wasted in short life span, low valued or poorly designed ones.

5.2. PANEL-FORMED PRODUCTS FROM RECYCLED OR LOW VALUE TIMBER

The housing standards in Poland are considerably lower than those in the other European countries. In the upgrading and remodelling of existing home units and the construction of new houses, a variety of wood based building products is being involved, including softwood and hardwood lumber, sheathing, millwork, doors and windows; all produced from solid timber. In the United States and EC countries in Europe, where construction market scale is much larger than in Poland, demands are being placed on forest ecosystems, attracting the attention of their industries to alternative wood based products to be developed. Using recycled wood-based waste for building board products offers a rational solution for a better forest resource preservation scheme and provides the way to reduce disposal coasts concerned with landfills and waste utilization facilities.

The full spectrum of wood based waste that might be converted into housing non-bearing and structural products includes a variety of waste sources, among which the most important are full-sized used lumber, small-sized wooden pieces, wooden pallets, timber scraps from construction sites, old wooden utility poles and railway ties, wastepaper, yard trimmings and paper mill sludge wood fibres. To ensure a relatively homogeneous raw material for manufacture processing, also the substance of fast mass growth trees is often needed to be combined with the recycled material. This additional raw material is renewable and can be grown either in plantations or in conjunction with holistic agricultural practice close to the manufacturing facilities.

Well known processes used for the production of such board elements as plywood, chipboard, particleboard and OSB, are based on a similar technology as needed for recycled wood based wastes to be converted into structural panel products. The fulfillment of increased demands for wood products in housing developments would only be realistic if the wide range of recycled wood waste based products are being in stock and could be used in a bulk amount. Since the technology refinements required for the manufacture of such products are relatively straight-forward, it is rational to start with mass production exercise in Poland as soon as possible. It would require a know-how technology transfer and the financial assistance made available to Polish wood industry by international development organizations involved in the global conservation program.

Several existing processes have the potential for manufacturing of housing panel-formed products from recycled wood waste. The most attractive solutions for both low and high performance products are reconstructed wood composites, wet-formed structural fiber products, wood-plastic composite products and finally wood-inorganic composite products. To be successful in mass application of the above engineered products in housing development, they have to be economically feasible, competitively priced and acceptable to builders and customers.

5.3. NEW GENERATION OF ADVANCED SOLID-FORMED TIMBER PRODUCTS

Housing construction, especially requiring the arrangement of larger unobstructed spaces, utilizes glue-laminated solid timber sections. By the application of gluing process to join long in span but small in sectional size sawn timber pieces, the degrading effect of natural wood defects and knots on strength and stiffness of long span structural elements is diminished, if not practically eliminated. Semi-engineered products of this type can however only be manufactured out of original sawn timber; therefore the input raw material has to yield from harvesting of forest resources. In this context, investigations should be intensified into the development of wood-waste based substitute technologies to balance the original glulam technology.

Two basic technologies used for panel-formed products have the potential to be applied also in the case of solid sections. The end products are reconstructed wood composite solid beams and wet-formed structural fiber solid beams. The term "reconstructed" describes a process in which low density wood of fast mass growth trees or wood waste are broken into strands, flakes or other chip-like pieces in order to be reinstated into new solid form with the aid of an adhesive. The term "wet-formed" describes a three-dimensional pulp molding process in which a pulp slurry is drained through a resilient mold. The mold is then hot-pressed in order to densify and dry the product. This molding technique has the potential of a great enhancement of house design flexibility, since elements produced in this technology may be arbitrarily shaped, even in curvilinear manner. It is expected that the advancement level of material engineering science will result soon in other timber high performance composites to be developed and used in practice.

6. Conclusions

A fully predictable and stable situation on the market of wood raw material and wood waste engineered products would facilitate the realization of socio-demands for the mass housing development in Poland. The utilization of forest resources is therefore a necessary condition for the accomplishment of the two basic purposes of the State forest policy:

1) satisfying the necessary demands of the national economy for the wood raw material;
2) maintenance of the sustainable development of the multi-functional forest enterprise.

The concluding remarks concerned with the timber building industry development in Poland are as follows:

a) A better national economic conditions and the creation of pro housing credit policy are required if the private housing market is to be seen as a permanent booming sector of national economy.
b) An intensive use of timber in modern structural application of mass housing production would help to solve Poland's most urgent socio-demand as well as to achieve the balance between the accessibility and affordability of housing products.
c) Prevailing restricted attitudes against timber among the lay public and the professionals working for the construction industry must change.
d) Timber has to be wisely marketed as a competitive material that can win over the concrete and steel because is sustainable.
e) It is important to prefer economical use of wood and its recycling as an environmentally friendly material; its lasting and renewable production ensured thanks the proper forest policy.
f) We should attempt at increasing the production capacity of Poland's forest resources; it is still significantly lower as compared with many other countries. The achievement of this purpose will facilitate both economic and environmental profits, e.g. improvement of CO_2 balance.
g) The quality of timber products, therefore also the housing products are big challenges to ensure a long-term reputation of timber in Poland to the similar extend which is existent in the EC countries.
h) Significant will be the influence of the principles of wood raw material certification as being introduced in the countries of European Union, both for the system and methods of satisfying the demand for wood raw material in Poland. A number of activities should be undertaken directed at coping with the challenge.
i) It is necessary to permanently assess the sources of wood raw material, and especially so those from import. To achieve this, drastic changes in the wood raw material price policy in Poland are necessary; the price policy aims should be defined so that they react to the changing situation in the European market.
j) A special concern should be devoted to modernization of the wood processing industry. This can be achieved through the technical modernization of production plants, the promotion of small-to-medium-sized production facilities situated near the sources of

wood raw materials. Undertaking an activity undertaken gives an opportunity to create new work places in the regions of structural unemployment.

The concluding remarks concerned with the management and utilization of forest resources are as follows:

a) The basic source of wood raw material is still the public purse forest and the privately owned forest. Other sources of wood like, e.g. tree plantations or urban greenery, play a marginal role. The latter fulfil, however, a very significant social function.
b) The current level of wood raw utilization in Poland (about 20 mln m^3 annually) places our country at the end in Europe with regard to *per capita* wood consumption. In order to reach the level of economic development of the West European countries it is necessary to increase the consumption level from 0.5 m^3 (today) to about 0.8 m^3.
c) The prognosed demand for wood raw material in Poland will be determined by the following branches of wood processing (data in mln m^3)

Sawmill industry	9.0 - 9.5,
Paper industry	10.0,
Board industry	2.0 - 2.5,
Extractive industry	1.0 - 1.5,
Plywood industry	0.25 - 0.35,
Energetic wood	3.0,
Export	2.0.

7. Observations from the ARW Sessions

The ARW sessions were conducted in a certain hierarchic way, starting from the meetings of individual Working Groups, followed by the joint meetings of every two Working Groups and finally ending with the meeting of the Whole Delegation. The organizational pattern applied allowed for:

a) detailed discussions related to the area of individual group expertise;
b) identification of a large number of issues that are of a great mutual concern;
c) determination the areas of overlap and complementarity existing in the individual and joint group discussions.

It has been clearly demonstrated that the multi-disciplinary body of experts is able to tackle successfully all the environmental issues related to the saving and better utilization of Central and Eastern Europe's forests. The linkage between such factors as industrial processes and pollution, technological aspects of energy storage and consumption, forestry and agriculture management balance, forest resource conservation and utilization, and finally economic and social development is very complex. This complexity results in conflicts between different society interest groups and lobbies. The aim of interdisciplinary workshops, held both on the regional and global levels, is being seen in avoiding the escalation of further conflicts created by polarization of arguments. The ability of mutual understanding and the atmosphere of tolerable acceptance of different lobbies is the pre-requisite for an ecosystem oriented model of social and economic development to be worked out and followed by policy makers. All these issues need the consensus of countries and wide-scale action urgently undertaken. The NATO ARW has significantly contributed to this challenge.

Regarding the constant decrease in reserves of fossil energy and raw materials made available through excavations, the role of solar energy utilization and renewable raw material resources will grow significantly in the near future. Forests of Central and Eastern Europe are important suppliers of wood raw material for industrial sectors, including pulp and paper, sawmilling, and finally furniture and building industries. The sustainability of our ecosystem can only be achieved by:

a) saving the regions subjected to the heavy industrial pollution and to a change in climatic and biological conditions; it can be done through human activities aimed at rehabilitation of regional forests and improvement of industrial processes;
b) better utilization of forest resources and forest value-added products; it can be done through the human activities aimed in multifunctional use of forest resources for the social and economic development.

The nature has to adopt itself to the changing conditions. We are interacting with a number of other factors in the whole ecosystem. The degrading effects of world industrialization and the human effort aiming in rehabilitation of forests in the area of the so-called black triangle, on the border between Poland, Czech Republic and Germany (formerly the German Democratic Republic) were seen during the study tour organized for the delegates of Post Communistic (PC) countries.

The multifunctional utilization of forest resources should be demanded even to the greater extend in order to achieve the objectives of civilized societies, namely sustainable forestry, industry and economy. Both foresters and engineers are interested in sustainable timber industry and timber construction market. This can be achieved by the introduction of computer techniques in the forestry planning, by the application of mechanized procedures in the forest management and finally by the development of highly efficient automatic sawmilling processes and building technologies utilizing timber based products. The workshop delegates had an excellent opportunity to visit the Thurn and Taxis forest resources and the modern sawmill. All this might be shown as a good example of well conducted activities which are market oriented and simultaneously friendly to the environment.

Expanding the use of timber as a structural material requires to remove the barriers existing on the national level and in the macro-region co-operation. Restrictions imposed on timber structures are typical for all the PC countries and result from:

a) lack of adequate know-how knowledge among professionals, including the phases of design, assembling and maintenance of timber structures;
b) specifications existing in the present building by-laws, for instance in fire regulations, which are in favor of traditional massive technologies;
c) psychological barriers and prejudices with regard to quality assurance, degree of durability and fear about safety in fire conditions.

To be successful in the creation of sustainable timber construction market in the PC countries, the following actions are needed:

a) transfer of technology advancements and educational programs;
b) reorientation of university engineering curriculum and continuing education of engineering staff in the area of timber structures;
c) general informative and promotional activities delivered to lay public and to potential end-users of timber products.

In the macro-scale, the removal of the trade barriers for building products and technologies would facilitate the international co-operation. Trading methods between the countries of the European Union (EU) and of the European Free Trade Association (EFTA) are being changed and the PC countries seeking the EU membership will soon be subjected to the same exercise. The concepts of self-sufficiency and open border trade require the constant assessment at least from the employment policy point of view; it requires also economic analysis and probably intervention measures undertaken whenever a pathological situation occurs. The direction towards the economically unified Europe seems to be unavoidable. Model European Standards are therefore needed for the assessment of raw material and end-product characteristics, identification of testing procedures, unified design methodology, and finally for the application of common quality assurance and certification requirements. The relevant package of Eurocodes (ECs) is being released and expected to replace the present National Standards. The adaptation of European Standards, relevant to timber engineering, in the PC countries will work for the reinstatement of the position of timber structures and for the development of timber housing industry. This has however to be associated again with educational activities.

The university education and continuing training of engineers for timber structures need to be a priority. The timber structure related education is not yet considered as important as the structural engineering education in concrete and steel structures. Regarding steel structures, the European Steel Design Education Programme (ESDEP) has been completed in order to facilitate the use of Eurocode 3. The package was prepared with the assistance of European experts, also from PC countries; the ESDEP launch was held at the Warsaw University of Technology in January, 1995. The program helped to update the university curriculum in the area of steel structures and to harmonize the knowledge level with that of the EU countries. In the area of timber structures, the similar project, called the Structural Timber Education Programme (STEP), is underway. However, experts from PC countries are not involved in the exercise. An urgent action is therefore required for the creation of conditions enabling a permanent co-operation also in this area. Otherwise, the PC countries would remain acting only as the timber raw material exporters, since the lack of knowledge transfer is being recognized as a main source for the depreciatory treatment of timber structures in Poland and other PC countries.

The knowledge transfer in the area of timber structures has to be combined with the government support for education, training and research programs as well as for providing the loans for better industrial development. An additional opportunity is being seen through the development of forestry trade association facilitating the monitoring of costs and setting a green peace approved certification of forest products.

8. References

Budownictwo; Monitoring budowlany (1993...1995). Instytut Gospodarki Mieszkaniowej, Warszawa. *Building Industry; Monitoring Bulletin,* Housing Research Institute, Warsaw.

Construction Market in Poland. Foreign Investors' Guide (1994). The Ministry of Physical Planning and Construction, Warsaw.

Draft papers (1996). NATO ARW Workshop on *Science and Technology to Save and Better Utilize Central and Eastern Europe's Forests.* Straubing, Germany.

Hajduk, H. (1992) *Housing in Poland in early stage of transformation period (1990-1991)*, Housing Research Institute, Warsaw.

Hajduk, H. (1992) *Restructuring and privatization of the building industry in Poland.* Housing Research Institute, Warsaw.

Informacje o mieszkalnictwie; Wyniki monitoringu (1992...1995). Instytut Gospodarki Mieszkaniowej, Warszawa. [*Information on Housing Development; Monitoring Bulletin (1992-1995).* Housing Research Institute, Warsaw].

Paschalis, P. (1994a) Zmiany srodowiskowe warunkow zycia czlowieka na poczatku XXI wieku. Stan lasow i prognoza zmian. *Nauka* [Environmental changes of human life conditions at the beginning of XXI century. Forest state-of-the-art and change prognosis. *Science*], 2, 115-116.

Paschalis, P. (1994b) Current status and potential of utilization of wood material as energy carrier in Poland. International Symposium *Energy for Central and Eastern Europe to the Year 2010 given the Need for Energy Efficiency Improvements.* Nov 23-25, Les Diablererets, Switzerland.

Paschalis, P. (1996) Uzytkowanie lasu wielofunkcyjnego [Forest harvest in multiple use forest]. *Sylwan,* 1, 5-11.

Rocznik Statystyczny (1990...1995). Glowny Urzad Statystyczny, Warszawa. [*Statistical Yearbook (1990..1995).* Sate Main Statistical Office, Warsaw].

TIMBER BRIDGE CONSTRUCTION - A VALUE ADDED USE OF FOREST PRODUCTS

R.M. GUTKOWSKI, Ph.D., P.E., Professor
Department of Civil Engineering
Colorado State University
Ft. Collins, Colorado, 80523 USA

1. Background

In 1989, the author made a three week Professional Development Visit (PDV) to Wroclaw Technical University (WTU) in Poland. This PDV was sponsored by the National Academy of Sciences in the United States and the Polish Academy of Sciences. The purpose of the visit was to merge the author's expertise in timber bridge construction with the counterpart expertise of Polish researchers experienced in bridges of other materials. The intended outcome was to be a cooperation in research in bridge infrastructure, connected to timber bridges. By way of that visit, a university level formal cooperation agreement in bridge engineering and management was signed by WTU and Colorado State University (CSU) in 1991.

Through the cooperation agreement, several exchange visits of faculty and graduate students have occurred as well as cooperative research activities. The author was also involved in the organizing of an international bridge conference in Wroclaw, lectures at WTU and at other universities (Warsaw, Krakow). Cooperative or mutual interest research activities include knowledge-based (expert systems) methods for computerized bridge management , portable computer inspection capability, non-destructive testing of wood members, field testing of bridges and experimental timber bridge technologies.

Later, the Colorado Department of Transportation (by its formal linkage to CSU) and the counterpart General Directorate of Public Roads in Poland (by its formal linkage to WTU) became involved. In October 1994, the author and Prof. Jan Kmita of Wroclaw Technical University, co-directed the NATO Advanced Research Workshop on "Regional Challenges to Improve a Deteriorated Transportation Infrastructure Within Poland and Its Neighboring Countries." It was organized in conjunction with the several national Ministries in Poland and Germany and held at WTU [1]. At CSU the cooperation now also extends into its inter-disciplinary graduate research program in International Studies in Timber Engineering. Interest exists in expanding this cooperation into environmental aspects of decaying infrastructure.

Through these ongoing professional activities in Poland, the author has developed an understanding of the bridge infrastructure situation in Poland and needs in that country vis-a-vis the United States. This paper focuses on the role of timber as a

R. M. Gutkowski and T. Winnicki (eds.), Restoration of Forests, 271–292.
© 1997 *Kluwer Academic Publishers.*

valued added forest product in timber bridge construction in the U.S.A. and the degree to which there is potential for such in Poland. It is believed the same or similar circumstances and scenario exist in other, maybe all, NATO Cooperation Partner countries and independent states involved in this workshop.

2. State of Bridges in the U.S.

For the past 15 years in the United States, extraordinary national attention has been given to the need for bridge repair and replacement of highway bridges. Prior to this period of bridge infrastructure needs were virtually neglected for about 20-25 years. This is because during the forty years of construction of the Interstate Highway System, beginning in 1956, most of the Federal highway funding was concentrated on that project. Bridge construction declined greatly during that period of time. Thus, many aging bridges in the United States are of pre-1940 vintage. With intended service lives of 35 - 50 years, the condition of bridges reached a crisis point as expected service lives were reached and passed. Failure (during service) of the Silver Creek Bridge in West Virginia precipitated public attention. Failure of a main span of the Mianus River Bridge in Connecticut outside New York heightened the attention. The bridge is located on a very heavily traveled major Interstate Highway and had been recently inspected.

The U.S. bridge problems and needs have been documented by the American Association of State Highway and Transportation Officials (AASHTO) [2]. The report indicates that 571,308 bridges over 20 feet in length (the official lower limit for a bridge) existed in the U.S. at that time. Of these, 371,041 (65 percent) were located on rural roadways. The National Bridge Inventory (NBI), maintained by the Federal Highway Administration (FHWA) classifies bridges as being either structurally and functionally adequate or structurally and/or functionally deficient. According to the AASHTO report [2], 131,248 existing bridges were structurally and/or functionally deficient and another 141,857 bridges were projected to be deficient by the year 2020. The FHWA's Eighth Report to Congress [3] gave an even higher number of 220,000 currently deficient bridges. In 1989, the United State Department of Agriculture (USDA) - Office of Transportation reported 48.3 percent of bridges nationwide as meeting only minimum standards or less [3].

Considerable work is ongoing for the improvement of existing bridges. Through the Highway Bridge Repair and Replacement Program (HBRRP), federal monies are made available for bridge reconstruction needs. Since the inception of the HBRRP, a few years after the Silver Creek Bridge failure, about 2000 - 5000 bridges have been removed from the deficient list in the NBI each year. Some states augmented the federal funds through state monies. For example, Colorado implemented a state gas tax for bridge repair and reconstruction needs.

3. U.S. Timber Bridge Initiative

In the United States, the use of wood materials in housing is extensive. Perhaps 90% of all residences are comprised of light frame wood construction. In that mode of usage, wood has provided low cost and durable housing with a service life of 50 years and more. Construction of residential housing is bench mark industry for the U.S. Since so many jobs are tied to the industry itself and the manufacture of needed products, the state of housing construction is paramount interest to government economists. Developments in the cost of wood products and housing influence the American stock market almost daily. Summary stock market reports in the daily news typically include the category of housing construction.

Use of heavy timber in construction has had considerable less presence in the market place, principally in low rise buildings. Though steel and concrete have been the predominant materials of bridge construction since the beginning of the 20th century, there is a renewed interest in timber for short- and medium-span bridges. In the past decade extensive efforts have been made by U.S. researchers and industry to introduce modern technologies into timber bridge construction for highways. This interest has been generated by the intensive national effort to rehabilitate or replace deteriorated bridges.

About 14% of highway bridges (71,200 bridges) involve timber as the main structural material mostly on secondary roads. The HBRRP only distributes between 20 and 35 percent of its funds to secondary roads. Since local municipalities and counties in rural areas lack the tax base to augment federal monies and most do not have state funds available, bridges on secondary roads are tightly constrained by cost. Typically, simple, unimaginative bridges must suffice on a bare minimum cost basis.

Recognizing this need and opportunity, a National Timber Bridge Initiative (NTBI) began in the late 1980's. It focused on construction of modern timber bridges as demonstration projects for secondary roads [5]. This federal initiative was motivated by a Timber Bridge Technology Transfer Plan developed by the USDA - Forest Service and the American Institute of Timber Construction. That plan had the goal of increasing the use of timber bridges on highways. This goal was to be fostered by technology transfer, organizing workshops and identifying needed research. The NTBI has principally focused on federally subsidized demonstration projects in which hundreds of recently developed bridge systems have been put in place as trial installations.

4. Modern Advances in Timber Bridges

4.1. ADVANTAGES OF TIMBER BRIDGES

Properly designed timber bridges provide numerous technological advantages. Compared to concrete and steel, the energy of production of wood products is measurable less for timber. The advantages of light weight, ease of handling, lower skilled labor for fabrication and assembly, renewability of the tree resource and natural appearance are also evident. High strength to weight ratio and impact energy-absorbing properties are two of the important performance characteristics of timber. Service lives of 50 years or more can be expected when treated timber is used and field cutting or boring is prevented. Protection of connection details from moisture also greatly enhances longevity. Treated timber has a favorable durability against chemical deicing agents versus concrete and steel.

Traditional concerns about wood have been the longevity in exposed environments, fire resistance, susceptibility to decay and an image as the low technology material. Decay and longevity are primarily aspects of poorly conceived connection details which entrap moisture. Concealing metal parts between paired members greatly deters such problems and reduces fire spread concerns and improves fire ratings. Large timbers do not support combustion and charred members can be refurbished and reused. Fire ignition and spread in timber structures is essentially related to the contents of the structure and heat transfer through metal construction and ducts. Mixed concrete-wood composite construction is feasible and affords additional protection against exposure to moisture, fire and sound vibration. Correctly designed and detailed wood construction has been approved in parking garage circumstances, too.

New concepts allow high strength connection details to be made using only ring shank nails, small screws and lag bolts with thin plates, even sheet metal. Compatibility of wood to the use of fiberglass connection details, in lieu of corrosion sensitive steel details, offers promise, too.

4.2. CONTEMPORARY TIMBER BRIDGE TECHNOLOGIES IN THE U.S.

For many years, beginning in the late 1960s, the only modern timber bridge system being fostered was the glulam stringer bridge. However, its high cost kept it non-competitive with other materials. So no dramatic market developed, even when industry developed a panelized version featuring prefabricated components for ready assembly.

The structural analysis of timber structures is now highly advanced. Systems analysis is possible for the structures involving the complexities of wood. Factors such as slip behavior of connections, inclusion of partial composite action, statistical simulations and probabilistic analysis are possible. In the U.S. the starting point for improvement of timber bridge technologies, was research to better understand, quantify and utilize the partial composite action between wood elements [6-9].

Bridge decking has also been improved with the evolution of glulam deck panels. They are used in both transverse and longitudinal orientations over glulam or steel beams. "Stress-laminated" timber decks were developed in Ontario, Canada, as a repair method for longitudinal deck bridges. Using this system, vertical sawn lumber laminations are clamped together on their wide faces by high-strength steel post-tensioning rods bearing on a bulkhead. Attempts to stress laminate new longitudinal deck bridges as well as parallel truss configurations and to form box girder bridges of timber are under study as demonstration projects. However, these have not been proven economical or, in some cases, technically successful.

5. Experimental Timber Bridges in the United States

5.1. BACKGROUND

Considerable research has been conducted at CSU to perform conceive and develop new timber bridge concepts. It has been aimed at putting innovative experimental bridges in place in Colorado. The U. S. Department of Transportation recently funded the author to conduct analytical, laboratory and field studies to develop efficient timber bridge technologies for rural and roads. These studies (done in cooperation with the Institute for Wood Construction at the Swiss Federal Institute of Technology, Lausanne, Switzerland) involved conceptual and analytical design and reduced-scale and full-scale load testing of innovative timber bridge technologies. In 1993, two experimental bridges were placed in service in the state of Colorado. Work is continuing on concepts for wood-concrete bridge construction and for improving railroad bridges.

5.2. ANISOTROPIC GRID TIMBER BRIDGE CONSTRUCTION

Presently, in the U. S., in order to have a chance to implement new bridge constructions for secondary roads, one must consider the economics and "mind set" of public bridge owners. Almost without exception, rural bridges are standardized systems. Commonly, they are either "off the shelf" steel girder, systems or prefabricated precast concrete systems or, for very short spans, simple culverts (either corrugated metal or ordinary concrete). One must be able to develop a cost effective bridge system that has the same repeatable use (is not "site specific") and simplicity. The following bridge achieved that and is innovative and attractive in its low profile for a deep system.

Figure 1 shows a 57 ft (17.4 m) long, single span experimental highway bridge incorporating an "anisotropic grid" floor system. It was built in 1993. The width of 32 ft (9.75 m) enabled a through girder construction, whereby the girders and abutments permit a two way action (interior system supported on all four sides) of the interior system. The deck (longitudinally oriented panels of dimension lumber) suspends from the abutments and the floor beams from the through girder main elements via special steel hangers suspended from the tops of the girders. As a consequence of the lightweight design, the deteriorated concrete abutments (originally scheduled to be replaced) were retained. Laboratory and field load tests have also been conducted (Figures 2,3).

The hangers also provide for an I-beam behavior of the entire bridge under lateral loading of the bridge guardrail. A key aspect is the moment resisting connection of the hanger to the transverse floor beams. Each floor beam is actually a double beam, whereby two side by side members frame (spaced slightly for ventilation) connect to the hangers. The hanger was designed to be connected to the top and bottom of the floor beams by outstanding legs fastened to them by numerous, closely spaced nails. This avoids large concentrated bearing forces encountered when bolts are used which generally lead to either seasoning cracks or splitting under load. While proven successful in Europe as a general connection method, the bridge owner was

Figure 1. Anisotropic Grid Bridge Construction

Figure 2. Laboratory Reduced Size Model of Anisotropic Grid Bridge

apprehensive about nails and an alternative connection was used.

Subsequent to the research on the "anisotropic grid' bridge superstructure, full-scale laboratory testing has been ongoing to the originally designed guardrail configuration. Load tests have shown that a conventional all timber guardrail system (which was successfully subjected to a standard vehicle crash test) is considerably weaker than the experimental system. Static loads that demolished the conventional system caused no evident damage in the experimental guardrail. In static and dynamic tests (Figs. 4,5) of a 25 ft (7.62 m) long, full scale section of the bridge configuration failure was not achieved, despite very high loads being applied. It was evident the I-beam behavior was occurring and the load was shared by all transverse members, somewhat approaching equally. Although not yet tested to destruction, it is anticipated the alternative system will provide up to ten times more load capacity than conventional systems. This is evident from comparative laboratory tests of each system [10-12].

5.3. TRUSS-FRAME TIMBER BRIDGE IN ROUND POLE CONSTRUCTION

In 1993, the Scott Lancaster Memorial Bridge was constructed in Idaho Springs, Colorado. The bridge was constructed by supervised community members as a memorial bridge for a local youth killed by a mountain lion. It too was conceptualized in Switzerland and the subject of research studies done at CSU. This experimental bridge is in distinct contrast to the anisotropic grid highway bridge. It is neither standard nor simplistic nor low profile. It is the only one of its type in the U.S. Basically it is a modern, covered bridge for a rural bike path. It has a 75 ft (22.9 m) span and was designed in round pole, truss-frame construction (Figure 6).

The bridge is on a bike path, immediately adjacent to a major Interstate highway, and prominently visible to passing traffic. A high profile was desirable for scenic visibility of the structure. Two parallel, pitched trusses, each framed of six round poles and three vertical steel tension rods interconnected by innovative curved steel plate connection detailing, comprise the primary system. The detailing of the six joints creates continuity of the chord members at four of these joints and semi-rigidity at the top and bottom chord midspan joints. The eccentricity of the curved plate nailing and the longitudinal bolt of a special triangular end bearing plate plates counteracts the inherent eccentricity of internal forces at the joints.

Wind is resisted by a transverse frame at midspan combined with diaphragm action of the roof and floor systems. The outriggers of the wind frame help support the long overhang of covered roof, thus sheltering the superstructure and its connections. The flooring is round pole floor beams supporting a grid of 1 in. x 2 in. (25.4 mm x 50.8 mm) lathing and 2 in. x 6 in. (50.8 mm x 152 mm) planks. Metal splices and tie rod ends are recessed inside the underside of members for moisture protection. Curved plate fastening is accomplished using multiple ring shank nails.

5.4. MIXED WOOD-CONCRETE CONSTRUCTION

One of the critical functional concerns in the external use of concrete for bridge decks, is the concern about the intrusion of chlorides into the reinforcement. Being weak in

Figure 3. Field Load Test of Anisotropic Grid Bridge

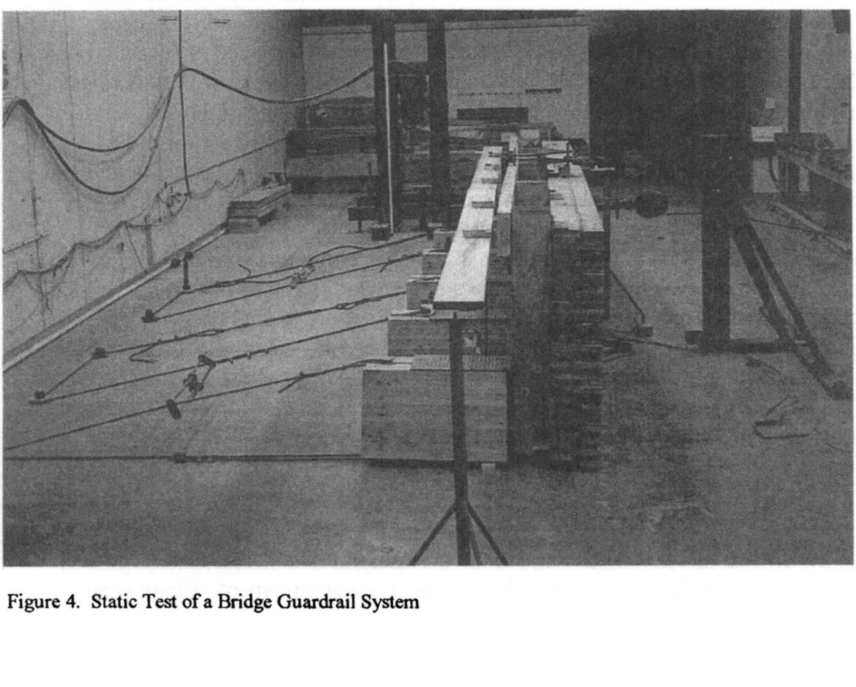

Figure 4. Static Test of a Bridge Guardrail System

Figure 5. Dynamic Impact Test of a Bridge Guardrail System

Figure 6. Truss-Frame Bridge Construction

tension, concrete cracks in the tension zone and is the reason for placing steel reinforcement. However, the micro-cracking (hairline cracks) exposes the steel to the elements to chemicals during salting or exposure to any contaminated atmosphere.

Besides the exposure problems, concrete decks have three major cost inefficiencies. First, the cracked concrete (extending through half the thickness or more) serves no structural purpose itself, except to hold the reinforcement in place. Second, the expensive steel reinforcement is needed. Third, cast decks have to be held in place by metal or wood form work for shoring. These costly aspects can be eliminated by the use of composite wood-concrete decking.

Figure 7 shows the configuration of a mixed construction slab bridge concept being tried in Europe. Half logs are interconnected to the concrete overlay by use of a special doweled shear key. The post-tensioned dowel maintains vertical interconnection of the layers. The upper end is sleeved to accommodate that need. The interlayer shear transfer puts the wood in bearing (at a small angle to grain) and the concrete in bearing and shear. The concrete overlay also protecting the wood from moisture and fire. The structural wood system serves to shore the cast-in-place concrete. Full-scale laboratory tests at the Swiss Federal Institute of Technology (Lausanne) have shown the success of this concept.

Studies are in progress at CSU to utilize mixed construction in layered bridge decks, employing dimension lumber or small timbers in place of logs (Fig. 8). A funded program has just begun. A wide array of shear key specimens (different cut dimensions, two connector types and embedment, etc.) have already been load tested for interlayer force transfer. They demonstrate that the shear key can transfer 3-4 times the force conventional mechanical fasteners (bolts and lag screws) can transfer. A large number of rectangular beam specimens have been prepared for initial determination of preferred placement of shear keys along the member. These will be followed by load tests of reduced size laboratory bridge deck specimens. The bridge specimens are available from the development work on the anisotropic grid experimental bridge. With a background in analysis and testing of partially composite layered wood system [6-9], CSU has available rigorous computer based mathematical models for studying incomplete composite action. A pilot test program on some simple laboratory T-beam specimens (concrete flange, wood stem) showed the software is applicable to wood-concrete layered beams.

5.5. RAILROAD BRIDGES

Historically, in the U.S., the predominant material used in railroad bridges was timber, specifically the timber trestle bridge. At the turn of the 19th century, the development of steel focused attention on longer spans and the use of timber trestles diminished in subsequent construction. However, the railroads have numerous existing timber bridges which have been in place for 40-75 years, or more. Maintenance has been good and they are in various states of condition but still in use. In recent decades train car loads have increased significantly. Presently, the typical design axle load is 66 kips, but the requirement is to be increased to 78 kips, soon. Consequently, the railroad industry has developed interest in field testing existing timber trestle bridges to determine capacity, field testing bridges strengthened with sleeper members, and

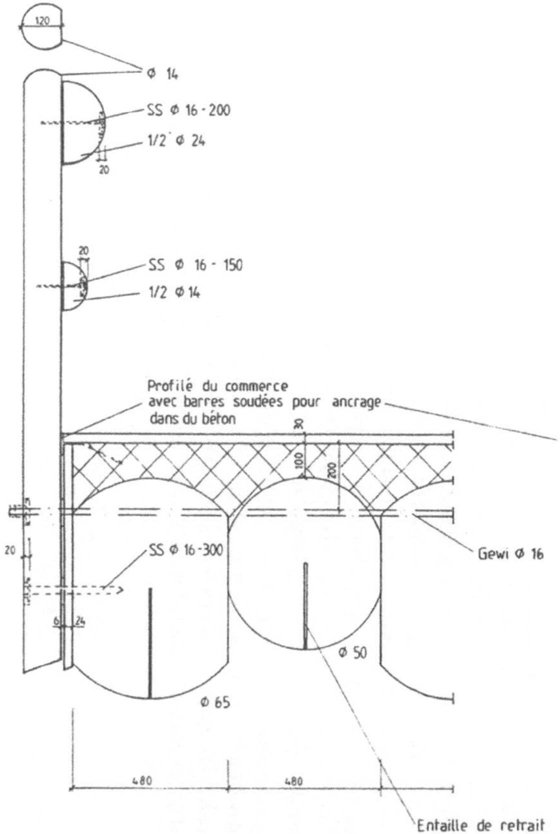

Figure 7. Composite Wood-Concrete Slab Bridge

Figure 8. Composite Wood-Concrete Beam Specimen

controlled laboratory testing of full scale bridge specimens. As the availability of large timbers has diminished or been controlled, interest in strengthening methods or new construction methods relying on smaller timber exist, too.

In conjunction with the Association of American Railroads, CSU has just initiated a comprehensive load testing program. In summer 1995, exhaustive field tests were conducted on three carefully selected open deck, timber trestle bridges in Colorado. One was a three span right bridge, one was a 4 span skewed bridge and one was a 31 span right bridge (Fig. 9). Each was extensively instrumented for deflection, strain, accelerations etc. A special track loading vehicle (traveling laboratory) was used to provide the controlled loading (Fig.10). The load axle was capable of applying load simultaneously in three directions; statically, by ramp or dynamically. Controlled vertical ramp loads and sinusoidal dynamic tests were conducted. In addition, the three car train unit (total weight over 880,000 lbs (3,916 kN)) was used statically itself in over 100 load positions. Finally, a set of moving train test were done at various speeds. The objectives were to study existing load paths under static and dynamic loading. Results are being analyzed and are yet to be published [13].

In fall 1996, the 3 span bridge was strengthened and will be retested for comparison. Later, a full size 3 span bridge will be constructed in the laboratory and tested at each stage of construction, so as to examine various construction and retrofit options.

6. Trends and Future Needs Timber Bridge Construction

6.1. OVERCOMING MISCONCEPTIONS ABOUT PERFORMANCE

Development of timber bridges for even long span bridges is technically achievable and a numerous case studies exist in the U.S. [5] and western Europe to attest to this reality [14]. The greater challenge is to foster broad-based desire and engineering capability to achieve such bridges and to solve the "real" and "perceived" constraints of wise utilization of the tree resource. In countries where modern timber bridge technology would be a new initiative, many misconceptions may be in place or encountered. A certain backlog of bias against wood as a material exists everywhere, too. For some engineers it is considered a plain, ordinary material for which the technological development has been minimal. Specifically, for some, wood is considered a low technology material whose performance is perceived from a track record of many years ago. Wood is mistakenly thought of as having low strength, requiring massive sizes, being subject to rapid decay and sustaining fire. This misconception occurs despite the reality that wood is indeed our most complex construction material and the state of the art of wood science and technology is markedly advanced. Modern materials have overcome these limitations of the past. Wood products are indeed in a high" tech" arena and provide significant advantages.

Figure 9. Field Testing of a 31 Span Railroad Bridge

Figure 10. Loading Bogey of the Track Loading Vehicle

6.2. THE PROBLEM OF INEXPERIENCE

With proper engineering, the upper limit of application of wood in bridge construction is considerably higher than typically thought by society or by architects and engineers either unfamiliar with its real potential or lacking full technical understanding of the material. Inefficiency in behavior and cost primarily occur from the use of improper or unwise framing configurations for wood. Detailing which overlooks the special aspects of wood or are left unprotected from moisture, will result in poor performance of a timber bridge. Some inexperienced engineers tend to adapt steel bridge or concrete bridge concepts to timber systems. This can lead to some poorly conceived or implemented projects and resulting bad image. Thus, the availability of special education focused on modern timber construction is prerequisite for architects and engineers interested in utilizing timber as a bridge construction material.

6.3. QUALITY CONTROL

Quality control is a vital aspect of production of any material. For wood, it is necessary to effectively control the manufacturing process for grading base lumber and laminate materials. For economy, such sorting is critical before assembly or into composite materials (such as laminated beams) or reconstitution (e.g. laminated veneer lumber). It is advantageous to place stronger or stiffer materials where they are most needed, for example, laminated beams should have the strongest material as it outermost laminations and weaker ones in the interior.

The jointing methods for achieving large member size (length and depth) need proper control, too. Experienced researchers would concur that finger jointing has the likely consequence of introducing the weakest point into a glue-laminated member. Failure in bending is most likely to initiate either at a finger joint or at a knot close to one. Unless proper joint strength is confidently achieved, some low strength laminated members will be present in supply. Proof loading (loading the joint beyond the tensile or bending design level) finger joints is presently used in production, but it is cumbersome and time consuming. Highly rigorous studies of finger joints has been evident in recent years and they are much better understood [15]. Automated, non-destructive methods to monitor joints and identify defective ones in laminating stock assembly of members are possible and would help.

In the recent decade, extensive research on non-destructive evaluation of wood members has been conducted. Non-destructive evaluation permits accurate determination of solid wood member strength and rejection of inferior material. Machine stress grading via production line flexural tests based on correlating modulus of elasticity to strength) have existed for two decades. However, this can be expensive and not readily affordable in developing plants. More advanced technologies for examining wood exist. Sonics, ultrasonics, acousto-ultrasonics, acoustic emissions, tomography, X-ray imaging, thermography, nuclear techniques among others are being used in various ways. Primarily these have been for condition assessment such as examining moisture content, density, detecting deterioration and decay, evaluating stiffness, monitoring surface roughness, locating defects, and other physical characteristics. However, ultrasonic methods exist to accurately determine solid member strength so that better sorting can be done after production. Currently, such methods are being used to allow acceptance checks of delivered material and to determine some properties of in place material.

Tests have shown that the strength is higher and variability less for natural logs than for large pieces (lumber) cut from them. In essence, nature generally produces a better product than mankind. Cutting and assembly introduces synthetic complications, that weaken the material too. Thus, round poles tend to have an inherent advantage over sawn lumber and timber. So applications in which they can be employed would tend to be more economical.

6.4. RESEARCH AND DEVELOPMENT CHALLENGES

Some needs in non-destructive evaluation of wood are apparent. The abilities to determine strength of laminated beams and deteriorated material are paramount. The former is being done approximately now and compounded by the related issue of determining finger joint resistance. The second requires the ability to detect location and extent of deterioration and relate it to a quantified indicator of the consequence to member strength. Success in the latter can help in assessing capacity of existing timber bridges more dependably than by analytical rating methods. A more advanced need in that regard is to detect degradation and approaching brittle failure during load tests of in place material.

Species, genetics, climate, exposure, harvesting, milling to sizes, grading procedures and mechanical jointing and assembly into products are key factors affecting the mechanical properties of wood and it end products. In total population, wood and wood products are extremely variable in their physical properties. Traditionally, design stiffness and strength values for wood and timber have been, and still are, established by consideration of sampling complete populations, be it by species, by grade or other artificial separations.

Two major directions in statistical codification of wood design properties are "in-grade testing programs" and "load and resistance factor design" (LRFD) formats. Each reflects a striving to statistically characterize a complete sub-population of the wood products and offset the lower tail end values as a mathematical point of demarcation between unacceptable and acceptable resistance. In effect, the larger proportion of the medium and high resistance material is penalized by the acceptance of the probability of having low resistance occur.

Whether by the past use of lower 95% exclusion limit (5% fractile) of strength or by emerging use of a safety index in LRFD for setting the base resistance for codification of measured resistance into allowable stresses, the implication is weakest members in a sub-population control the permissible stresses. This is in contrast to steel wherein the industry has developed a wide range of material grades with modest variation about the mean. Thus, designers can dependably "select" steels from a range of "resistances" (types and grades of steel) to match the strength and exposure requirements. In contrast, wood is penalized by conservative lower limit strengths. The preeminent challenge is timber bridge construction is to be able to identify, with high dependability, the specific properties of the individual pieces.

6.5. ENVIRONMENTAL CONSIDERATIONS

The environmental and ecological aspects of accessing forests for the needed tree resources is of increasing importance. Attention to and debate about the wise and sustainable utilization of wood fiber and protection of the forest and eco-systems is universal. Some compelling and critical constraints on forest access are sharply debated. Some present trends are reducing access to the forest and diminishing availability of the tree resource. Fear of excessive exploitation exists. Concerns about endangered species have closed forest land and sharpened the and polarized the political process affecting that consequence. Despite their renewability some strong critics wish to prevent all such access to forests. These strong to extreme environmental constraints are very limiting to the tapping of the opportunities and benefits of timber construction. They tend to place limitation on even logical, rational forest management.

On the other hand much more is known about the science and silviculture of forest growth. Extensive research on the control of damage due to infectious insects is also ongoing. Computer tools provide better means to monitor and manage forest growth (or depletion). Clear cutting of trees, which is contested as being both visually imposing and ecologically damaging to the forest and eco-systems, is being more closely examined and debated. Managed harvesting of trees is imperative to regrowth of the forest and approaches to this are improving.

Virtually every fiber of the tree can now be utilized. Reconstituted products utilize veneer peeled from even low grade logs. Material that is otherwise wasted, such as peelings and sawdust, are used in panel products such as particle board. Logs with numerous defects can be reconstituted into laminated veneer lumber (in population, stronger and less variable than solid sawn dimension lumber), wafer board, oriented strand board and other "high tech products." "Waste" is also usable in methanol production for developing efficient fuels. In contrast to other materials, energy of production and construction are low since steel producing furnaces, welding, flame cutting, production of cement and other energy demanding processes are avoided. The deleterious environmental effects of producing alternative products is of concern, particularly when plant modernization, efficiency and containment of pollutants are lacking.

Standing trees with older wood hinder new growth and are susceptible to biological attack which can spread throughout a forest. In reality, expansion of the use of timber construction fosters forest conservation and new plantings needed to insure regeneration and life of the forest. Ability to selectively identify (by NDE) the stiffness and strength of wood members has been progressing via NDE. Achieving this would allow designers to use them "selectively" putting strongest pieces where most needed. The beneficial outcome would be dramatic since a) less of the "better" material would be wasted, b) marketing less production of "better"material at value added premium prices could result from the benefit justifying the cost and c) demand for older growth trees would be reduced or more controllable without consequence to industry. Similar rationale applies to NDE assessment of standing trees for potential harvesting and production planning. There is a compelling need to be able identify the physical characteristics of standing trees. This capability would foster selective harvesting of trees for assigned (graded) use in different end products. It would also provide a means to non-destructively monitor the consequences of different forest management approaches. This infers the need to develop practical means to use such high technology.

7. State of Bridges in Poland

Poland's bridge infrastructure has experienced dramatic changes this century. The evolution was described by the Deputy General Director of Public Roads [16,17]. The following summarizes the main developments.

During World War I most of Poland's roads and bridges were destroyed and had to be rebuilt. In between the two World Wars, about 137 miles (221 km) of bridges were reconstructed. During the World War II, about 46% of the bridges were destroyed. Ten years after the war, the recovery from the losses had taken place. Presently, there are about 227 miles (365 km) of public roads in Poland. This network is divided into four categories, national roads (12.5%), Voivodship roads (35.2%), county and communal roads (47.2%) and industrial (service) roads (5.1%). About 30,000 bridges of 334 miles (538,000 km) total length exist on the public road systems. Only abut 90 of these are longer than 656 ft (200 m). About 22,000 are concrete, 4,000 are steel and 3,000 have timber deck on steel girders.

The General Directorate of Public Roads (GDPR) has responsibility for construction and maintenance policy for all public roads, allocates the State Budget for National Roads (national funds) and establishes technical specifications and regulations. Regional Directorates of Public Roads manage and maintain the National Roads and its 7,000 bridges and Voivodship roads and its 10,500 bridges. From 1982-1991, the ADT on main international roads increased by 100% and is much higher now. However, in the period of 1986-1992, the State Budget for National Roads decreased by 75%. As a consequence, condition of roads had deteriorated significantly. The condition of bridges does not meet the traffic requirements, either. About 75 miles (120 km) of bridges officially qualify for reconstruction. Many recent bridges were concrete bridges, built in the 1970s or later during a period of a forced fast pace of development of the construction industry. The result was a diminished quality of the materials and the structures. The traffic density, vehicle weights and air and water pollution have increased, too. Thus, many bridges only 15 years old are under design capacity, defective and corroded.

In 1983, Poland ratified the European Agreement on Main International Traffic Arteries AGR. One consequence is that a road network model was developed, in which construction of three motorways and a system of expressways would be built. One roadway is part of a North-South Trans European Motorway, extending from Poland to Greece. Another is part of the Via Baltica System. Thus, the international roadway has primary priority for the diminished limited national funds. To add funds, market economy relationships between employer and contractor, toll roads, service industry tax income etc. are being fostered. In 1986, the GDPR in collaboration with WTU began development of a comprehensive, modern bridge management system (BMS). The BMS is advanced (based on expert system format, fuzzy set theory and neural networks and logic), meeting or exceeding standards of many western European countries. This BMS is operational now and provides the means to make sound fiscal decisions regarding efficient use of limited bridge construction funds. An adaptation of the BMS is also being developed for railroad bridges.

In effect, Poland is pursuing the model used for building the Interstate Highway System in the U.S. The immediate goal is apparently to meet a national need to improve national transport avenues, as part of increased tourism travel and movement of goods to and from neighboring countries and evolving market economies. A secondary, but important need is the movement of goods from source to market. For example, a need exists for numerous short to medium span bridges over Poland's waterways. The existing motorway from Krakow, through Katowice and Wroclaw to eastern Germany is an ongoing upgrading project. Access to it from communities, industry sites and farms requires shorter bridges across the Oder River. Similar "farm to market" routing is needed throughout Poland. However, Poland's economy is such that at present funds are so inadequate even its major motorway goals are not achievable for some time.

8. Training and Development Needs in Poland

It is clear that in the future, a need and an opportunity exists for timber bridge development in Poland. Poland has the necessary forest resource to develop the needed industry support. It also has the need for lower skilled jobs that a forest products industry requires. However, its industrial technology for material production is not modern and its universities lack expertise in timber bridge technology [18].

In 1993, the interest of the World Bank was derived by the author to begin to address Poland's "access bridge"needs. In conjunction with the author's PDV (arranged because of Poland's interest in timber bridge technology transfer), a verbal arrangement was discussed to provide World Bank funds to Poland for enabling CSU to provide advanced training in timber bridge technology to faculty at Polish universities. In addition, U.S. experts would evaluate the needed development of the timber industry in Poland and economics of the material supply needs. A comparison of the U. S. wood industry/products situation vis a vis the Poland situation would be made to assess that potential. Coincident with these steps, via the cooperation agreement between CSU and WTU, the partners universities would augment the BMS of the GDPR to include timber bridge technology. Unfortunately, due to the priority needs of the National Road System, the GDPR elected not to pursue the World Bank loan option.

9. Conclusions

Interestingly, the present track of road and bridge rebuilding and expansion in Poland is similar to the much larger scale development of the Interstate Highway System in the United States. Having just begun to tie into the major trans-European network, the timing is the present for Poland. The priority in Poland is on the primary national road and bridge system, but secondary roads are an inevitable need. Modern timber bridges appear to have a potential for meeting this second progression of bridge needs.

In the U.S., the emergence of timber bridges is recent and after the completion of the Interstate System. Timber bridges have become competitive and economical due to careful, individualized design predicated on in depth knowledge of wood as a material; creative conceptual design, framing and connections; attention to details; emphasis on site-specific considerations; accessing the variety of wood based materials available in the market, awareness of the limitations in present grading and allowable stress values, and insistence on leading edge NDE/NDT technology for quality control of the specified materials both in the fabrication and on the site.

Poland has a natural resource base and labor force, but the forest products industry is not well developed. Training, research and development and establishing of academic research specialists the area are prerequisite to making this opportunity possible. Some lead time exists to pursue this path, since the development of the primary road system is moving slowly. Situations like Poland's probably exist in other NATO Cooperation Partner (CP) countries and independent states in central and eastern Europe.

10. Pertinent Observations from the ARW

The author's observations principally consist of a number of "surprises" learned as a consequence of the activities at the ARW. The most important surprise is that despite the consequences of a forest situation impacted by heavy industrial air pollution, the predominant unifying viewpoint and opinion was that their was indeed plentiful forest resource available in most CP countries represented. This sharply contrasted the prevalent descriptions he had read in the U.S. literature of the late 1980s and early 1990s. Many papers told of a crisis of dying forests throughout central Europe, to the point of possible extinction. While environmental damage is serious in the "Black Triangle" region and must be abated, this grave scenario does not appear true elsewhere. In more distant areas, the situation does not appear to be hopeless, but is, in fact, becoming more manageable and improvable.

In the context of discussions at the ARW, the meaning of "resource" was consistently expressed as the multi-purpose use of the forest ranging from social value to economic value. This was not surprise as it is becoming almost universally accepted and fostered. However, the fact that wood utilization for construction was seen as a positive theme in all four Working Groups was a striking surprise.

In the context of this paper, valued added utilization is presented as a major component of the forest resource and was so for the delegation. Predominantly, individual delegates expressed the importance of expanding, not constraining such forest utilization. Considering the range of fields represented from genetics to botany to environment to ecology to environmental engineering to timber construction, this is was enlightening. The common setting for this viewpoint was two-fold: a) utilization must be within a setting of maintaining forests as a sustainable natural resource and 2) "market conditions" were the primary factor preventing expanded usage in construction. The latter point was particularly striking for the author, a delegate from in the United States where broad "protectionism" (rightly or wrongly) principally hinders access to public (even private) forests by the forest building products industry.

It became evident to the author that prior to and even after rule by communist governments (and, for some delegates, presently during) the pricing structure was set by the state officials (or as some delegates stated, the "central government") with either conscious or blind bias. One deduces some industries were favored over others and still are. On one hand, wood and timber products were and remain priced artificially much higher than production costs would dictate in some countries. Apparently, without the factor the end products could be provided at considerably less disadvantageous costs relative to other construction materials. On the other hand, the demand for wood building products is low in many CP countries due to a cultural preference for non-wood based housing. The dilemma is that delegates opined that wood products could be provided for substantially less end cost if government controls and pricing were removed. If so, they related, the extensive need for modernized, low cost housing and low rise building infrastructure could be more readily met with wood construction.

A "pent up" desire to replicate the successful use of wood construction to provide affordable single or multi-family residences was evidently compelling within some CP countries. The author deduced that the if the cultural preference (tradition) for other more massive building materials was overcome, a freer pricing system would lead to affordable solutions via wood construction. Thus, the means exist but the incentive within government does not.

A second surprise in the context of this position paper was the evident negative situations of cross-border customs and trade policies. Related to this was the particular surprise that Hungary was, in contrast to the other CP countries represented, primarily an importer of wood products from the western countries. It is the author's recollection, from his visit to that country in 1980, that at that time Hungary was heavily forested in hardwoods and was exporting furniture products heavily, primarily to the U.S.S.R. Evidently, the reverse was true for building products, perhaps due to the above factors and lack of softwoods.

In regard to the larger cross-border issues, the need to have freer trade between CP countries would allow mutual benefits to evolve. One delegate expressed (author's paraphrasing), "We buy our own poor quality window products when our immediate neighbor CP country can supply a much superior product at a better price. On the other hand, we make better lumber than theirs at better prices. Pity we cannot trade for mutual advantage." The striking thought is international trade associations might induce paths to achieving mutual benefits. However, customs duties defeat the cost advantage. On the other hand, the importance of providing products that meet the products standards of unified western Europe was considered equally critical, should cross-border barriers diminish or disappear. This would be particularly true for any CP country which might gain admission to NATO as a member country.

To close, the above might be self evident to some readers but the author was unexpectedly surprised that his paper, written in part to promote the valued added potential of wood construction materials, was philosophically unnecessary. The belief in value added potential already exists and was convincingly exhibited by a delegation with extreme differences in their fields of endeavor and responsibilities as related to the broad "value" of the basic resource, the forests themselves. What seems most needed is the technology transfer and indeed, lobbying from western countries via experienced "wood engineers," and "wood architects" and economists to overcome centuries of hesitation (in-grained social "bias") against a potentially standard of life helping material. The resulting inertia to adopt wood construction would also contribute to solving immense housing and infrastructure needs in CP countries, as well as generating badly needed jobs in wood related professions and trades.

11. References

1. American Association of State Highway and Transportation Officials, Inc. (1988) *Keeping America Moving: The Bottom Line, A Summary of Surface Transportation Investment Requirements 1988-2020*, Washington, D.C.

2. Favre, P.-A. (1994) Development of an Experimental Skewed Anisotropic-Grid Timber Bridge M.S. Thesis, Colorado State University, Fort Collins, Colorado.

3. Federal Highway Administration (1987) Highway Bridge and Rehabilitation Program, *Eighth Annual Report to Congress*, Washington, D.C. 1987.

4. Gutkowski, R.M., and Robinson, G. (1996) Field Studies of Timber Railroad Bridges draft report to the Association of American Railroads, Pueblo Colorado.

5. Gutkowski, R.M., and Kmita, J. (1994) *Transportation Infrastructure - Environmental Challenges in Poland and Neighboring Countries*, Springer, Heidelberg, Germany.

6. Gutkowski, R.M. and Favre, P.A. (1993) Ultimate Capacity and Mode of Failure of a Timber Bridge Guardrail, MPC Report No. 94-19B, Mountain Plains Consortium, North Dakota State University, Fargo, North Dakota.

7. Jauslin, C. (1992) Tension Behavior of Fingerjoints in Wood Laminates, Ph.D. Dissertation, Colorado State University, Fort Collins, Colorado.

8. Ko, M. F. (9172) Layered Beam Systems with Interlayer Slip, M.S. Thesis, Colorado State University, Fort Collins, Colorado.

9. McClain, T. E. (1975) Curvilinear Load-Slip Relations in Laterally Loaded Nailed Joints, Ph.D. Dissertation, Colorado State University, Fort Collins, Colorado.

10. Mistewicz, M. (1993) Planning, Financing and Managing of Roads and Bridges in Poland, Invited Paper and Lecture, Colorado State University, Ft. Collins, Colorado.

11. Mistewicz, M. (1993) Bridge Inspections and Management Activities in Poland, Invited Paper and Lecture, Colorado State University, Ft. Collins, Colorado.

12. Natterer, J., Herzog, T., and Volz, M. (1991) *Holzbau Atlas Zwei*, Institut für internationale Architektur - Dokumentation, GmbH, Munich.

13. Pault, J. D. (1977) Composite Action in Glulam Timber Bridges, M.S. Thesis, Colorado State University, Fort Collins, Colorado.

14. Pellicane, P.J., Gutkowski, R.M., and Czarnocki, J. (1996) Poland: Threatened and Neglected Forests, *Journal of Forestry* **2**, 29-32.

15. Shigidi, A. M. T. (1995) Static and Dynamic Testing of a Timber Bridge Guardrail Specimen, M.S. Thesis, Colorado State University, Fort Collins, Colorado.

16. U. S. Forest Service (1993) The National Timber Bridge Initiative, *United States Department of Agriculture.*

17. Vanderbilt, M. D., Goodman, J. R., and Criswell, M. E. (1994) Service and Overload Behavior of Wood Joist Floor Systems, *Journal of the Structural Division,* ASCE, Vol. 1, ST1.

18. Walzer, N. and Chicoine, D. L. (1989) Rural Roads and Bridges: A Dilemma for Local Officials, *United States Department of Agriculture*, Office of Transportation, Washington, D.C.

THE RELATIONSHIP BETWEEN FOREST RAW MATERIAL AND THE CONSTRUCTION OF SLOVENIAN HAY RACK

J. KUSAR
Faculty of Arhitecture
University of Ljubljana
Slovenia

Abstract

Keywords: Timber (Wooden) Construction / Hay-Rack / Framework / Heritage

Forests have always been the most important pillars of the Slovenian farmer's rural economy, particularly in the mountainous regions. The tradition of the ecological approach and the principle of exploitation of small surfaces in Slovenia has existed for a whole century. The Slovenian hay rack represents a very interesting Slovenian achievement in the field of national joiner architecture.

1. Introduction (Some facts about Slovenia)

The young Slovenian state has a rich, over a thousand year history and a cultural background tightly connected to Western Europe. Some of the most significant milestones of Slovenian culture are: church and monastic architecture from Romanesque period onwards, castles and towns from the period of 12th to 15th century, Slovenian Renaissance sculpture, painting and music masterpieces, the first book printed in Slovene in 1550, the first scientific academy operated in Ljubljana at the end of 17th century, Baroque architectural creativity, the 19th century poetry and architecture and literature of the last century. Science has a long tradition too. From the 15th to 17th century humanists of Slovenia were active at European universities and high school studies were introduced in the 17th century by Jesuits. Today Slovene scientists significantly contribute to the creation of world-wide knowledge.

The Alps, the Dinaric mountain range and the Panonian plain spread over the area of 20.256 sq. km where 2.020.000 people live. About 91% of them are Slovenian, though Hungarian, Italian and other national minorities also contribute to the Slovenian cultural variety and character. Climatic conditions are influenced by the geographic configuration. The average temperature in January is around 2 degrees C and in July around 22 degrees C.

R. M. Gutkowski and T. Winnicki (eds.), Restoration of Forests, 293–300.
© 1997 *Kluwer Academic Publishers.*

2. Forests in Slovenia

Forests are an indispensable element of Slovenian country with important ecological and Social parts. They cover 53% of Slovenian area and are an important natural wealth. Wood cut in 1994 amounted to 1.411.275 m3 of conifers, 843.606 m3 of deciduous trees.

The relatively well-preserved natural environment and its beauty with numerous natural attractions represent an important developmental potential of Slovenia. At the same time, however, they also impose the duty on the country (cf. Article 73 of the Constitution of the Republic of Slovenia) to manage this heritage responsibly in accordance with up-to-date knowledge and with professional and ethical criteria which will ensure the preservation of the heritage that has been passed on to us and the formation of new heritage to be passed on to the next generations.

Slovenia lies in the temperature zone where the temperature and humidity conditions are favourable to the development of good soil and the growth of forests trees. The forest topsoil is, on the average, very fertile. In comparison to the European and global situation, the Slovenian forests are preserved well above the average, especially as regards the diversity of natural tree species. Owing to the intensive care devoted to the growth of forests in the last thirty years, there are more cultivated systems in Slovenian forests than in other European countries.

The ecological management of renewable natural resources has become the basic condition for the sustainable development and quality of life, part of the philosophy of life that can ennoble human beings. In the management of forests, this means that nature can be preserved, as well as our living environment and our national identity, and that a sustainable development of our countryside is made possible. The rural economy, forestry, and all other activities in the environment should have an adequate basis for a sustainable development.

Under present conditions, the Slovenian natural heritage is endangered not only by sheer ignorance, unawareness and accidental events but more and more by privatization and marketing attempts and accompanying impacts of planned interventions in the environment. The use of these resources is (has been) in fact free and unrestricted, as is their management and the management of their direct surroundings. The natural heritage will be safe only when the awareness of its values as part of the general culture penetrates into all activities involved in the natural heritage in some way, either by managing or exploiting it commercially, protecting or endangering it, or making any other decision on its fate.

3. Slovenian Hay Rack

Forests have always been the most important pillars of the Slovenian farmer's rural economy, particularly in the mountainous regions. The farmer soon recognized the importance of the forest's protective role. This sustainable ecological approach to the forest has always been, and will also be in the future, a good basis for a successful co-operation between the foresters and forest owners in the management of forests.

In the past, the ecological management of forests was well known to the Slovenian forest owners. The tradition of the ecological approach and the principle of exploitation of small surfaces in Slovenia has existed for a whole century. Essay on hay-racks in Slovenia is therefore an apotheosis of the wood and its beauty, and of the creativity of numerous anonymous carpenters whose Memory is preserved by and within their "living" Creations.

The Slovenian hay rack is undoubtedly a very interesting Slovenian achievement in the field of national joiner architecture. It is also an important contribution of the Slovenian architecture to the global knowledge on timber skeleton design and construction.

The present description tends to render a picture of the hay-rack as a part of the Alpine and European territory. However, within this territory we are linked not only by hay-racks, but also by the Slovene folk culture as a whole, since our ethnic territory is a meeting point of the Mediterranean (Roman) civilization from the south, the middle-European from the north, and the Pannonian from the East. This special position of Slovenia in Europe also reflects in the culture of the hay-rack and its types, draws its borders, and sheds a most vivid light on the transformations of the hay-rack as it was adapted to various cultural environments. (T. Cevc, J. Cop, Slovenski kozolec, Agens; Zirovnica, 1993)

4. Hay Rack Construction

Hay rack design and construction combine an aesthetic feeling of static functions of the hay rack bearing construction structure and simple and clean lines of design structure.

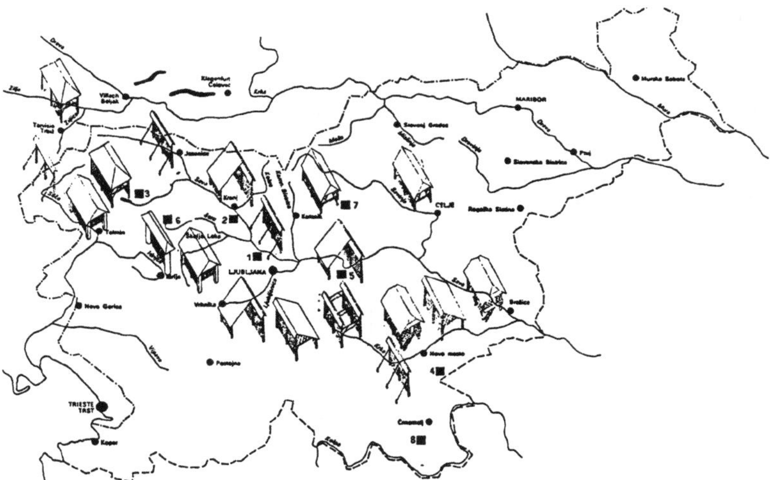

Figure 1. Spread of the Slovene "kozolec" (hay-rack), with its full territorial extent and its characteristic shapes (M. Music, Arhitektura slovenskega kozolca, C. Z.; Ljubljana, 1970)

Timber (wooden) skeleton construction has a long tradition in Slovenian joinery. We have found archaeological findings of wooden skeleton construction from the Hallstatt period from the 8th century B.C. at Most na Soci.

For a Slovene soul, a hay rack means a unique monument of the Slovenian joinery. In the opinion of architects, hay rack construction is a perfect harmony of construction and functionality, the author is convinced that with its specific design forms a hay rack will remain a unique monument of the Slovenian timber construction and an integral part of the Slovenian landscape architecture.

Timber rack system construction is known all over the world. It is known since the beginning of the joinery when metal tools were developed.

Figure 2. Characteristic shapes of Slovene hay-racks
M.Music, Arhitektura slovenskega kozolca, C. Z.; Ljubljana, 1970

A
Simple streched "kozolec"

B
Simple streched "kozolec" with stone columns

C
Hay-rack with "cloak"

D
Double streched "kozolec"

E
Coupled "kozolec" - one-tree "stog"

F
Coupled "kozolec" - two-tree "stog"

G
Coupled "kozolec" with stone columns

H
Coupled "kozolec" with corridor

I
"Dog kozolec" or "Goat kozolec"

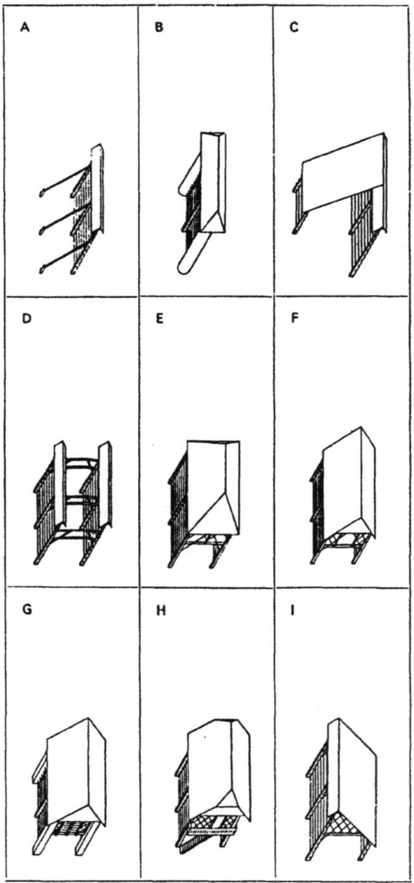

Statics of a coupled hay rack is performed with its framework construction system and its functional structure. Its structure is divided in three levels referring to its height: bottom or floor level, middle level and roof construction. The bottom level serves for storing of agricultural implements, and hay wagons in rainy weather. Walls are vertically positioned columns with transparent walling of thin boards which function as an intermediate between inner and outer space (interior and exterior of the hay rack).

Middle level of all four walls functions as hanging store for holding of hay, and is inserted in columns. Hay dries quickly due to good airiness enabled by the hay rack construction. Middle level walls are made of harrow of the crossed joint timbers which are put in the upper and lower connected beam.

Figure 3. Coupled hay-rack, two-tree "stog" (photo by J. Kusar)

Figure 4. Coupled hay-rack, details (photo by J. Kusar)

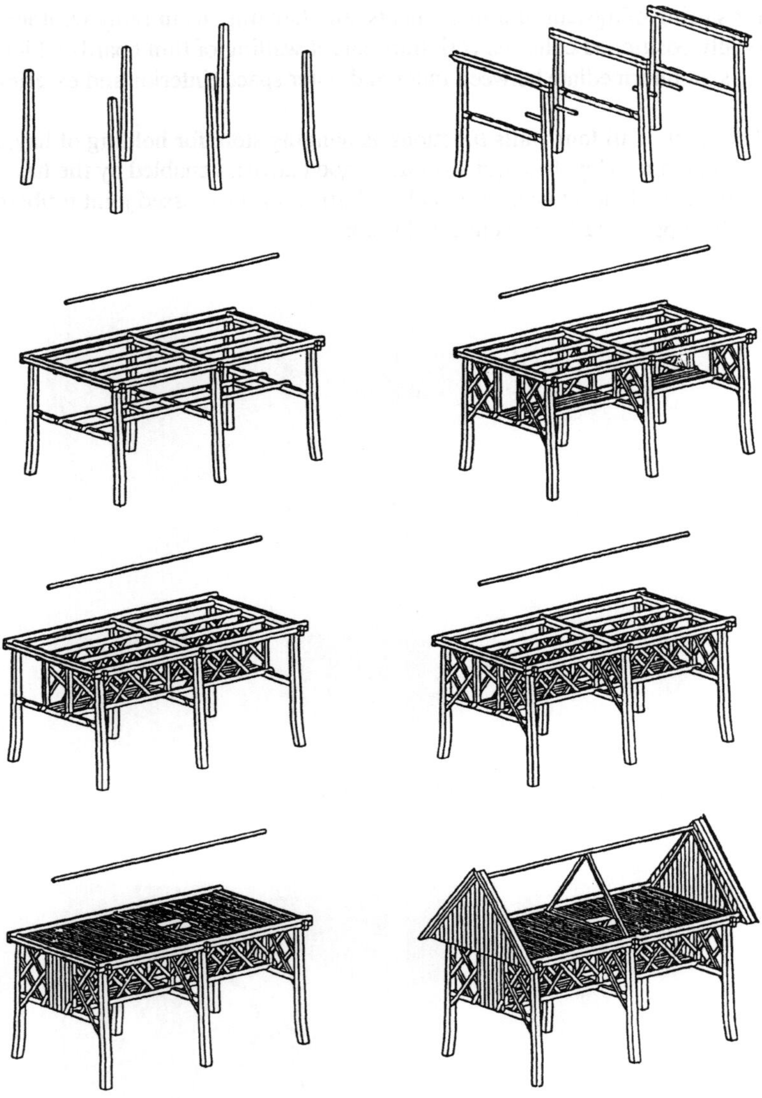

Figure 5. All necessary columns, beams and timbers are cut. In these elements all connected bungs and holes are made and marked.

Figure 6. Coupled hay rack construction is clearly seen from the construction technology and mounting assembly system.

Front crossed harrow is linked with connected beams and columns on the floor. The wall is then mounted vertically on stone stands and temporary supported at sides. All other frameworks are also put vertically. At the end, another front wall is mounted. Timbers are assembled in prepared bungs and secured with horizontal cross inner harrow.

Slovene culture is intimately linked to the hay-rack. Although the hay-rack is definitely a utilitarian structure, as a characteristic element of the Slovene countryside, it has been given its due place among the most precious monuments of folk architecture. The hay-rack is widely thought of as a Slovene speciality.

5. References

Cevc, T. and Cop, J. (1993) Slovenski kozolec, Agens; Zirovnica.

Fister, P. (1985) Umetnost stavbarstva na Slovenskem, C. Z.; Ljubljana.

Kusar, J. (1995) Influence of earthquake on the construktion of slovenian hay-rack. - COST TC Urban Civil Engineering Conference, Ljubljana.

Melik, A. (1931) Kozolec na Slovenskem, Ljubljana.

Music, M. (1970) Arhitektura slovenskega kozolca, C. Z.; Ljubljana.

THE LITHUANIAN TRENDS FOR FOREST UTILIZATION

A.K. KVEDARAS
Head Department
of Steel & Timber Structures
Civil Engineering Faculty
Vilnius Technical University
Saulėtelio alėja 11
LT-2054 Vilnius, Lithuania

Abstract

This paper discusses the main trends for forest utilization in Lithuania. Shortly the present state of Lithuanian forests and the main needs of timber industry are described. It is shown that during the last five years the situation in forests and in the timber industry was sharply stressed to an unsatisfactory one. The main reasons of such a situation in the timber industry are: domestic market becoming weak because of a great decline of building activity; export of raw logs and rough sawn timber products which for the State economy is harmful; insufficient investment into forests and aforestation and into timber industry either from the State or from foreign investors. Subsequently the paper discusses conversion of wood as raw material into various products including building products and application of its in building practice of Lithuania. Finally the paper deals with main trends to develop new value-added building products by the means of increase of load bearing capacity and fire resistance of timber structures and development of new composite members as most efficient value-added products enabling more effective utilization of raw materials and waste from the forests and timber industry permitting to maintain the forests in a better health.

Keywords: State of Forests, Timber Industry, Conversion of Wood, Building Products from Timber, Load Bearing Capacity, Fire Resistance, Mixed Composite Members, Stress and Limit States.

R. M. Gutkowski and T. Winnicki (eds.), Restoration of Forests, 301–314.
© 1997 *Kluwer Academic Publishers.*

1. Introduction

1.1. PRESENT STATE OF LITHUANIAN FORESTS

According to the announcement at news agency ELTA the present area of the forest land in Lithuania is equal to 1,975,500 hectares. The forest is growing on 1,860,000 hectares. The indices of Lithuanian woodedness are constant from the 1980. Present woodedness of the country constitutes 30.1 percent of the land.

At all types of cutting areas of state and private forests, 5,980,000 cubic meters of trunk wood have been logged during 1995. It is calculated that only 5,000,000 cubic meters of trunk wood can be logged in Lithuanian forests now, and about 5,500,000 of it after a decade because usual amount of cuttings from 1961 till now was only 3,000,000 - 3,200,000 cubic meters of trunk wood.

The situation in the forests of Lithuania is very complicated, but according to the view of specialists of Forest Ministry it is still not a catastrophy. The forest is under heavy damage from storms, squalls, insects, mould diseases, fire, adverse natural circumstances. At the end of 1995 the concentration of the forest pests and mould diseases have been over an area of 65,000 hectares (for comparison during 1994 it was over an area of 74,000 hectares). The needle gnawing pests have covered an area of 21,000 hectares and the mould diseases - 2,000 hectares of the forests. To improve the status of trees in the State forests during the 1995, preventatives have been used on a total area of 215,500 hectares (during 1994 - on 122,500 hectares). During 1995 trees have fallen in an area of 9,432 hectares, most of them (92 percent) because of damage by insects.

In Lithuania, public opinion is widespread that, during the last five years of re-established Independence taking place, wholesale destruction of forests occurred. Heavy machinery was extensively introduced into forests. From the signs of labour left in forests, many people held the opinion the foresters hunted only after profits. They aspired to cutting down of excellent wood and to capitalize on it. So the devastated areas have left with deep tracks turned up by caterpillar or wheeled tractors, which destroyed vegetation of underwood. Stubby places and devastated surroundings is not a very good inheritance for the future Lithuanian generations.

The old population remembers the independent period of Lithuania before the Second World War, when forests were looked after more carefully. Then timber cutting was going on only in winter time, when soil was frozen and there had been snow fall because the vegetation would not suffer from that.

The forest areas are on a continual renewal. During 1995 aforestation had been fulfilled on the area of 7,000 hectares, for silviculture 5,500 hectares of field had been prepared. But all measures of renewal are carried out in the State forests. In the nearest future they would have to constitute about 70 - 75% of all Lithuanian forests. The remaining part would have to be private. It is foreseen to carry out a privatization of equipment and enterprises for forestkeeping and wood cutting.

Privatization is foreseen as one of the cardinal means to improve the conditions in Lithuanian forests and timber industry. But problems involved are very complicated because of very spontaneous character of economic reforms in Lithuanian forestry as well as in agriculture and industry. On the other hand, private possession in the West countries had been increased at the end of nineteenth and at the beginning of twentieth centuries. It was useful for the forests. In Lithuania it is also necessary to create a private forest economy. Unfortunately, Lithuania has no traditions of such an economy. But it is necessary to create special programs to establish and enlarge co-operative forest estates with serving foresters structures. The owners of forest estates have also to know a little about management of forests. But now for realization of all measures in private forest economy, there is a great lack of means.

Export of raw logs and timber products during 1995 came to 10% of total export of whole State (during 1990 the 5.3% of whole Lithuanian industrial product was produced by enterprises of the timber industry). Such timber export is useful for enterprises, but for State economy it is harmful because there is growing export of raw logs and products of rough sawn timber. The State receives no revenue from the export of timber products or from the state forest property. Therefore the State has no money to support care under forests and aforestation. Help may come from foreign investments, but that requires a thorough analysis of total natural forest resources.

A great danger is now springing up in the Lithuanian forests because of difficult ecological and financial situations and because of quick spreading of ticks infected with Russian (Rosslag) encephalite viruses.

Most people were infected by this virus when they visited territories of increased risk and were stung by infected ticks. Nevertheless now more and more infections arise through the milk products of cattle.

In Lithuania during the latter years the number of people falling ill with Russian encephalite constantly is increasing. In 1994, 284 people were ill with this disease, and in 1995 - about 500 people. There is widespread opinion about being left infected or increased risk territories each year. But the Lithuanian situation is one of the very fast spreading of forest areas infected with Russian encephalite ticks. Now about 30% of all Lithuanian forests are so infected. The greatest density of infected ticks is in the forests of the central part of Lithuania. The means to stop the spreading of this virus still do not exist. Vaccination of people against the Russian encephalite is complicated and expensive, but treatment of infected people is ten times more expensive and not always successful. Therefore the international serious attention and investment is necessary to investigate this also very important problem and to find ways to localize and overcome the fast spreading of ticks infected with Russian encephalite.

1.2. THE MAIN NEEDS OF TIMBER INDUSTRY

The timber industry is one of the priority matters of all Lithuanian economy branches. During the last five years many very important changes have taken place in this industry. The domestic market became quite weak because of a great (some times) decline of building activity. Most of production units, seeking foreign markets, have changed not only owners but have started restructuring actions, reduced capacity, and sometimes changed the profile of production. At the same time many very small private production units have been established.

The greatest decrease of production is taking place in the large factories which converted timber into products for the building industry: doors, windows, furniture, plywood, laminated veneer lumber, paper and cardboard. There is less production of cellulose, veneers, wooden single family houses, and fodder's yeast.

Unfortunately the sawmill industry suffers too. The level of manufacture of sawn timber has not changed, but the largest production units have lost the market and customers. Their place was occupied by new small units and therefore without good prospects. Thirteen of the largest sawmills reduced their production during the five year period more than twice.

Now enterprises use about 5,000,000 cubic meters of wood: about 4,000,000 cubic meters of it are manufactured by industry, and 1,000,000 cubic meters of it are used as biofuel. Hopefully during the next five year period production of sawmills will increase up to 1.3 million cubic meters of sawn timber, up to 25-30 thousand cubic meters of plywood, up to 150 thousand cubic meters of particle boards, and up to 20 million square meters of wood wallboards each year. The previous level of production has to be restored. It is expected to have an increase of sawn timber but not of woodwork (timber structures, windows, doors, furniture, etc.). Incidently, sawn timber and plywood have now and will have at the nearest future the largest and fully warranted demand in the local as in the foreign market. However, for the State there exists a great difference between the export of round wood and products of preliminary manufacture (sawn timber, plywood) or woodwork.

The export and local use of woodwork means creating new jobs and more incomes. The most important development would be the widening of utilization of woodwork in construction of buildings or their most important parts instead of steel or concrete elements and structures. Then a possibility to minimize the import of large amount of steel products and structures, which are not local resourses, for construction of buildings would become more real. There is also the possibility to limit the manufacture of a large quantity of cement, leading to great energy savings. Timber structures are coming into construction sites more widely because a growth of local building market. Being an excellent renewable resource, are a very important part of the means of protecting our limited global resources.

The activity of timber industry in Lithuania may be stimulated by foreign investments, and by more serious building projects. Presently some joint venture companies exist, but their contribution to modernization of timber production is still paltry. The main attention still usually is paid on extension of export of round wood but not to expansion of local production capacities to manufacture the high quality and value-added wood products for export and local market.

2. Conversion of Wood

2.1. PRINCIPLE TRENDS OF CONVERSION

In Lithuania, for better conversion of the wood chain from silviculture and wood supply to high quality products and wholesale operations, it is necessary to create one or more major forest industry organizations that would be able to market its products world-wide. The practice of forestry has to be based upon the sustainable utilization of renewable natural resources and assuring its subsidiaries of a permanent supply of round wood. Subsidiaries may be specialized for different types of wood conversion into products of the sawmill industry, plywood and building products, pulp and linerboard, printing paper, board, packaging and soft tissue and chemical products, etc.

For successful and profitable operation of corporations, considering the difficult market situation, strategic investments have to be done namely:

* to improve customer service;

* to a restructuring program aimed at adapting corporate units operations to the prevailing weak market situation;

* to supply according to restructuring program the plants with most advanced technology, improving the characteristic and quality of products significantly also by carrying out projects for developing quality assurance systems;

* to aforestation and permanent care causing sustainable renewal of natural forest resources.

2.2. UTILIZATION OF ROUND WOOD

Lithuanian forests still have possibilities to assure sustainable utilization of raw material because of renewable natural resources. But volumes of preparation of sawn logs today are such that the situation requires a very careful sustainable aforestation and planned management of all forests. In the Lithuanian forests a constant renewal of forest land is occuring. In State forests young trees were planted on an area of 7,000 hectares. For silviculture the 5,500 hectares of forest land were prepared. In forest land about 21,500,000 plants of trees were planted. There were 35.6 tons of trees' and shrubs' seed produced. Problems arise with re-forestation of private forest lands. The private sector, as in other countries, is presently interested more in consumption than in long term renewable resource policy. In some areas, illegal deforestation becomes a problem, too.

A problem exists for conversion of wood waste into value-added products and simultaneously solving environmental tasks. Wood waste - branches, tops, stubs, barks and dead-wood - often are left in Lithuanian timber cutting areas to decay.

That often becomes a cause of concentration of forest pests and mould diseases and impedes the process of sustainable natural renewable resources and unfavourable losses instead of utilization into the value-added products. Even bark, as well as sawdust, is used as a biofuel for energy production. But for the Lithuanian situation, again, it is necessary to have special investment to reopen the pulp industry and develop the production units for manufacturing from all possible wood waste into value-added products including various building products and to gain a large economic benefit from the effort.

2.3. ORDINARY BUILDING PRODUCTS

Different load bearing building structures, parts and elements are usually manufactured from different sawn timber products, laminated veneer lumber, plywood, and particle boards. Part of it is coated, wood woolboards, cement-wood fibre boards, cardboard, paper, eco-wool, etc. One direction of manufacture of building products in which experience exists is production of windows, doors, furniture components for kitchen furnishings, furniture, office furnishings, etc. The low building activity in Lithuania and sudden decline of purchasing power of significant part of the population during the last five year period has created a real hardship for the Lithuanian door, window and furniture industry. Also, the ongoing restructuring is likely to reduce capacity. But a positive side of such restructuring is modernization of this industry and its products.

The Lithuanian practice of wooden (as well as brick-stone) construction has lasted through the centuries from the Middle Ages and earlier and during this long period considerable experience was gained. Modern wood construction in Lithuania has periods of rises and falls. About two decades ago, capital investment was made into the construction of two modern factories for manufacturing load bearing structures. One, in Jure, used laminated veneer lumber and another, in Alytus, used wood panels and components for single flat houses. At the same time, some large sawmills supplied with modern equipment have been constructed, too. Applications of veneer lumber (well suited to use in load-bearing structural solutions calling for durability, light weight, resistance to warp and dimensional accuracy) included public buildings, large prefabricated buildings, components of manufactured products, single-family houses for farmers, and renovations.

General trends of above mentioned applications were towards increasingly spacious premises requiring long structural spans: beams, girders, trusses, arches, frames, domes, space grid structures have been usually between those structural systems used for such long structural spans. Straight and curved structural members were manufactured. In some buildings very efficient load bearing timber structures and claddings have been used in mixed structural systems, as rule with lightweight steel structures. Timber members sometimes have been used in composite load bearing structural systems, again usually in combination with steel and composite elements, or sometimes with concrete structures, too. Timber beams and panels have been used in roofing, flooring, wall sheathing and renovation.

At that same time, research in various fields of timber structures was fulfilled, including involvement of timber industry and construction [2-9]. The valuable collection of data from theoretical and experimental investigations has been useful for further development of the most progressive and efficient timber structures and structural systems. On the basis of such research, attempts have been made to develop non-traditional and more efficient structural solutions using timber structures.

The present situation with development of manufacture of building products is again tightly connected with very low building activity in Lithuania. The overhaul of the entire building products industry may be foreseen only after re-structuring. Modernization of it requires solid investments which now are impossible to collect from very poor Lithuanian resources. On the other hand, growing social demands for low cost housing, and restructuring of national infrastructure are the threat for the realization of rational forest resource utilization policy. The virgin area concerned with the application of modern timber structures in building industry in Lithuania also is still to be developed. For better forest resource utilization the relevant education activities, informative courses and technology transfer will have a tremendous importance in successful realization of programs of protecting our limited global resources.

3. Trends to Develop New Value-Added Building Products

3.1. INCREASE OF LOAD BEARING CAPACITY AND FIRE RESISTANCE OF TIMBER STRUCTURES

Increase of load bearing capacity and/or fire resistance of timber structures would lead to saving of wood products and ultimately to saving of forests as a source of sustainable renewable resources and raw materials. There are some examples of such possibilities arising after innovations of load bearing members and technique of their manufacture. Reference [10] deals with the increase of load bearing capacity and fire resistance in the connections of glulam structures. Laboratory research results are presented for 28 timber beams, made of wooden logs of square cross-section, glued plank packages and multilayered beams, and produced by biaxial pressurization method, tested until rupture under the action of transverse short and long term loading. The research revealed the strength and crack propagation of laminated veneer beam being dependent on the preliminary pressurization quantity σ_y. At the value of σ_y equal from 1.5 to 2.25 MPa the beam strength and crack resistance increased by an average of 10%.

The research resulted in those two methods of the axial pressurization being given a patent (SU 1716035A1) [11] useful to strengthen connections of such important structural members as large span beams, frames, arches and trusses. This allows not only the improvement of the resistance of such connections under lateral forces but also reduces or completely eliminates the effect of tensile lateral stresses on their bearing capacity. The essence of the method given in the patent

SU 1719586A1 [12], is such that it enables to set the steel parts of the joint of complicated configuration into a wooden mass of member so that the holes of corresponding shape and size are made the same in each separate plate. After the holes and setting of steel parts of any size and shape into a wooden mass are performed, the elements have to be glued. That makes the process of joint setting easier and results in increased fire resistance.

Similar problems are being solved with stretched wood joints glued in steel reinforcing rods [13]. The conducted tests prove the strength of above mentioned joints is ten times higher and deformations about twenty fire times less than corresponding parameters of regular pin joints.

The results of longevity study [14] carried out with hard wood fibre boards internally sized with phenol-formaldehyde resins are interesting. It has been established that flexural strength is decreased after 30 year natural aging in the atmospheric conditions at an average: after pressing of the boards - by 23.7%, after it heat treatment - by 37.1%. The water absorption of the boards after have decreased by 16.9% after its pressing and this index has increased by 4.5%- after heat treatment of the boards.

3.2. ON DEVELOPMENT OF COMPOSITE MEMBERS

3.2.1. *Determination*

Composite structures are those having elements made of different structural materials usually with different mechanical properties, especially with different values of Poisson's ratio. Because of that, the total strength of members or whole structure becomes significantly higher. Sometimes, under load, an interaction occurs between the separate parts of composite member made of different materials which leads again to the significant strengthening of the whole member. One example of such a type of composite members is so called "concrete filled tubes" [15,16], also composite steel-concrete beams, consisting of steel beams with concrete slabs on jointed together through special anchors, so called "shear connectors", which have to resist longitudinal shear.

Many different types of composite structures combining different materials exist. Those involving timber have to be classified so:

* mixed timber-steel structural systems in which the compression members are made of timber and tension one of steel i.g. timber-steel trusses, strutted beams, etc. Many such structures have been in application for decades, other new such structures are in development now, including these consisting of timber;

* reinforced timber structures in which reinforcement steel rods, strips, perforated or tootbed fishplates, and glass or carbon strips, fabrics or wires can be used;

* multilayer timber structures made from plywood, laminated veneer lumber, wood fibre boards, etc.;

* timber composite members, consisting of external plywood shell and internal solid or hollow value-added wooden core;

* composite metal-timber members, consisting of external metal (steel, aluminium or super lightweight alloys) shell and internal solid or hollow value-added wooden core. The external shell may be done only from metal or it may-be multilayer including metal as well as timber (say plywood, woodfibre boards or so on).

3.2.2. *Mixed Timber-Steel Structural Systems*
This type of composite structures can be considered as "traditional" because it has been in application for a long time. New structures now under development differ more not by structural system but with materials used: instead of solid timber elements more often glulam is now used, instead of steel members - lightweight or super lightweight materials are used.

The typical structural schemes of mixed timber-steel structures may be demonstrated by such examples given in Figure 1.

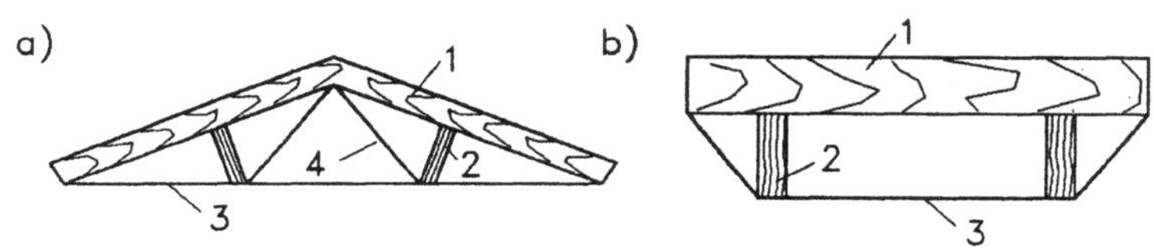

Figure 1. Triangle mixed timber-steel roof truss (a) and strutted beam (b) 1 - upper compression timber chord (a) or timber beam (b); 2 - timber compression posts; 3 - steel tension chord (a) or steel tension stretching wire (b); 4 - steel tension diagonal.

As one of the new mixed timber-steel structures may be mentioned low floor or roof trusses made of upper and lower saw timber chords connected from both sides by toothbed thin-plated steel diagonals. All solutions, traditional and new, allow saving of building materials. For greater savings the super lightweight metals, instead of steel, in mixed timber-metal structures can be considered. In Lithuania for such purposes it is realistic to consider production of super lightweight metals, based on magnesium because a great quality raw material - dolomite - exist in great quantity. That may stimulate more rational and efficient application products of timber industry, too.

3.2.3. *Reinforced Timber Structures*
Some aspects of development of timber structures reinforced with steel rods were mentioned above. Great possibilities for more efficient and rational development of timber structures can come from other reinforcing means - toothbed steel strips or plates, glass carbon fibre strips, fabrics located in glue layers of glulam structures. Techniques of development of mixed timber structures are enabling to use wood and other structural products and materials in more efficient way.

3.2.4. *Timber Composite Members*
Timber composite members are usually made of glulam where weaker wood can be located in less stressed parts of the member. Correction for the influence of diffusion property of timber is needed in the manufacture of such products as plywood and efficient use of it in structural members. Many timber composite products such as plywood, wood wool boards, cement-wood fibre boards, hard wood fibre boards are used also for flooring, roofing, claddings, and insulation elements.

For load bearing structures it is very rational and efficient to have to be timber composite members, formed from external multilayer (timber or mixed) shells and internal cores from weaker material.

For such cores the timber waste can be very successfully utilised as value-added product. That can happen because under load action different parts of composite member get a different properties of deformability and an interaction occurs. The composite member is able to resist larger forces. Some examples of possible timber composite members are shown on Figure 2.

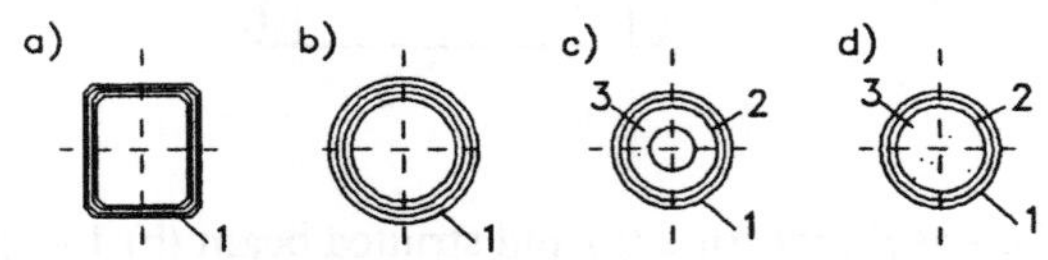

Figure 2. Some possible examples of timber composite members.
1 - plywood external shell; 2 - internal metal (steel, lightweight metal or super lightweight metal) shell; 3 - hollow core of value-added wooden product; 4 - solid core of value-added wooden product.

In the case of a multilayer composite shell, involving internal metal tube, one can consider additional layers of the external wood (plywood) shell - yielding fire protection of the metal tube and core. External composite shells can also be used for strengthening of timber structures (columns, piles, etc.) worn and torn during the long maintenance the buildings [16].

3.2.5. *Stress and Limit States*

Stress and limit states of timber composite members under uniaxial actions have to be very similar to that of concrete filled steel tubular members. Behavior of concrete filled steel tubular members has been discussed [16].

Theoretical investigations of the strength, stress-strain behaviour and limit states of composite members usually are based on behaviour of short uniaxially compressed members with solid cores encased with circular external shells. Stability investigations can be based on strength results of short elements. For description of stress-strain behaviour the presumptions of the elasticity or plasticity theories usually are used. Application of the plasticity theory as a rule leads to the analysis of solid cylinders loaded separately with coaxial force and internal pressure. The yield relationships in such cases are ordinary expressed by one of two possible equations:

$$f_r^2 + f_r \cdot f_t + f_t^2 = k^2, \qquad (1)$$

or

$$0.5 \cdot (f_1 + f_2) = k, \qquad (2)$$

where $k = f_y / \sqrt{3}$ for the relationship (1), which expresses the beginning of plastification due to the Huber-von Mises-Henckey criterion, or $k = 0.5\, f_y$ for the relationship (2) which expresses the same due to the St.Venant - Tresca criterion; f_r and f_t = radial and transversal stresses respectively; f_1 and f_2 minimum and maximum stresses respectively; f_y – yield stress of material. Some of the plasticity theory fundamental relationships have been used to solve problems of strength of different composite members.

Taking into account the law about equality of the axial strains of both media - internal core and circular external shell - on contact circle

$$\varepsilon_{zb} = \varepsilon_{zs} \qquad (3)$$

the radial stress of interaction between the above mentioned both media occur

$$f_r = \frac{f_b \cdot E_s - f_y \cdot E_b}{E_s - 0.5 \cdot E_b}, \qquad (4)$$

where f_b and f_y - the characteristic strength of value-added product of internal core and the yield strength of shell (steel, lightweight metal or multilayer plywood) respectively; E_b and E_s - the moduls of elasticity of each material respectively.

This stress f_r gives an opportunity to estimate the increased stiffness of composite member and, as a result, the degree of increase in load bearing capacity of compressed short composite members, what is expressed by ultimate longitudinal stresses f_{zb} · and · f_{zs}.

Then the ultimate load bearing capacity of the composite member N_u, may be expressed by equation:

$$N_u = f_{zb} \cdot A_b + f_{zs} \cdot A_s, \qquad (5)$$

where A_b & A_s – the values of cross-sectional areas of each medium.

Similar ways exist to develop an efficient composite timber members also under different loading conditions: columns, beam-columns, beams, etc.

4. Summary

The situation in Lithuanian forests and timber industry is not satisfactory and requires permanent care and very large investments. The main products of timber industry still are the rough sawn timber, plywood, mainly assigned for export although the experience of production of other timber products is quite great, but for efficient it realization investment also are necessary. The background and experience to widen production of ordinary building products and new innovations based on trends to develop new value-added building products exists in Lithuania. But for realisation of that the investments it is necessary to stimulate building activity which now is very weak.

5. Pertinent Observations from the ARW

During the ARW it was more strongly evident that the existing situation in a great number of European countries including the Cooperation Partner countries can be expressed as insufficient understanding of the scale and impacts of air pollution on forests growth and performance. Excursions into the degradated areas of forest ecosystems in the Polish and the Czech parts of the Sudety Mountains left a very sad impression. It is a very clear proof of the necessity to encourage cleaner technologies and to not allow the forest damage to continue.

A very important demonstration of optimized biological production combined with modern technology for private forestry was carried out during the field trip through the Thurn and Taxis forests in Regensburg. Such technology is one of very important means to modernize the forestry in Eastern European countries.

Of great importance was identification of many of salient issues relating to the saving and better utilization of the forests of Central and Eastern Europe carried out through a professional dialogues between key interdisciplinary specialists.

6. References

1. Briediene R. (1996) Timber industry needs owner and investments. Diena, 106 (532). 12.

2. Cechavicius R. (1983). Calculation and reinforcement of curved glued timber beams subjected to transverse tension stresses in G.Marciukaitis et al. (eds.), *Transactions of Lithuanian Universities "Constructional Structures,"* Mokslas Publishers, Vilnius, 50- 57.

3. Cechavicius R.,Valiunas B. (1983). Calculation peculiarities of ridge joints in sagittate three-hinged glued timber arches in G.Marciukaitis et al. (eds.), *Transactions of Lithuanian Universities "Constructional Structures,"* Mokslas Publishers, Vilnius, 58-62.

4. Cechavicius R. (1989) Manufacturing method of glulam bearing elements. *Patent bulletin*, 8 (RP 135 SU 1716035 Al E04 C3/12). 8-11.

5. Cechavicius R. (1989) Manufacturing method of glulam bearing elements. *Patent bulletin*, 10 (RP 155 SU 1719586 Al E04 C3/12). 15-18.

6. Cechavicius R. (1993) Methods to increase for glulam bearing elements knots bearing capacity and fire resistance in R.Cechavicius et al. (eds.) Proceedings of International Conference Modern Building Materials, Structures and Techniques 3, Technika Publishers, Vilnius. 108-113.

7. Kvedaras A., Sapalas A., Sapalas K., Valiunas B. 1990) The possibilities of repair and bearing capacity increasing of structural elements from composite members in Z.Kamaitis et al. (eds.), *Transactions of Vilnius Technical University "Structural Repair and Strengthening,"* 2, Technika Publishers, Vilnius. 100-108.

8. Kvedaras A., Sapalas A., Valiunas B. (1995). The outlook and survey of research and applications of spun composite structures in G.Marciukaitis et al.(eds.), *Transactions of Vilnius Technical University "Constructional Structures,"* 20, Technika Publishers, Vilnius, 19-30.

9. Rolicius J. (1996). Stretched wood joints with glued-in rods in R.Zilinskas et al. (eds.), Proceedings of Conference Civil Engineering and Architecture, Technologija Publishers, Kaunas. 167-169.

10. Valentinavicius A.Valiunas B. (1983) Rational design of the three-hinged glued timber portal frames in G.Marciukaitis et al. (eds.). *Transactions of Lithuanian Universities "Constructional Structures,"* Mokslas Publishers, Vilnius, 7-10.

11. Valentinavicius A. (1983). On the horizontal displacements of double tapered-curved timber beams in G.Marciukaitis et al. (eds.), *Transactions of Lithuanian Universities "Constructional Structures,"* Mokslas Publishers, Vilnius, 33-38.

12. Valentinavicius A. Valiunas B. (1984) Influence of cross-section weakening of bending timber rods on their strength in G.Marciukaitis et al. (eds.), *Transactions of Lithuanian Universities "Constructional Structures,"* Mokslas Publishers, Vilnius, 34-38.

13. Valentinavicius A. Valiunas B. (1988). Some problems of designing joints of braced frames in G. Marciukaitis et al. (eds.), *Transactions of Lithuanian Universities "Constructional Structures,"* Mokslas Publishers, Vilnius, 136- 143.

14. Valentinavicius A. Valiunas B. (1989). The relationship between the load capacity of solid-sawn timber beams and their fire-resistance in G.Marciukaitis et al. (eds.), *Transactions of Lithuanian Universities "Constructional Structures,"* 16, Mokslas Publishers, Vilnius, 86-91.

15. Valentinavicius A. (1990). Determination of fire resistance of bending glulam elements with account of its dimensions and actions on its. *Vilnius Civil Engineering Institute Repor*, Nr. 16-28 t., 355 p.

16. Vilkas R. (1996). Longevity study of wood fibre loads internal sizing with phenol-formaldehyde resins in R.Zilinskas et al. (eds.), Proceedings of Conference Civil Engineering and Architecture, Technologija Publishers, Kaunas. 211-215.

RESEARCH AND DEVELOPMENT GOALS FOR WOOD TECHNOLOGY IN AUSTRIA

H. RESCH
Austrian Forest Products Laboratory and
University BOKU
Gregor-Mendel-Strasse 33
A 1180 Vienna
Austria

1. Resources

Austrians have been cultivating and using their forests for centuries on the basis of professional knowledge, tradition and love for nature. The idea of sustainable use was born centuries ago and cast in forest law. The latest version of the Austrian forest law dates back to 1975 with an amendment in 1987. It aims at protecting the forest for its 4 functions of multiple use, i.e. commercial production, protection for human settlements, protection of the environment and use for recreation.

The forested area has been steadily increasing for decades. Today, about 3.9 million ha, equal to about 46.3% of the total area of Austria, are under forest cover. Austrian forests are primarily in private hands as public forests comprise only 20.4% of the total. A large portion, about 3.33 million ha, are maintained as commercial forest area.

The growing stock amounts to about 972 million m^3 with an annual increment of 31.4 million m^3. Thus the annual cut of 19.8 million m^3 falls far short of annual growth. the conversion of timber by sawmills amounts to about 11.1 million m^3 and by the pulp and paper industry to about 3.3 million m^3. Residues are converted to chipboards and fibreboards. The semi-manufactured products are converted to structural products as well as windows, doors, flooring, skiers, furniture, etc. About 2/3 of the lumber, pulp and paper, composition boards, and finished products are exported mainly to countries of the European Union.

Thus, Austria finds its place in the center of Europe with strong commercial linkages not only to the surrounding countries, but also, as far as its forest products are concerned, in a world market.

2. Incentives

Research and development of forest products is extremely important for the future of Austria, however, need to be considered within different industrial groupings (here exclusive of pulp and paper). Because of its material specific and ecological

R. M. Gutkowski and T. Winnicki (eds.), Restoration of Forests, 315–318.
© 1997 *Kluwer Academic Publishers.*

advantages, wood must be subject of further technological development to position the forest as a major natural resource for the 21st century.

Incentives for development of new and improved technologies spring mainly from the desire to save raw material and energy, reduce production costs, increase productivity and product quality as well as compete with other materials, for instance in the construction and residential sectors. Products like steel and concrete do not only use high amounts of energy, whose generation causes pollution, but also emit pollutants of their own from their processing plants. The energy consumption for the production of bricks, cement, plastics and aluminum amounts to about 500, 1,000, 15, 000, 72,000 kWh/t. In contrast, wood products are manufactured of a material produced by nature with relative low energy input. Structural timber, for instance requires about 100 to 150 KWh/t. However, not all projects aim at economic improvement, but increasingly at environmental protection and occupational safety. There is a need to increase the magnitude of uses for wood and to position it in contrast to other material on the basis of its aesthetic and ecological values and advantageous energy contents. The energy required to construct living space within a wood frame building amounts to about 1,270 kWh/m^2 and compares well with that in a brick building which requires 52% more energy or 1,921 KWh/m^2 (Weller, K. and Rehberg, S. 1989). Similar, the energy consumption for production, use, and removal of a storage facility with a floor area of 1,000 m^2 and an enclosed space of 6,000 m^3 constructed with wood, steel, and pre-stressed concrete amounts to 1,480, 1,827, and 2,223 kWh/m^2 respectively (Baier, B. 1982). Wood follows a natural cycle from its growth in the forest, to harvest and production of goods, to recycling of the products and, in a final step, to utilization for energy. While wood is in use, approximately 255 kg of carbon, equivalent to 935 kg of CO_2, are sequestered in one cubic meter (assuming a specific gravity of 0.5). When wood is finally converted to energy, that cubic meter contains about 6,000 MJ of energy.

3. Non-destructive Testing (NDT)

The assessment of varying wood quality characteristics by means of NDT methods and recognition systems will make it possible to segregate materials and thus allow for their better utilization (Proceedings, 1990). Possibilities already exist for round timbers of smaller dimensions when using penetrating radiation or sound transmission. However, greater opportunities for quality improvement and assurance may be found in almost all processing steps for smaller pieces from lumber all the way down to flooring strips. Here, NDT assesses the physical characteristics and properties in accordance with important criteria of utilization and market demand. This certainly includes improved computer vision for the evaluation of surfaces.

4. Material Savings and Quality Improvement

A large potential of savings on raw material still exists in the manufacture of lumber. Much of the material is converted to sawdust because of unnecessarily thick saw kerfs.

Especially for circular saws, the theory already exists for tensioning the saw blades with the help of vibration analysis. Saw blades thinner than used today, could be tensioned to render a more stable cut and thinner saw kerfs without much reduced speed of sawing (Szymani, 1993).

Not all wood products should be dried using the same technology of compartment kilns with air as a means of transferring heat to the wood and, at the same time, removing its moisture. Special drying techniques should be developed for special applications and products that would allow for a reduction in the amount of often ignored drying defects and/or in actual drying time. For certain products, an optimum should be obtained through a combination of different drying technologies. For instance, some research is now focused on vacuum drying with various means of heat transfer, among them heating with microwaves and high frequency current.

5. Combining Smaller Pieces to Improved Products

One of most difficult tasks is the development of better and more cost-effective adhesives for the production of panels and of structural members of various forms. some of the glues that have been used for decades are now thought of as not being "environmentally friendly" enough, mainly because of the use of formaldehyde, and substitutes should be found. Another goal is the conversion of batch processes to continuous manufacturing such as the continuous pressing of wet formed high-density fibreboard. Also the production of glued-laminated products and similar structural elements should be automated using more computer decision making, if not specialized roboters. Laminated products can also be improved through better assembly using either pre-sorted materials for best placement in tension or compression zones, but also through the use of other, non-wood materials to be strategically located such as in zones of high tension stress.

6. Value Added Products

Research and development are especially important for the production of finished, high value products. Windows, doors, flooring and pre-manufactured buildings, all made of wood, compete head on with similar products made of alternative materials. Especially windows, where the natural beauty of the wood should remain visible, are desired to have minimal maintenance requirements. The goal of research should be two fold on the one hand, radically different window frame designs should be developed to meet higher insulation and ecological standards, while on the other hand, an improvement of the existing product lines should aim at market demands. Another aspect is the development of burglar resistant doors for which the demand increases especially in urban centers with increasing crime rates. In the manufacture of furniture, it will become necessary to correlate innovative design with requirements for increased strength and stability conforming to the new European standards for testing.

7. Environmental Requirements

All the above mentioned products from boards and panels to furniture, windows and doors, are in need of surface coatings to be applied with a minimum of organic solvents (Republik Osterreich, 1995). A newly released government regulation pushes paint manufacturers toward formulating either high solid or mainly water based coatings.

In this connection, it is important to point toward the need for new, ecologically friendly wood preservatives, especially those that may not pose an environmental hazard when, as a final step of wood utilization, the wood products are burned and converted to energy. With all these manufacturing steps, questions about air and water pollution must be answered. To assess the many parameters important to environmental safety, new or improved chemical, but especially biological test methods must be developed.

8. Timber Construction

One of the basic human needs is that for shelter. Coupled with the desire to own property, there is a strong need for affordable homes. Wood construction is well suited for many countries with good supply of wood as a natural and advantageous building material. In addition to single story homes, there are opportunities to construct multi-story buildings for living and for offices.

Structural design for pre-fabricated as well as on-site constructed dwellings and industrial buildings must assure safety. The EURO-code encompasses probability based design. Fire safety is to be enhanced not only for single, but multi-story buildings. However, meeting the multitude of building codes across the European Union, seems to be a difficult, illusive goal. Nonetheless, the technical advances in material science and innovative approaches to manufacturing promise a renaissance in the use of wood and wood products.

9. References

1. Baier, B. (1982) Energetische Bewertung luftgetragener Membranhallen im Vergleich mit Holz-, Stahl- und Stahlbetonhalen, Verlagsgesellschaft R. Muller.

2. Proceedings of many authors (1990) Process control/production management of wood products: technology for the 90's, University of Georgia, Athens, USA.

3. Republik Osterreich. (1995) Verordnung des Bundesministers für Umwelt über Verbote und Beschränkungen von organischen Lösungsmitteln, Bundesgesetzblatt Nr. 872, December 29, 1995, pp. 9083-9121.

4. Syzmani, R. (1993) Sawing technology update, Proceedings of the International Conference of Woodworking Technologies, LIGNA '93 Hannover, Germay, pp. 12.1-12.11.

5. Weller, K. und Rehberg, S. (1989) Lösungsansatze für den energie-und rohstoff-sparenden industrialisierenden Wohnbau, Technische Unviersitat Berlin, Fachbereich Architektur.

BUILDING WITH TIMBER, A POSSIBLE CONTRIBUTION TO SOLVE EASTERN EUROPE'S ECONOMICAL AND ECOLOGICAL PROBLEMS?

W. WINTER
Professor at the Technical University of Vienna,
Karlsplatz13/254, A-1040 Vienna,
Austria

Abstract

Looking back to the finishing 20th century we realize that Europe changed considerably the basic orientations of the societies, of politics and economy and of the relationship between man and nature. Some keywords of these changes are: Colonialism, Nationalism, Faschism, Socialism, Communism, Democracy; Industrialisation, Post Industrial Period; Communication, Globalisation; Environment, Pollution, Ecology.

Building has always played a key role in a community's economic and social life, and will continue to do so even in modern, highly developed industrialized societys. In the countries of Western Europe 10 to 20 percent of gross domestic product is directly or indirectly created by the building industry. During the 20th century the use of timber for construction in Western Europe went down to 5 %, in Eastern Europe even to less. A completly different development can be observed in North America and Scandinavia were timber could hold a 50 to 90 % market share. As a consequence, the question of "How do we build and which materials do we use?" is a vital issue.

The paper discusses the main reasons for these different developments and works out some arguments for a wider use of timber in Europes construction industry.

1. Building Materials in the Course of Time

The past centuries were characterized by a continual change in the use of construction techniques and building materials, depending on which raw materials, energy sources and labour forces were available at a given time, and on economic and social structures as well as prevalent systems of cultural and spiritual values. Since the time of the French Revolution and the social upheaval it brought to Europe and the world a radical change in dominant building techniques and materials has been observed. Essentially, every century has embraced its own building material.

Up to the 19th century wood and natural stone were the predominant building materials. The universal material wood, in particular, had a large variety of uses, from dwellings to furniture, tools, ships and bridges. The associations of carpenters and

R. M. Gutkowski and T. Winnicki (eds.), Restoration of Forests, 319–330.
© 1997 *Kluwer Academic Publishers.*

stone masons with their closely organized structure and elite thinking, monopolized the skills and know-how of their craft, thus acquiring a high social status.

The 19th century, i.e., the period of industrialization, created new social structures; medieval guilds and organizations were not able to supply the labour potential required by industry. New schools were opened to the public, providing general education and access to knowledge for everybody.

The artisan's timber was replaced by steel, the most prominent product of the newly established industries. This new material was manufactured by the capitalist big industry in unlimited quantities and processed by unskilled labour forces working for wages much lower than those demanded by the carpenters' guilds.

The necessary practical know-how was gathered by a new profession, the construction engineers, and spread by the newly created Technical Colleges. Such democratization of knowledge was necessary to permit large-scale selling of the new material. Railway construction finally acted as a motor for the development of building materials made of steel. New technologies were tested to permit the construction of railway bridges and stations. The big world exhibitions gave rise to spectacular prestige structures, such as the Crystal Palace and the Eiffel Tower, offering opportunities for identification to the new bourgeoisie.

The 20th century, finally, has had little use for the "soft and combustible" natural material of timber, and even steel has lost its undisputed position in view of the possibilities of reinforced concrete.

This new building material resulted from combining cement, the traditional man-made stone material, with steel as reinforcement. Invented by the French gardener Monier and long fought by cement manufacturers who did not believe in the durability of steel in concrete, reinforced concrete experienced an unprecedented triumph and advance throughout the Western world.

After World War II the giant task of reconstruction was undertaken with the use of reinforced concrete, especially in the European countries which were directly afflicted by the trauma and sufferings of war. This young building material, which had acquired an image of "eternity" after being on the market for a mere thirty years (see, for instance, the trade name "Eternit"), reduced the market share of steel construction to some 15-20 percent. The understandable desire of the people in Western Europe for protection and safety, and the production structure of the concrete industry developed in a wartime setting, led to the 70-80 percent share of reinforced concrete in the construction industry.

In the industrialized nations of North America and Scandinavia, which were less affected by the World Wars, structural steelwork retained much of its former importance in industrial and commercial buildings. In the private home sector timbere structures increased their impact on the market to 60-90 percent in these countries, whereas their share dropped to 5-10 percent in Western Europe, giving way to "single-family fortresses" with stone walls and floors and ceilings of reinforced concrete.

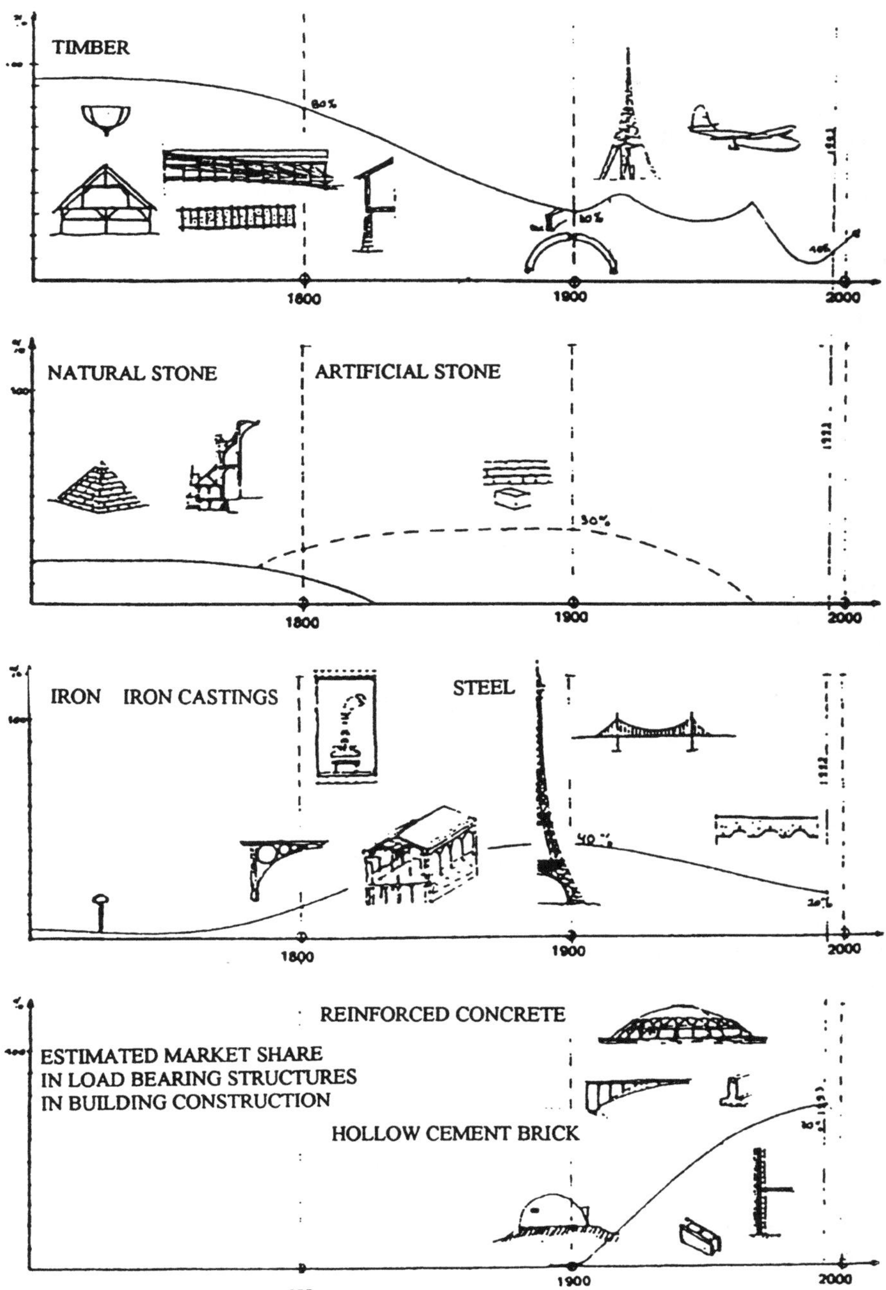

Development and importance of building materials in the course of time

2. Speculations on the Material of the 21st Century

2.1. REINFORCED CONCRETE IN A DOMINATING POSITION

Can there be any doubt that reinforced concrete will continue to be the leading building material in Western Europe?

Over the past decades the cement industry has continuously raised its production; large monopolistic industrial groups have developed. The Swiss "Holderbank" controls some 50 percent of the cement production of the entire western world, including the U.S.A. In the East European countries the monopoly of the states' factories producing cement and prefabricated concrete elements is even more dominating the whole market. Combined with the - usually state-owned and subsidized - steel industry, which supplies reinforcing steel of a value amounting to approximately fifty percent of all basic materials required, the cement industry has production capacities at its disposal that were developed for the high demand of the seventies and eighties.

The production volumes of operating steel mills and cement kilns cannot simply be reduced, as the technology calls for a 24 hour manufacturing cycle (blast furnaces cannot be switched off when there is a lack of demand), and the money invested constantly demands its return. If the share held in the market is threatened, the industry's market behavior is likely to become more aggressive, the profits made during the fat years of 1960-80 permitting it to defend its monopoly in the construction market.

Today's marketing strategies show the first signs of such developments (for example, the slogan, "concrete - living material"). Regulatory bodies such as the standards committee for fire regulations, are most attentively courted by the concrete industry. As a consequence, other building materials will find it far from easy to conquer a place in the market that is anywhere near the one occupied by the domineering reinforced concrete.

Architects and engineers whose professional training reflected an 80 percent market share of reinforced concrete, will hardly be in a position to propose materials and construction techniques other than those for which they were trained, in particular since the technical guidance and service offered by the reinforced concrete industry is unmatched. Industry-dependent research and development institutions whose main customer over the past fifty years has been the reinforced concrete industry, will of course concentrate on solving the current technical problems of this material, such as steel corrosion in concrete, carbonization and a somewhat ambiguous long-term behavior.

In view of its mature state of development and position of power, reinforced concrete is not likely to encounter any serious rival in the 21st century. The decisive question rather will be whether it can adjust to the global changes to be expected in a more general setting. Is it conceivable that reinforced concrete will experience the fate of the dinosaurs?

2.2. THE IMPORTANCE OF ENVIRONMENTAL ISSUES

The 21st century will become the century of the environment, though not in the sense of a romantic call "back to nature," but as an existential imperative for human life. Scientists are mainly concerned about two developments:

Depletion of fossil energy resources: Oil will be exhausted within the next 50 to 100 years, coal within 100 to 200 years. The developed industrial societies, which to this day meet 70 to 80 percent of their energy demand by consuming fossil energy, will have to change their habits completely in order to escape the fate of the dinosaurs.

Global warming as a consequence of the greenhouse effect: The developed industrial societies produce incredible amounts of gases which accumulate in the atmosphere and thus influence the temperature balance between the earth's surface and space. In the greenhouse, Earth's temperature is expected to rise by 2 to 4°C within the next 100 years. In view of the fact that the average temperature of the planet has never changed by more than one degree, even during the Ice Age, we can imagine (or possibly cannot) the radical changes in vegetation and sea level which may be caused by such a massive increase in temperature.

Carbon dioxide accounts for some 50 percent of all greenhouse gases, and is released during the combustion of fossil energy carriers. Policymakers concerned with the medium-term future try to react accordingly and to convince society, or force it,

- to consume fewer non-renewable fossil raw materials, to reduce energy consumption in general, and to utilize renewable energy sources to a greater extent;

- to drastically reduce emissions of greenhouse gases, especially of carbon dioxide.

How do the major building materials, concrete, steel, and timber, compare in view of the absolute necessity to reduce the consumption of fossil fuels and energy in general, and as regards the emission of greenhouse gases?

3. How to Compare Building Materials

3.1. "CLASSICAL" ENERGY BALANCES

In the literature of the eighties we find numerous attempts to estimate and compare energy consumption for manufacturing, processing and recycling (so-called grey energy) of building materials. One method of calculation is to add up consumed energy in kilowatt-hours, and to compare the added-up amounts per unit volume (cubic meter) or weight (kilogram) of the building material manufactured. A typical energy balance for the manufacture of 1 cubic meter of saw timber (ST), glue-laminated timber (LT), concrete (C), reinforced concrete (RC) or steel (S) is shown in Table 1.

Table 1: Energy used for manufacturing building materials

	ST	LT	C	RC	S
Kwh/cu.m	660	1200	500	1600	50000
Kwh/kg	1.5	2.7	0.2	0.67	6.4

More detailed studies compare the energy expenditure for the supporting structure of one and the same building. Table 2 shows the values obtained for an industrial hall of some 25 m span and 1,000 sq.m floor area, with a frame of glue-laminated timber (LT), or reinforced concrete prefabs (RC), or steel (S).

Table 2: Grey energy for the supporting structure of an industrial hall

	LT	RC	S
Total weight (tons)	18.5	183	24
Grey energy (Kwh/t)	42.6	180	192

Interpretation of these widely known figures did not give rise to any principal questioning of current market positions of the individual building materials. The amount of energy saved if timber is used seems to be comparatively small, while the reinforced concrete industry (rightly) points out that the amount of energy needed for heating the building during its lifetime, which amounts to an annual 20 to 80 kilowatt-hours per square meter floor area heated, will by far exceed the amount of energy required for building it.

Accordingly, reducing the amount of heating energy was given first priority, by an energy-efficient design and improved thermal insulation. Stricter heat insulation regulations throughout Europe have not brought about any dramatic advance of timber construction, however, but rather resulted in the plastering of reinforced concrete walls with insulating material, much to the pleasure of the relevant industry.

3.2. "LIFECYCLE ASSESSMENT" OF BUILDING MATERIALS

There is a growing awareness today as to the complex consequence of human dealings with the biosphere. It will not suffice to evaluate good and bad behaviour byapplying one-dimensional criteria. Energy use in kilowatt-hours is such a one-dimensional criterion. One of the current questions is, for instance, whether it is meaningful to use enormous quantities of petroleum-based mineral fibers to meet heat insulation standards in the "k-factor Olympics." And other relevant questions are: What is the ratio of heating energy saved and production energy used? How scarce are raw materials? How are the materials disposed of when their useful life has come to an end? Is there a chance of reusing them? How much money and effort go into a manufacturing process? Science attempts to deal with these complex issues and is about to develop a new tool, the "lifecycle assessment."

Lifecycle assessment are primarily concerned with materials which are felt to be an endangered species as a consequence of their poor performance in classical energy balances. Manufacturers of PVC windows will prepare lifecycle assessment which give a clear proof of the superiority of their product over non-recyclable windows made of impregnated wood. Their line of argument is on the point of embarrassing the timber industry on the other hand. Well-founded studies and analyses are required to corroborate the intuitive arguments set forth. Below are indicated some facts which should enter into a more holistic evaluation of timber as a building material.

All forms of energy are not equal. A comparison of energy requirements in the manufacture of building materials should not be based purely on the calorific value of the consumed energy carriers, but should also take into account which types of energy carriers are used. Although the same amount of energy (some 600 KWh/cu.m) is needed for producing one cubic meter of saw timber and one cubic meter of concrete, it does make a difference whether the cement kilns are fired with CO_2 emitting oil, or whether the energy needed for the drying of wood - a process which accounts for two thirds of the energy consumed during timber production - is mainly gained from wood wastes. If the wood is dried in the open, i.e., with the use of solar energy, this would have to enter the balance as energy gained, similar to the use of solar equipment for the preparation of hot water.

Producing timber for construction means gaining energy. One cubic meter of dry wood has a calorific value of some 2,300 KWh; ten times the energy needed for cutting, sawing and transporting it (with outdoor drying). Compared to the manufacture of other building materials, in which scarce energy sources are used up once and for all, the manufacture of timber does not only yield a useful building material but also a surplus of energy to be utilized in future times.

All raw material are not equal. Even if we assume that the mineral raw materials used for making cement and steel are available in inexhaustible quantities - which is not so certain, at least as far as gravel and iron ore are concerned - we should bear in mind that extracting these minerals leaves "wounds" in nature which must be tended. Timber, on the other hand, is a raw material that renews itself if carefully harvested, and is not easily depleted. The extraction of this material does not leave harmful traces in the environment (provided forests are managed for sustained yield), and is indeed necessary for reasons of forest hygiene, to prevent overmaturity of the trees.

Trees serve as solar collectors, air filters and CO_2 storages. Biomass, in particular trees, is a major storage medium for CO_2, thus contributing to the CO_2 balance in the atmosphere. In order to form one ton of dry wood mass a tree draws 1.5 tons of CO_2 from the atmosphere. At the same time it generates 0.6 tons of water vapour and 1.1 tons of oxygen. Although the stored amounts of CO_2 are released upon natural decay or combustion (yielding large amounts of useful energy), this will have no degrading effect on the atmosphere as the same amount of CO_2 was withdrawn from the atmosphere during the lifetime of the tree, i.e., only 100 to 200 years ago. Fossil energy sources, however, release CO_2 upon combustion that was bound billions of years ago, thus effectively increasing the CO_2 content of today's atmosphere and adding to its burden.

In the course of the past fifty years the CO_2 content of the air has risen by 20 percent; at present an annual five million tons of CO_2 are released world-wide by the combustion of fossil fuels, which is added to the constant CO_2 amount of the atmosphere of 110 billion tons. Another 1.5 million tons CO_2 are released per year in clearing tropical rain forests by burning. The trees are destroyed rather than used for building, in order to obtain pastures for (more lucrative) cattle ranching. The developing countries hardly use timber for construction purposes, but prefer concrete instead, after the European model, thus creating a two-fold CO_2 burden by burning wood on the one hand and producing cement with the use of oil on the other hand, to replace the burnt building material timber.

4. Forests as Resource of Raw Material

4.1. FOREST MANAGEMENT AND ECOLOGY

To act ecologically means first of all to minimize unavoidable damage in order not to saw off the branch one is sitting on (maybe some other branch will do, if there really is a need for firewood), nor to slaughter the cow whose milk one wants to drink. Eco"logic" behaviour also demands a moral attitude, self-restriction and modesty. Making use of what is available and feasible, and resorting to "unfamiliar" resources only after everything else has been exhausted and no further alternative can be found - such behaviour has a long tradition in the relationship between man and nature.

Forests, like humankind, are part of the biosphere. They have been cultivated and domesticated, at least in Europe, and are coexisting with human societies under a system of mutual giving and taking. Man has learned how to tend them and profit from them without inflicting damage. It is eco"logically" imperative to take advantage of this symbiosis, i.e., to fully utilize the forests, both as CO_2 neutral energy sources and as suppliers of raw materials saving outside energy.

On the other hand we are under the obligation, for moral reasons and out of self-interest, to give them optimum care in order to make sure that they can fulfill their function of improving climate and providing intermediate CO_2 storage.

In an extreme formulation, this implies that - regardless of ambiguous, one-dimensional energy balances and market-oriented profitability arguments - wood should be considered as the first and foremost building material, provided enough of it is available from responsible forest management, and no grave technical drawbacks are involved in its use. Only after available supplies have been consumed or if technical requirements forbid its use, man-made building materials such as concrete and steel should be resorted to.

4.2. THE UTILIZATION POTENTIAL OF FORESTS IN EUROPE

In Europe, forests have not been fully exploited since the end of World War II; only 60 to 70 percent of their annual growth is being harvested. They have been left to regenerate for a century, and have expanded in Central Europe by an average of 30 percent. Today there is a growing tendency towards even further increase of forest areas, as less territory is needed for agriculture.

Parallel to the expansion of forestland the demand for building timber and fuelwood has dropped. Petroleum and reinforced concrete have conquered the traditional markets of timber. Today's forests are underexploited and overmature on average. For this reason they are more susceptible to injuries (either inflicted by natural pests or man-made damage, such as polluted air, acid rain, etc.), growth will decline (and thus the amount of bound CO_2) and unused wood will decay, emitting CO^2. Although it is not possible to quantify the useful potential in detail, some general considerations will permit rough estimates.

- Annual growth of today's overmature and underharvested forests totals 6 to 8 cubic meters per hectare. Forest experts assume that the new growth of forests which are exploited fully (100%), will rise to 8 to 10 cubic meters per hectare, and that growth figures for "forest plantations" with optimum "layout" can reach 12 to 15 cubic meters per hectare under the climatic conditions of Central Europe.

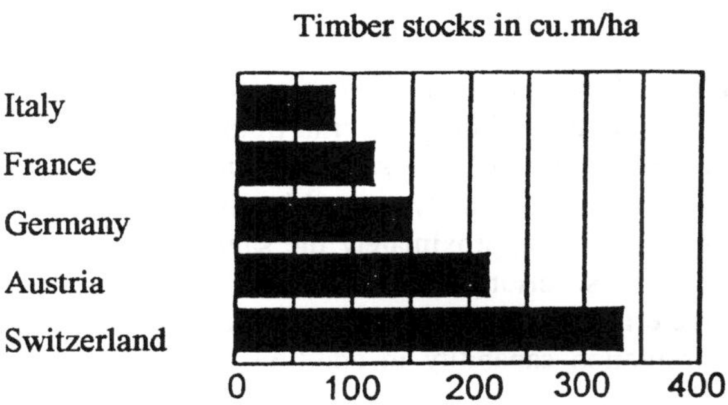

Switzerland has the largest timber stocks per hectare forest in Europe, totalling 333 cubic meters per hectare.

- Uncut timber growth of the past 50 years totals 50 to 100 cu.m/ha on average. This amount may be regarded as a stockpile ready for consumption, in addition to the normal amount of new growth per year.

- The vast forestlands of the thinly populated countries of Scandinavia (and, possibly, Russia) represent a natural supply base for Central Europe. It may be assumed that the currently imported amounts of "Nordic wood" (about 30 percent of today's timber consumption) will also be available in the future.

Model calculations based on these assumptions show that the amounts of timber ready for cutting in the forests of Central Europe would suffice, if import volumes from Northern Europe remain unchanged, to permit a substantial increase in the current market share of timber structures (an estimated 10 percent of the total amount of building materials used in construction, and 50 percent of the entire timber consumption going to the construction industry).

In general we may draw the conclusion that the amount of timber produced by an ecologically sound forest management would just be absorbed if the current share of timber construction in the building market were increased by a factor of two. The question otherwise arising is, who will pay the costs of such forest management if timber construction retains its present share in the market and cannot buy the amount of timber produced?

Table 3: Potential rises in the share of timber construction in the building market by increased utilization of forests in Central Europe

Current share	10%
100% cutting of new growth	18%
Additional cutting of overmature stands	21%
Increased yield by thinning	28%
Ten-percent increase in forest area	32%

5. Timber Construction and its Technical Potential

A comparison of the major technical performance data of wood, steel and concrete reveals the superior characteristics of the natural building material timber. Timber is a composite material reinforced by cellulose fibers, with a considerable amount of hollow space. Knotless timber has the same load-bearing capacity as steel but is lighter, it can absorb tension and compression forces and has approximately the same compression strength as concrete while featuring far superior tensile strength. In contrast to crystalline, mineral-based building materials, timber, on account of the hollow space contained in it, has good thermal insulation capability in addition to its ability of storing heat and moisture. It has the best load-bearing properties of all thermally insulating materials.

Each material has its weak points. Steel is sensitive to corrosion and heat (burning and heat expansion), concrete is susceptible to temperature fluctuations and cracking, and timber is sensitive to moisture and has low fire resistance. Timber is the ideal "decathlete" among building materials, with a well-balanced performance in all major "events." In light construction, timber structures have proved their efficiency even if the high standards of developed industrial societies are applied (Scandinavia and North America, for example, where timber structures account for 60 to 90 percent of all private buildings).

In engineering construction, i.e., the building of supporting structures and bridges with high loads, large spans and multiple stories, examples of the efficiency of timber are more difficult to find than in the building of homes. This may be due to innate weaknesses of timber construction, or to halting efforts on part of the civil engineers to translate the qualities of the material into efficient supporting structures.

A number of exceptions, such as the timber radio towers of the twenties (rising as high as 160 meters), the "Spruce Goose" of the forties (the biggest aeroplane of the world, made of spruce and mold-laminated ribs), and some large-scale buildings of the eighties seem to prove that the small number of objects actually built of timber is not so much a result of essential technical weaknesses of the material, but of non-technological factors from outside.

6. Future Prospects

In Central Europe timber construction will find itself in a market niche at the beginning of the 21st century. Reasons why it has lost its former position of power in construction are manifold, though hardly of a technical nature. The material's technical efficiency is undisputed; individual examples show the great variety of applications, including the most complex tasks undertaken by the highly industrialized societies. Taking an extreme position we might say: If steel and reinforced concrete had never been invented, everything, or almost everything, could just as well be built of timber.

The main reasons for the subordinate role timber construction is playing today are rooted in social developments and factors of economic history. Although industrialization, capitalism, world wars, industrial concentration and the forming of monopolies may have been necessary stages in the development of Europe, they have not been conducive to the use of timber.

Forests in Europe are community property, and it would be against the principles of capital-oriented industrial production to rely on resources that are not freely available. It is more interesting, certainly, to use raw materials that are not claimed by anybody (such as minerals) for the manufacture of building materials which cannot be produced by many.

It would be unwise, though, to decry such developments and call for a new beginning. The existing industrial infrastructure, the knowledge compiled, and the experience gained in using man-made building materials, are essential prerequisites for the tasks of the 21st century.

It would be just as unwise, however, to exclusively rely on artifical materials whose ecological disadvantages have become apparent. Timber construction should not be understood as an alternative, but as a complement and corrective against the potential abuse of the monopolistic position of man-made materials. For this purpose it should be led out of its niche; on its own, devoid of outside assistance, it will not be able to advance against well-established market positions and assume the role of a meaningful corrective.

Timber construction engineers are too modest. Too often do they feel guilty if yet another assembly hall is being furnished with a steel frame and sheet metal panelling, the timber structure proposed having been found too expensive. And if yet another footbridge is cast in concrete, many are those who believe that timber does not have enough load-bearing capacity and is not suitable for outdoor construction.

Current market prices have nothing to do with an eco"logical" use of building materials, and badly planned negative examples reveal nothing of the efficiency of a material whose application technology has only sporadically been revised over the last 100 years. The potential for development does exist in the trade, but if it is to be fully utilized, considerable efforts will have to be made in all areas, from training and research to the restructuring of businesses.

Position Papers

Working Group 4

"Socio-Economic Considerations for the Forest Resource"

SOCIO-ECONOMIC CHALLENGES OF CENTRAL AND EASTERN EUROPE'S FORESTRY NEEDS

M. ST. KOSTKA
Technical University of Bialystok
Institute of Informatics
Wiejska 45 A
15-351 Bialystok, Poland

Abstract

This paper is an attempt to present the main commentaries, findings, conclusions and suggestions concerning the fundamentals of management of forestry resources in regard to socio-economic aspects. The author strongly distinguishes between the natural, technological and economic sphere of subjects, which have been analyzed. A systems approach is presented. The basic concepts have been discussed. The matter of economics and economy, concerning the forest resources and in forestry, is stressed especially.

The purpose of the is paper is to outline the basic questions of systems approach to forests and to forestry, and to suggest some actions required to save and better use forests - not only as elements for the human environment, and not only in Central and Eastern Europe.

1. Introduction

Humanity is losing its forests at an exponential[1] rate. Europe, and in particular Poland, are a kind of exception,[2] but only in some criteria. There are still a lot of evidences and examples of a loss of Poland's forests quantity and quality. Different initiatives which have been directed toward changing these processes have failed to arrest the loss of forests.[3] Because of this new initiatives are needed.

The matter of saving and better using the forests is very complicated. There are a

1 Forestry Statistics for Tomorrow. FAO Rome, 1993

2 K.Rykowski: Sustainable Development of Forests in Poland: State and Perspectives. Ministry of Environmental Protection, Natural Resources and forestry. Agencja Reklamowo-Wydawnicza Arkadiusz Grzegorczykl p. 12.

3 "...numerous well meaning initiatives have failed to arrest the loss of forests...each (have) been launched with fanfare and high hopes, and have each proved disappointing, if not fruitless..." A.T. Durning: Redesigning the Forest Economy. In: L.R. Brown (ed.): State of the World. 1994. A Worldwatch Institute Report on Progress Toward a Sustainable society. W.W. Norton & Company. New York-London p. 23.

R. M. Gutkowski and T. Winnicki (eds.), Restoration of Forests, 333–344.
© 1997 *Kluwer Academic Publishers.*

lot of different aspects of the present and future forestry. Philosophical, ethical, physical, ecological, technological, economical, and political aspects are common topics of discussion. In Central and Eastern Europe the matter of saving and using forests is more complicated. This is due to fundamental and rapid changes in the structure of the entire socio-economic system, and in the basic principles of their functioning.

Socio-economic considerations about forest ecosystems, forest resources and about forestry involve a range of natural, technological, geographical points of view. In socio-economic considerations the focus of observations, analyses, estimations, commentaries, statements and conclusions should be on interactions among different subjects (units) of socio-economic systems: persons, families, households, enterprises, state and self government institutions, non-governmental organizations, etc. Moreover these units should be analyzed both individually and jointly as: possessors, disposers, and users of forest resources; and as: consumers, producers, distributors for forestry products and services. These interactions vary considerably. There are many differences in forest matters and questions dependent on the nature of the mentioned units.

The purpose of this paper is to outline the basic questions of systems approach to forests and to forestry - not only as elements of the human environment, and not only in Central and Eastern Europe.

2. Forestry in the Face of Global Crisis

Present processes concerning forest ecosystems and forest resources, are only the tip of the iceberg. Human society is at the deadly curve. Humans face a challenge, it is like Gordian's Knot - a Global Crisis. We simplify the seriousness of the present situation of the human being if we concentrate only on the environmental crisis. There are many other crises, concerning Homo Sapiens L.: human birth rate, human consciousness, human ethics, human system of values, human health, human safety, human rights, human consumption, human activity, etc. It seems that a crucial splice of the mentioned Gordian's Knot is the global crisis of economics.

A global crisis of economics with environmental aspects was signaled very early (T. Malthus, K. Marx, H. George, W. Kapp, and some others). In 1969 it was mentioned by the then Secretary General of the United Nations - U. Thant. In 1972 evidence was given by the authors of I Rome Klub's Report.[4] In Poland known settlements of W. Bienkowski from 1965, concerning the suggestions of the universal economics.[5]

4 D.H. Meadows at al.: Limits to the Growth. 1972.

5 W. Bienkowski: Dzialalnosc czlowieka w przyrodzie. Problem ekonomii uniwersalnej. (Human's Activity in nature. Problem of Universal Economics). In. W. Szafer (ed.): Ochrona przyrody i jej zasobow. Krakow. 1965.

In spite of the newest findings of very prominent authors,[6] concerning the economics as a cause of global crisis, the Global Gordian's Knot is growing, and growing.

The main reasons for the mentioned phenomenon seems to be the existing Model of Euro-American Civilization. The main processes of this model are: exponential Growth of Consumption, exponential Growth of Production, exponential Growth of Exploitation of the Whole (Global) System of Gifts of the God, exponential Growth of Wastes. There are many failures of economic and political institutions which have been created and developed to supporting the existence of the mentioned model. Distribution of property rights, of economic and political power, of benefits and profits, of inputs and costs, etc., are not acceptable to a great majority of human beings. All this is focused in different Systems of Life, in which forests, wetlands and aquatic ecosystems are of crucial importance.

More detailed attention on linkages between Man and his Environment exhibits linkages between Man and Forest Ecosystems. Global crisis pervades all this in the same degree. One of the crucial splices of the Gordian Knot is dramatic situation concerning the Earth's forests. In both cases all is penetrated by economy, and its theory - economics. All is penetrated be technics, and its theory - technology. There is a large disproportion between economics as a whole, and economics of linkages between Man and Environment. In all this the forestry economics is really unseen. If forests are very important parts of human environment, than forestry (forest economy), and first of all forestry economics, must be important, too. If not, the crises must be. Technics and technology are too weak.

In the face of global crisis Poland is an especially surprising situation. This situation is very visible through forestry's problematics. In detailed analysis of Polish forest ecosystems; in Polish relations between forest and other ecosystems; in Polish forestry and its linkages with other units of Polish socio-economic system; in many another examples where something forest is existing; much of Poland is very visible.

Poland undoubtedly is a country of very interesting paradoxes. There is a paradox concerning the Poland's forests:

-Poland is a country in which the mostly valuable European forest ecosystems still exist: Bialovieza Primeval Forest, Knyszynska Primeval Forest, Roztocze's forests, and many others - about 2 million hectares of smaller forest ecosystems in different regions of Poland.

---

6 J.K. Galbraith: The Affluent Society. 1958. Houghton Mifflin Company, Boston

"...The hard truth is that our economic system is partially blind. It "sees" some things and not others. It carefully measures and keeps track of the value of those things most important to buyers and sellers, such as food, clothing, manufactured goods, work, and indeed, money itself. But its intricate calculations often completely ignore the value of other things that are harder to buy and sell: fresh water, clean air, the beauty of the mountains, the rich diversity of life in the forest, just to name a few..." **Al Gore: Earth in the Balance. Ecology and the Human Spirit. Plume. New York 1993. pp 182-183.**

-Poland is a country which has experienced basic examples of forest ecosystems in terms of reduction of forest quality and quantity; The European Triangle of Dead Forests in Sudety Mountains, forests adjacent to heavy industry centers, comprised of pine (Pinus Silvestris L.) and spruce (Picea Excelsa L.) monocultures have about 2 million hectares which have been permanently attacked by fungi, insects and stormy winds.

Because of this paradox, Poland is very interesting as a forest field experiment for solving the challenges of the Global Crisis.

3. Syndrome of Babel's Tower on the Example of Forestry

Another splice, which is a main cause of the Global Crisis, is Syndrome of Babel's Tower. It seems that one factor in many different crises, both in nature and in social life as well, is language i.e. difficulty in correctly describing the facts of physical reality. In many cases people cannot understand each other and cannot co-operate because of syndrome of "Babel's Tower".

Languages of naturalists, technologists, economists, and other specialists are considerably different. Specialists who have been involved in very important matters of human existence and maybe development, are speaking different languages. The concept of a forest differs considerably depending upon whether it is treated from physical, from ecological, from technological, from economical, from political, or from many other points of view. There is a need to have agreement about what we understand under the main concepts we use, and must use in putting partnership into action.

There are two concepts, which are of crucial importance: forest ecosystems and forest resources. Is a forest only "...a thick growth of trees, etc. covering a large tract of land...," and forestry only "...the science of developing forests; the management of growing timber...," what we know from Webster's definition?[7] The detailed analysis of definitions and describing of forest and forestry, which are known from many handbooks, cannot satisfy them, who have been involved in solving strategical problems in continental and centurial scala, too.

Everybody has his/her own original consciousness, knowledge, and experience concerning the reality, and abstract concepts which reflect different parts of this reality. They are results of elaboration of the information we receive, if we use our five senses, and it we use our mind as a natural computer. Moreover all of this changes with changes of human civilization.

Forest ecosystems are crucial parts of the Earth's Nature. There are forest parts of matter and energy. There are forest parts of atmosphera, lithosphera, hydrosphera, and biosphera. There are forest air, forest minerals, forest waters, forest soils, forest plants, forest animals,...forest substances, forest atoms. A forest ecosystem is one of these concepts, which, in one word, joins many physical things, phenomena

7 Webster's New Dictionary and Thesaurus. Concise Edition; Gedded & Grosset LTD., New Lanark, Scotland, 1990 p. 223.

and processes. A detailed knowledge concerning the forest ecosystem is a main subject of interests of naturalists, of ecologists, and of growing number of other specialists and of common people. A science will never achieve the finite knowledge about forest ecosystems, about its elements, and about the relations among them. What we know under the concept forest ecosystem is mostly complicated system of physical things, phenomena, and processes in a human environment.

Only some parts of forest ecosystems have been the subjects of people's interests. These parts are forest resources. We consider something a resource when we can use it as a physical mean of satisfaction of our needs, desires, wants. The term "good" is often used as a synonym of the term "resource."

Since the start of human civilization forest resources were crucial part of the System of Resources of the Nature. A real development of human civilization was always, and in every part of the Earth's space, based on use of forest resources. There are a lot of misunderstandings concerning forest resources. These misunderstandings are visible in more detailed analyses of results of different uncoordinated actions in forestry and concerning the forestry from outside.

The forest resource should be understood both, as a physical, and as an abstract concept. As a physical concept the forest resource is mostly a subject of interest of consumers, producers and technologists, who could be possessors, disposers or users of its products. In this sense forest resource is a physical good - something what we can measure in physical indicators of quality and quantity - measures of: volume, area, length, weight, color, sound, taste, fertility, productivity, etc.

It is also important to understand the forest resource as an abstract socio-economic quantity (economic concept, economic term), too. In this sense the forest resource is additionally an abstract good - something that is a subject of given demand, something of a given supply, and something that can be measured in monetary terms. As an abstract concept, the forest resource is a subject of interest of business people, economists and decision makers, who could be possessors, disposers or users, too.

Dealing with the whole and parts of the forest resource is the essence of very complicated theoretical and practical activity, it is called forestry. In the essence up to now, forestry is simplified because of the impact of economics and economy on material goods, and on short term benefits and profits. This is the main danger of Euro-American Civilization. But this matter is more complicated. For detailed explanation, we need a systems theory approach.[8] Forestry is only a part of the whole human activity.

Basically, we still have the syndrome of Babel's Tower. No angry God had "mixed their languages," but people themselves loss their ability to understand each other. It was because the "building of the Babel's Tower" was a very complicated matter, that people needed to divide their actions on many detailed parts: theoretical, intellectual, practical, physical; production, distribution, accounting, controlling; masonry, carpentry, transportation, etc. Everybody had been concentrating only on his own interest, and did not recognize the indispensable necessity of

---

8 L.V. Bertallanffy: The General System Theory. Human Biology. No. 23. 1951.

contact with all the others. The Biblical analogy seems to be a one of the great challenges of the next century and millennium. The challenge is to create understanding among naturalists, technologists, economists, politicians, etc.

4. Economic System's Approach to Forest Resource and Forestry Output

There are many references concerning the economic approach to forest resources. There are also many considerations besides economic options. The most controversial approach is a monetary valuation of so called "non-consumptive" uses of forests, such as forest beauty, forest landscapes, forest recreation, and many other forestry services. There are also unresolved problems of "economics:" definitions, qualifications, quantifications of many forestry variables, like the (gene pool), absorbing rainwater and releasing it gradually into streams, preventing flooding, keeping soil from eroding into rivers, protection fisheries, stabilizing climate, etc.

Within given limits of time and space, forest resources have been varied considerably. The oldest forest resources were water, wildlife, fruits, timber, hunting areas, places of settlements. Much later the forest resources became minerals, fodder, soils, areas of colonization, military objects. In the past century the forest resources became monuments, recreational places, open educational "rooms," filters of falls, garbages.

Forest resources could be parts of material things, energy, phenomena, and processes. There are thousands of detailed forest resources, which in given limits of time and space create the given systems of forest resources. This system could and, in many cases, should be divided on the basis of very large number of criteria on a very large number of detailed, partial elements. As forest ecosystem can be divided on differences in rocks under the soils, soils, minerals, organic substances, biocenoses, plants, animals, populations, organisms, ...atoms; so system of forest resources can be divided on millions of partial forest resources, too. The problem is that, in the last case, there are too many simplifications.

A forest resource is something that could be either physical or abstract. To be a resource, it must be something of people's interest, no matter what kind of interest. These elements of forest ecosystems, which are of interest of people, have a least one kind of use value. Because of this they become and are the forest resource. Every physical forest resource has many different use values (utilities). Each use value creates a different forest resource, and different exchange value of this resource.

A forest resource could be, of either natural or human origin. In first case they are natural resources (gifts of the God, natural capital). In the second case they are products. The natural are very often mixed with free.

Forest resources could be either free (free good) or limited (scarce good). If it is free, if there is no competition among people concerning its use - the demand for them is lower than their supply. It must be emphasized that free or limited could be both natural resources and products. Moreover free resources have no exchange value - no one wants to buy them, no one could sell them.

The limited (scarce) forest resources are called economic goods. In this case there is a competition between (among) people, they want (need, desire) to use, to possess, to dispose of them. There are people who want to buy them, and people who want to sell them, because the demand for them is larger than their supply. They have an exchange value. Because of this they could, and should, be priced.[9]

Up to now the focus considerations of the most economists and decision-makers are maximization of particular and short term forest benefits and profits, and minimalization of inputs and costs. In the existing model of Euro-America civilization there is still the treatment of the essential part of forest resources, forest products and services as free goods. This occurs despite the reality that, for years, there has been very scarce or very short supply and because of this they are actually economic goods. Even in such modern market economies the viewpoint favors utilization. For example, in the USA the, "...Forest Service...favors commodity extraction over recreational nonconsumptive uses because peculiarities of its legal structure allow it to retain most of the proceeds from timber sales. Its budget does not benefit, however, from camping, fishing, hunting, hiking, skiing, or any other nonconsumptive use. Nor does the Forest Service budget benefit from the water its forests provide the farmers and cities downstream, or from the fish and wildlife that depend on its forests for habitat..."[10]

Other opinions are exceptions: "...O'Toole contends that the Forest Service will behave as a prudent manager of all the national forests' functions only when it can itself benefit from them. To do that..., the agency reorganized to charge user fees for everything from camping to mining in national forests. The Service's own budget would come exclusively from those user fees. Because the nonconsumptive benefits of forests greatly exceed the consumptive ones in value... Forest Service would allow little timbering, mining, or other high-impact activities on the lands it controls. Charging visitors as little as $3 a day would generate more revenue than selling timber now does, leading forest managers to favor nonconsumptive uses over consumptive ones..."[11]

5. Valuation

There are many misunderstandings concerning of the concept "valuation." The question: "what is a valuation?" is open. This concept is used in many natural and technical sciences, and in many practical activities. It is very important to be conscious of what a valuation is as a concept in economics and economy. This question is still a subject of fundamental discussion among economists.

In economics and economy the concept of valuation very often has been substituted or mixed with other economic concepts: accounting, appraisal, assessing,

---

9 "...Ecological pricing is the second necessary condition of a sustainable forest economy. Under prevailing prices, the world economy works against integrity, making forest conservation an uphill battle..." A.T. Durning: op.cit. p. 30.

10 A.T. Durning. op. cit. p. 27.

11 R.O'Toole: Reforming the Forest Service. Islan Press, Washington 1988. Citation from A.T. Durning...op...cit. p. 27.

assessment, costing, estimating, evaluating, evaluation, pricing.[12] All of these terms have been used for quantifying economic variables. Some of them have been used behind the sphere of money.

A lot of misunderstandings in the Polish language are results of simplifications that occur in translations. For example in "The Great English-Polish Dictionary" "valuation" is translated as:

"1. oszacowanie (appraisal, appraisement, assessment, estimation); otaksowanie (appraisal, appraisement, assessment, estimation); ocena (estimate, estimation); wycena (pricing, valuation);

2. cena (price, value); wartosc (value, worth)"[13]

No wonder that in many cases the specialists who are very actively seeking economic solutions of different problems of growth and development, are like the builders of the biblical Babel Tower.

The author agrees with the settlements that "...valuation means money valuation...The justification for monetary valuation lies in the way in which money is used as a measuring rod to indicate gains and losses in utility or welfare..."[14]

"...By valuation we shall mean the assignment of a monetary value to economic goods and services and, in particular, to environmental resources. For most conventional goods and services, this valuation is done by the free market. For others, the valuation is done by pricing boards (for example agricultural pricing), by trade unions, by central banks etc...The purpose of valuation techniques is to identify the "correct" prices for ...goods and services."[15]

There are two concepts which must be clear to understand the economic aspects of the forest resources: use value (utility) and exchange value (monetary value) of given parts (elements) of forest ecosystems. Distinguishing these concepts is of fundamental importance. The author understands the concept of valuation as the act of reflecting something in categories of money; giving to something a monetary quantity; fixing the monetary value of; measuring the quantities of things, phenomena, and processes in monetary units. Every separate use value of a given physical form of forest resources implies a different exchange value, too. Exchange values of the same forest physical object as a source of timber and a source of mushrooms, differ

12 M.S. Kostka: -Ekonomiczne aspekty pozaprodukcyjnych funkcji lasu (Economic aspects of non-productive function of forest). Unpublished Ph.D.. dissertation. Warsaw Agricultural University. 1973. pp. 93-102, 129-148;
-Ochrona przyrody jako element procesow gospodarczych - uwagi do teorii ekonomii politycznej (Nature protection as an element of economic processes - remarks to political economics). Kosmos 1973, Nr. 3, 286 292;
-Teoria renty gruntowej a gospodarowanie zasobami przyrody (Rent of land theory and natural resource management). Dr. Sc.'s dissertation - Warsaw University 1987. Rozprawy Uniwersytetu Warszawskiego. 1987.

13 J. Stanislawski: The Great English-Polish Dictionary. WP Warszawa 1986, p. 464.

14 D.W. Pearce, R. Kerry Turner: Economics of Natural Resources and the Environment. London 1990. pp. 120-121.

15 Karl-Goran Maler: Multiple Use of Environmental Resources A Household Production Function Approach to Valuing resources. Beijer Discussion Papers No. 4 1992.

essentially. Exchange values differ greatly for the same forest physical object when viewed as either as a source of traditionally forest products or as a forest reserve.

An important question to answer is what do we want to value, i.e. to make the subject of valuation? D.W. Pearce and R. K. Turner have answered this question many times in different occasions. The conventional economists very carefully place value on these economic variables which are known from traditional handbooks of economics and those which are essential as the variables of National Accounting System (variables of GNP, GDP, National Income, economic affectivity of enterprises, etc.) such as products, commodities, service, means of production, capital-and labor-resources. "...Economic research into the monetary valuation of environmental commodities is still in a state of flux, although considerable progress has been made..."[16]

The essential broadening of valuation is in an output of environmental and ecological economists. Among new subjects, which have been valued by them are: externalities, ecological (environmental) losses, ecological benefits, ecosystem services, environmental services, non-market amenities, and ...the environment. In recent years the output concerning the environmental valuation has multiplied.

The author cannot agree with "environmental valuation" or "ecological valuation" at all. In economic sense, what ecologists define as an environment means "all," it surrounds MANKIND. ALL, implies all things, all phenomena, all processes, all people, that surround human organisms, human families, human groups, human nations. It is impossible to give a concrete monetary value to this ALL, taken as a whole. Economists can value all, but in given parts of this all - only the parts for which supply is not larger than demand. Above all the rational valuation has limits of the healthy mind.

The author believes that one cannot forget the unquestionable, fundamental principles of classic economics. Economists have their own world of economic concepts. One can transform the different concepts from other disciplines of science and practice into the system of economic concepts. But one cannot do it indiscriminately. At first one must use all possibilities of verifying the essence of form of old concepts. On the base of ecological principles one could, and should, verify knowledge about the place and functions of economic concepts in economics and economy; and knowledge about place and functions of economic concepts in natural sciences, in technics, in technology, in law, in politics and so on.

Valuation is not an "art for the art's sake" task. It is an integral part of economics and economy. Its goals, subject, methodology are parts of economics, and should serve to economy, to technology and to technics, and ultimately to the whole society. Valuation is an economic means (instrument), and an economic way (method) of resolving of very difficult economic (and not only economic) questions, goals and tasks. With the help of valuation we could better observe, analyze, reflect, describe, and change the economic reality. In rational forming the configurations of economic variables in given limits of space and time, valuation is indispensable.

If in market economy something is very limited, is not priced, it becomes

16 D.W. Pearce, R.K. Turner, op. cit. p. 22.

an object of waste exploitation, squandering, depreciation, depletion. Adequate valuation and pricing of detailed forest resource, of detailed forest products and services; is a basic methodology of action directed toward saving and better using of the forests.

In incorporating of the system of variables of ecological crisis into the whole system of economic variables, special attention is needed on those efforts which have been concerned not to environmental valuation, but to valuation of natural resources (e.g. land). But first we must answer the question "what do we understand about the concept of natural resources?" In our case we come back to the question "what do we mean by the concept of a forest resource?" Without an agreement concerning this question it is impossible to take a rational action for saving and better using forests, not only in Central and East Europe.

One main cause of Euro-American civilization's crisis is an inadequacy of the valuation of a large number of economic variables: goods, services, resources, commodities, inputs, losses, outputs, benefits, and many others. There are many examples of how we spend much time and money for placing value on the variables which have not been essential for the solutions of the mentioned crisis, and variables which we really do not want to value. But from the other side for decades one could not convince one's self about the necessity of monetary evaluation of fundamental economic variable, which up to now have not been valued, and because of this they remain outside the common economic accounting. Among these variables a very extensive understanding of natural resources (e.g. land) is most important.

The author is suggesting that a very complex economic valuation of all essential variables which are applied to forest resources as a whole, to forestry's products and to services - especially - are treated as nonconsumptive, nonproductive, social, ecological, and environmental. This is the main lock of the doors to better future of Euro-American Civilization. Without economic criteria and indicators a sustainable forestry is impossible. The concept of sustainable development is impossible, too.

6. Some Other Aspects Concerning Forests and Forestry

The economic forest resources are objects of property rights.[17] There are three main forms of these rights: possession (ius pessendi), disposing (ius disponendi) and using (ius fruendi). In the course of time, in relations between Man and Matter - this relation is the essence of the concept of a resource - at first a resource is the object of property rights, and later the object of economy, and last the object of economics (theory concerning economy).[18]

At first a given resource, once no longer a free good, has been treated as

17 "...Whether the resource is land, timber, floral supplies, or indigenous knowledge, secure tenure is the first necessary condition of sustainable forest economy. Without it, the people who actually manage the world's forests will have little reason, and less authority, to safeguard forest health..." A.T. Durning: op. cit. 27.

18 M.S. Kostka: Ekonomiczna interpretacja wlasnosci (The economic interpretation of property). In. T. Wawak (ed.) Wlasnosc prywatna (Private property). Jagiellonian University's Proceedings. Cracow. 1993

protected good (for example: the sankt trees and stones, protected plants and animals, protected species and populations, protected elements of cultural heritage: castles, pictures sculptures, etc).[19] This resource (good) is undoubtedly limited (scarce, economic), but business people, entrepreneurs, economists and decision-makers commonly haven't been conscious about this. In this case people, who have a great knowledge about this good, and about the whole environment (natural and cultural) of this good (specialists of natural and cultural heritage) at the end of their very hard battle for protection (conservation), receive support from lawyers, and from powerful decision-makers (kings, princes, lords, parliaments, courts); and thus these (resources) become and have been protected by law.

In every reality "...Designing tenure and price policies for sustainability is not difficult. The problem is meeting the third necessary condition of sustainable economy: political change..."[20] But it is obvious that in democratic, socio-economic systems political change is a result of the will of the majority of electors. So basic political change requires a change of consciousness, knowledge and actions of the mentioned majority. In democratic systems elected officials should be servants of electors. Because in "normal" society's populations the majority of electors are short term oriented, the whole problem must be resolved by the elected minority, the decision-makers. But they must be supported by specialists. If specialists have not been supported by democratic decision-makers, in the long term victims are both the decision-makers, and the majority of electors.

Many of Poland's outstanding forested areas have been granted official conservation status through designation as either a national park, a wildlife reserve, or some other protected category. In the essence this designation implies essential changes concerning property rights. Many of these areas are at serious risk.

The mentioned problem is very important. In Poland, as an example of Central and Eastern Europe, more and more forestry objects receive the status of a protected object (area). In the essence there is a revolution in subsystems of forestry property rights. This revolution causes a revolution in forestry economy, not only in a local scale. In this case a great paradox is that mostly valuable forest objects (forest resources), which first of all must be under the strictly economic control, have been losing by forestry because of the change of property rights. These changes have been made allegedly in the name of the whole society, and in the name of future generations. From system's approach to economics the matter is very controversial.[21]

---

19 M.S. Kostka: Ochrona przyrody jako element procesow gospodarczych (Protection of Nature as an element of economic processes). Kosmos No. 3, 1973 pp. 282-297.

20 A.T. Durning. op. cit. p. 38.

21 M.S. Kostka (ed.): Proceedings of Student's Scientific Camps: Wigry National Park (1992, 1993), Elk (1994). Unpublished materials in the course of edition in research papers of Technical University of Bialystok and of University of Finances and Management in Bialystok.

7. Conclusions

From general, from synthetical, from the whole set of options, the ways to rationality of saving and using forest resources are passing through many areas of a very large desert. For this travel scientists and technologists have no compasses or maps. They have not much of their own, (or from their predecessors) consciousness, knowledge, experience, good will, honest intentions. In many cases the ways and means they have been looking at, have been founded by other specialists and common people. Most valuable are settlements, commentaries, conclusions and suggestions of scholars. Because of this answers to the questions on the theme of our workshop are not only a matter for us alone. But the author believes that our actions are very important, nonetheless.

Unresolved problem of forestry management refers to monetary valuation of natural - not of environmental or ecological - forest resources; and to monetary valuations of, up to now, not valued "nonconsumptive" forest products and services. It is impossible to choose rationally between different options for the use of the limited forest resources without developing monetary valuations of these resources. It is impossible to choose rationally between very limited inputs and costs of forestry without monetary evaluations of all elements of forestry's outputs: products and services. It is impossible to value economic resources, goods and services, without previous assessment of them.

A lot of theoretical output exists. Practice is not far behind. The problem is, that they must be more united. There is a need of pilot, experimental projects. Central Europe, especially Germany, Poland and the Czech Republic, are very predesignated as pilot projects of forestry. It must be adequate to meet challenges of the twenty first century - for all the world. As a regional pilot project the best are is North-East Poland, as it concerned Lowland; and Bayern-Sumava, as concerned Mountains and Highland. Among some local pilot projects must be concerned Puszcza Bialowieska (Bialoveza Primeval Forest), as a whole - National Parks and Forest Promotional Complex.

A PERSPECTIVE ON ISSUES IMPACTING THE PROFITABLE AND SUSTAINABLE UTILIZATION OF FOREST ECOSYSTEM RESOURCES

P.J. PELLICANE
Professor of Forest Sciences
Department of Forest Sciences
Colorado State University
Fort Collins, Colorado 80523
USA

Vice-President
Center for the Advancement of Forest
Ecosystem Utilization and Management
Fort Collins, Colorado 80525
USA

Abstract

This paper reflects a position concerning issues related to profitable and sustainable forest ecosystem utilization. An attempt is made to extract from the author's extensive global experience in forest products and wood engineering an understanding of the key components in successful international forest products commerce. Experiences drawn from Australia, Finland, Honduras, Malaysia, Mexico, Poland, Sweden, U.K., and U.S. constitute the frame of reference for the discussion. General areas of consideration include: resource base, cultural issues, national and international policy concerns, national infrastructure, and educational levels.

1. Introduction

Earth's forest ecological system (ecosystem) resources and their distribution have played a fundamental role in human development. In particular, forest ecosystems have provided humanity with a variety of tangible, life-sustaining resources. These include: food, water, clothing, medicines, energy, and shelter. In addition, there are a number of less obvious, but equally important, contributions made by forest ecosystems to human existence. Among these are: weather moderation, soil stability, maintenance of the hydrological cycle, clean air, and providing a variety of aesthetic features that enhance the quality of human life.

Today, in more developed countries a large number of people directly earn their income from sustainable forest ecosystem use, while the remainder of the population utilizes the various products derived from forests and realizes the environmental benefits derived from responsible and sustainable forest ecosystem management. For example, in the U.S., the forest products industry is consistently among the top ten sectors in the national economy. In Finland, it is estimated that one in every four jobs is directly or indirectly tied to the forests (Pellicane 1984). In developing countries, especially those with extensive and diverse forest ecosystems, the opportunities to extract a multitude of wood and nonwood products, as well as a variety of services, for social and environmental benefit are substantial. Although the evolution of technology and societies over the centuries has changed the nature of the relationship between people and forest ecosystems, it is as true now as it has ever been that a sustainable, rationally-utilized forest ecosystem can provide numerous environmental

R. M. Gutkowski and T. Winnicki (eds.), Restoration of Forests, 345–355.
© 1997 *Kluwer Academic Publishers.*

benefits and a vehicle to improve the quality of life for society.

2. Species Availability

Although forest resources exist virtually everywhere on the planet that provides a reasonable habitat for people, the distribution of global forest resources is quite diverse and uneven. Factors such as latitude and altitude contribute to the species distribution and resource quantity available in a certain area. Although the forests of the north in Asia, Europe, and North America are vast, the species diversity generally increases with close proximity to the equator. In certain tropical regions, it is quite possible to find hundreds of different species on a few hectares of land. This biological diversity represents a variety of opportunities and challenges to those interested in sustainable forest ecosystem utilization.

2.1. OPPORTUNITIES

With the availability of a large number of species which demonstrate a wide variety of physical, mechanical, and appearance characteristics, it is possible to create a large variety of very diverse products. For example, in certain tropical regions, it is possible to find a tree whose wood has a specific gravity of about one-tenth that of water growing next to a tree with wood sufficiently dense so as to sink in water even when dry. This degree of variation covers the total spectrum of end use possibilities for wood. Also, it is possible to find species which are highly resistant to biological degradation near those that are not. Therefore, for some applications of certain species, treatment is desirable.

Among the more interesting features of wood grown in the tropical regions is the diversity associated with color and appearance. The various chemical constituents in the wood can produce colors that range from nearly white, through yellow, red, green, purple, brown, and black. The potential for making very high quality, and high value added, furniture and decorative items exists in these regions.

2.2. CHALLENGES

The challenges associated with species diversity are real and often formidable when technical information about the species is in short supply. Problems associated with machining and processing are frequently encountered. The permeability of species can be quite different making chemical treatment easy or nearly impossible. Surface features can influence the appearance of the material after machining. Perhaps the most troubling feature associated with material variation is the difference in drying rates and characteristics. Since all species (even those of near equal specific gravity) do not dry at the same rate, it is necessary that different drying regimes be used to create similar effects. The same magnitude of drying variation can also be found within species. In northern Mexico, it is possible to find approximately 35 commercially-valuable species of oak (*quercus*). Although much remains to be learned regarding their drying tendencies, preliminary information suggests that when certain combinations of species are dried in the same kiln, unacceptable levels of warp, collapse, and/or differential moisture content can be found.

Among the greatest challenges to using many species is the issue of material classification for marketing purposes. It is impractical to consider the possibility of having hundreds of species combinations available to the customer. A more rational approach would be to create a small number of species groups for marketing ease. For example, the U.S. species group, southern pine, is a collection of six yellow pines that grow throughout a very large geographic area in the southern and eastern U.S. These species have relatively similar appearance and mechanical properties and grow in the overlapping regions. The marketing efficiency that is gained by not having to sort materials by individual species more than compensates for the loss of benefit derived from identifying difference in properties. Douglas fir-Larch (western U.S.) and spruce-pine-fir (Canada) are two other popular North American species groups.

Another difficulty associated with species diversity is that of grading. Though similar to the arguments made in the preceding paragraph, design values for the many species and species groups need to be derived from numerous tests of either small, clear specimens or larger members. The number of combinations of species, sizes, growth characteristics, replications, and other factors makes the development of design values associated with grades a laborious and time-consuming effort.

3. Forest Products

It has become popular in recent years to discuss the management and utilization of forest ecosystems, as opposed to speaking of forests. This change in semantics arises from the now commonly recognized notion that wood is not the only valuable products derivable from forests. The commercial value of wood products from forest ecosystems is well known. Products, such as firewood, lumber, veneer, chips, etc., have been the focal point of forest ecosystem based economies throughout the world. Figure 1 illustrates the worldwide production of certain key wood products.

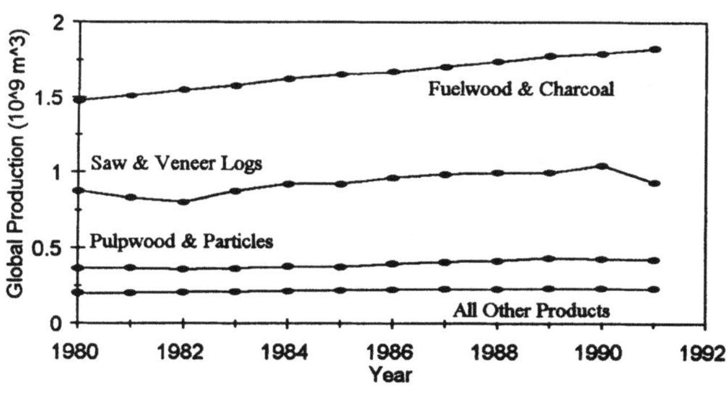

Figure 1.

Worldwide production of key forest products between 1980-1991.

3.1. CHEMICAL TECHNOLOGIES

With the advancement of chemical technologies, a large number of non-wood products (especially extractable chemicals) have been developed from forest ecosystems. These include primary botanochemicals (substances directly extractable from plants) and secondary botanochemicals (those derived from conversion of insoluble saccharides and lignocelluloses). Primary botanochemicals include products such as naval stores (pine chemicals), tall-oil products (paper-pulping byproducts), oils for industrial uses, waxes, tannins, and rubber. Traditional secondary botanochemicals include furfural and ethanol. Potential new botanochemicals include latex- or whole-plant oils, guayule rubber, polyphenols, specialty seed oils, and such potential lignocellulose conversion products as methane, fuel alcohol, and a variety of other fermentation-produced fuels and chemical intermediates. In the U.S. today, it is estimated that approximately 2500 products (clothing, preservatives, sweeteners, paints, etc.) routinely used by people in their daily lives are derived, at least in part, from forest ecosystems.

In recent years, the price of petrochemicals has remained higher than that of similar products from agriculture and forest ecosystems. As such, the latter have become more economically attractive. For about the past 15 years, various major U.S. petrochemical companies have been making substantial investments in new technologies for producing pine chemicals, guayule rubber, fuel alcohol (e.g. ethanol as an additive to gasoline) from cellulosic residues, whole-plant oils, and other botanochemicals. Clearly, the petrochemical industry has made a dramatic shift toward a renewable resource base.

3.2. MEDICINAL DRUGS FROM PLANTS

For all of the excitement that the botanochemicals have generated in the chemical industry throughout the world, the product area with the greatest potential for explosive economic growth and impact on the lives of people throughout the world is in pharmaceuticals and phytochemicals. Plants have always had an important role in medicine and public health (Akerele et. al. 1991). In developing countries, more than 80% of the population depends upon traditional plant-based medicines (Farnsworth and Soejarto 1985), and even in the U.S., 25% of prescription drugs are still based on phytochemicals (from only 40 species). Phytochemicals, which protect plants from the harmful effects of sunlight, are being found to have positive medicinal effects on people. In particular, these chemicals appear to be able to block the multiple processes that lead to cancer. In April 1994, researchers at Johns Hopkins Medical Institutions have confirmed the cancer-preventing value of sulforaphane (a compound found in broccoli). The National Cancer Institute is so interested in phyto-chemicals that it has launched a multimillion-dollar project to find, isolate and study their properties. Pharmaceutical companies are also extremely interested in phytochemicals for their enormous economic potential. It is estimate from available commercial data that in 1990, retail value of plant-based drugs in the U.S. was $15.5 billion (Principe 1989).

4. Perspective

Over the past 13 years, the author has had significant international experiences in numerous countries with very diverse resource, cultural, economic, political, and other circumstances related to forest utilization and management. These countries are: Australia, Finland, Honduras, Malaysia, Mexico, Poland, Sweden, U.K., and U.S. The forthcoming discussion focuses upon certain key issues that have been recognized as having an impact on the profitable and sustainable use of forest ecosystems. It is felt that an understanding of these issues in their global perspective can be useful in developing meaningful natural resource policies at the national and international levels.

5. National and Global Considerations in Forest Utilization

5.1. HUMAN AND TECHNOLOGICAL INFRASTRUCTURE

How a country chooses to participate in the global forest-based products market is often dependent on the nature of the national resource base and the markets in which they are capable of competing. For example, in lesser developed nations with access to substantial domestic forest resources and minimal access to even marginally sophisticated technology, it is quite common to see commerce dealing with a small variety of products that are easy to make, require little if any quality control procedures, and do not have many competitors in the global economy. In particular, Honduras produce a large quantity of products such as tomato stakes (for commercial growers), broom sticks, lobster traps, and lattice fences (for decorative purposes) (Pellicane 1995). These products can be made from any species, require little technology, do not rely on skilled labor, and return only a modest profit on the investment. This is in stark contrast with countries such as the Netherlands and Japan which have very little in terms of forest resources, yet are able to add value to commodity products, and become serious competitors in the world economy. By buying raw materials at modest prices and adding value; they can create significant economic activity. For example, a company in the Netherlands buys green iroko logs from west Africa (Pellicane et. al. 1985). They then use modern technology to process the logs, design, fabricate, and deliver entire bridges for sale in western Europe. Iroko is an extremely hard and durable species which is particularly well-suited for bridge applications. This process adds a great deal of profit to material. Of course, this can only be accomplished in a higher-technology environment by an organization which possesses the business skills to compete internationally.

5.2. INTELLECTUAL RESOURCES

Fundamental to the ability of any society to compete in global commerce is the presence of a human infrastructure capable of executing the myriad of activities surrounding product research, development, manufacturing, marketing, distribution, sales, etc. The type of people that are needed in this process must possess knowledge at a variety of levels. First, there must be a population of people that are literate both in language and basic mathematics. In Honduras and other developing countries, the absence of people capable of making even the simplest of calculations or measuring length with a ruler is a compelling obstacle facing industry. Above the level of the basic laborer are technicians and other technical specialists which provide a myriad of services. Trained professionals at the undergraduate and graduate levels are needed to take on leadership positions throughout their economic sectors and provide the intellectual resources to train future leaders in the educational arena and create policy in government.

In spite of living under a debilitating socio-economic system, the former communist block countries of Europe possess a cadre of well-educated people, and more importantly, an infrastructure to educate future generations. Therefore, the basic educational needs of the society can be developed. This is the missing component in much of the third world. The experiences of the rapidly growing economies of Asia (i.e. South Korea, Malaysia, Taiwan, Singapore, etc.) are examples of how investment in basic education can make a fundamental impact on socio-economic development in only one generation. Economic projections suggest that of the current seven largest economic powers in the world, only two of these will still be on the list 25 years from now. The new entries to this group will come from the emerging nations with substantial natural resources and the educated population to lead their country through the rigors of global economic competition.

5.3. GLOBAL AND INTERNATIONAL TRADE AGREEMENTS

Over the past several years, a number of extremely important global and regional agreements have been negotiated and finalized which are designed to facilitate and expedite international commerce. Regional alliances, such as the ASEAN, EU, and NAFTA, have created trading blocks which will make commercial intercourse between partners more convenient. The finalization of the Uruguay Round of the Global Agreement on Tariffs and Treaties (GATT) will strive to make the world one large common market while reducing tariffs and expediting the flow of goods and services throughout the globe.

For all countries that are participating in the new economic orientation of the world, the road is paved with both challenges and opportunities. Difficulties associated with harmonization of codes and standards, product modification for compatibility with domestic markets, and other issues associated with the entry into new markets will be weighed against the potential benefits of market expansion. For countries that are new to global commerce and the associated competition, there are additional considerations to be considered.

In Mexico, for example, producers of commonly used products such as lumber have been selling their goods in domestic markets. Without the stimulus of external competition, they have not seen the need to practice quality control methods that would satisfy more demanding foreign customers. As barriers to free trade start to disappear, the sudden and often overwhelming competition from abroad is having devastating effects on domestic economies. Higher quality products at competitive prices are displacing domestic goods of poorer quality. In addition, the lack of code acceptance of certain species for particular structural applications could form an impenetrable barrier to export possibilities. Furthermore, the lack of experience in dealing in international economics has kept participants from these out of the circle of individuals and organizations involved with that area. As such, the business community in these countries find themselves both wanting to participate in the larger market, yet frustrated by their inability to even see how and where to begin entry into the system.

At the moment, Mexican lumber and pole producers are anxious to exploit the commercial opportunities afforded by the NAFTA agreement and the weak Mexican peso to export their products. However, the problems that they are facing regarding the sale of ungraded materials made from species that are not recognized in U.S. codes, their lack of familiarity with customer preferences (combined with their lack of ability to satisfy the consumer desires for those preferences), and the lack of opportunity to make contact with potential customers seem to be overwhelming.

5.4. INTERNAL NATURAL RESOURCES POLICIES

The late 20th century has been marked by the proliferation of natural resources policy and legislation which is precipitated by social activism on environmental issues. In the U.S., legislation, such as the Endangered Species Act, have had major impacts on timber harvesting, mining, grazing, agriculture, and many other land use activities. Judicial decisions regarding the spotted owl have put severe constraints on timber harvesting in the states of California, Oregon, and Washington. Many other countries have followed the same path. In Australia, the national government has made a politically popular move by sanctioning the boycott of all tropical hardwoods for any government project. This type of decision has far-reaching negative effects on the tropical hardwood producing countries since the national government was the countries largest consumer of tropical hardwoods (Pellicane 1990, 1994). However, such decisions represent a commercial opportunity for producers of competing wood products. For example, producers of products from Australian forests can now compete for government contracts that once were beyond their reach. Producers of plantation-grown radiata pine from both sides of the Pacific Ocean have new commercial opportunities that previously did not exist.

In Thailand, the government has completely outlawed commercial harvesting on public lands in response to international public opinion. It is frequently the case that such policy creates uncontrolled illegal harvesting with far worse environmental consequences. Currently, producers of timber products, such as veneer and lumber, are procuring their raw materials from Cambodia which has far less rigorous environmental monitoring and does not respond as readily to the voice of international opinion.

In the lesser developed nations, there are a myriad of models for government participation in forest ecosystem management and utilization. Government policies related to management, administration, and utilization have evolved (often abruptly) over the past few decades. This is the case in Malaysia, Honduras, and Mexico. In Honduras, the government agency responsible for overseeing national forestry, La Corporación Hondureña de Desarrollo Forestal (COHDEFOR), took control of all public lands in 1974. In subsequent years, the timber output on public lands dropped precipitously from 943,000 m^3 (1974) to 94,000 m^3 (1992) (Pellicane 1995). This drastic change was due to a number of factors. These include: no rational forest management plan, lack of qualified forestry professionals, no accurate inventory data, corruption, etc. Furthermore, as the years past, COHDEFOR's role changed from that of a "mining" operator on forest lands to more of an integrated resource manager with a mandate to focus more attention on such matters as watershed management, wildlife, soil conservation, and other non-commodity issues. COHDEFOR also changed policy regarding land tenure so that land ownership was privatized giving control to people who had a vested interest in the productivity and sustainability of the resource. This act alone fostered an increase in the number of forest products companies in Honduras from 58 (1989) to 230 (1994).

In Mexico, at the federal level, forest policy was often a confused issue with several independent agencies making unilateral decisions on policy. This lack of coordination created difficulties in developing a unified and coherent forest policy in a timely fashion. Furthermore, the federal government's lack of ability to respond to local issues, together with interagency political power struggles, corruption, and other complications, often lead to slow-moving and counterproductive actions.

Recently, government policy in Mexico has become a stimulus to rational forest ecosystem management and utilization. The Mexican federal government has divested itself of a great deal of authority and returned many decision-making responsibilities to the states. This has manifested in the individual states forging their own economic and political directions. In northern Mexico, a major economic analysis (funded by the state of Chihuahua), entitled *Siglo XXI*, was conducted to determine the areas of future opportunity. Natural resources in general, and forest ecosystem utilization in particular, were identified as areas of potential economic growth and development. The action items contained in this document outlined the steps necessary to advance a sustainable and profitable forest-based economy.

In Malaysia, the government is an active participant in progressive forest policy. The ongoing support for the Forest Research Institute of Malaysia (FRIM) , an internationally recognized forestry-related research organization, has lead to the success and vitality of the center that has contributed to the Malaysian forest community. Funded programs to improve forestry and forest products education in Malaysia has been a priority in recent government strategic planning. As part of their most recent five-year plan, the Malaysian government identified both strategies and tactics for economic development. As in northern Mexico, sustainable forest utilization has been identified as an area of promise in Malaysia, and has stimulated a great deal of business, research, and educational activity in this sector.

5.4.1. *Land tenure*

In much of Latin America, the question of land tenure is particularly troublesome. Historically, the lands were owned by certain groups of indigenous peoples. Colonization of the Americas in the 16th and 17th centuries caused a transfer of land ownership to colonists. Of course, this arrangement was never acceptable to the indigenous peoples, who have claimed a moral ownership of the land. A series of populist revolutions in the 19th and 20th centuries caused a redefinition of ownership of certain lands. Today, in Mexico there are a large number of tracts of land, called "ejidos," upon which indigenous peoples live and having controlling interest. On other properties, ownership is a more contentious issue. Illegal claims, supported by forged documents, often constitute the case for an individuals assumption of land ownership. In Honduras, physical violence toward forest harvesting crews is the result of attempting forest operations under contract with an assumed owner on land that indigenous people believe to be theirs. These land tenure issues are often intractable situations for which institutional remedies offer no relief.

6. Cultural Considerations

In many parts of the world, it is confusing to an uninformed observer why certain materials are used in particular applications. For example, in Latin America, Asia, and much of the Arabic and Islamic world, it is difficult to find structures built of any material other than concrete and masonry. In parts of the world which are routinely plagued by earthquakes of devastating proportions, it is rational to enquire why construction methodologies revolve around as brittle and heavy a material as concrete. Note that heavy structures that lack flexibility and ductility are more prone to disastrous consequence in earthquakes. An analysis of the domestic natural resources, combined with the environmental boundary conditions of the region, suggest that timber structures would be a rational and practical alternative to current construction methods. However, a more detailed understanding of local culture reveals that timber structures are socially stigmatized. It is thought that residential timber construction is associated with the most extreme poverty and low social class. Therefore, it is unacceptable to a large majority of the society to live in timber housing.

In Europe, where one finds timber to be the material of choice for roofing but infrequently used in wall, floor, and other structural applications, one can find similar cultural biases against timber. Many of these biases are based on distant historical precedent and poor technological practices. In the U.K., timber is all but banned from most urban residential applications. This is due to concerns over fire danger that arose in 1607 when the famous fire of London destroyed a significant part of the city. However, modern construction techniques mitigate the situations that created the 1607 devastation. Yet, this mentality still survives. Furthermore, producers of competitive materials have been known to take a cynical view in product promotion, and deliberately feed on the fears and misunderstandings of the population to portray timber as a dangerous and problematic construction material.

To change people's conceptions of timber in structural applications is very time and effort consuming. Furthermore, the problem has even greater ramifications. In countries, where there is little or no market for timber construction, there is no significant market for products, and consequently, few if any suppliers of materials. In addition, with no market for timber construction there is no supply of carpenters and other trades people that make up the links on the chain of production from raw material to finished products. In Honduras, COHDEFOR, recognized the role of supporting industries and trades in a vital forest-based economic sector. To this end, and in cooperation with a variety of international funding agencies, a number of trade schools have been developed to create the workforce necessary to make the desired products (Pellicane 1995).

Although it is difficult to change the cultural propensities of a society. Work done in the U.K. in the mid-1980's demonstrated that it is possible to influence national market preferences. In 1985, the Timber Research and Development Association (TRADA) made a significant movement to stimulate the use of laminated members in structural applications. This effort resulted in a doubling of the volume of laminated timber used in the U.K. over a three-year period. Prior to this effort the volume of this product had remained stationary for many years. By the late 1980's, an economic down turn in the national economy caused the termination of efforts instigated in 1985 that generated earlier success resulted in the volume of laminated materials used in construction to return to the pre-1985 norm. This effort did demonstrate that short-term changes in marketing and promotion, research, and other activities have a positive impact on the forest-based economic sector (Pellicane et. al. 1985).

7. National Infrastructure

In much of the first world, communications and the movement of goods within most technologically-advanced countries is sufficiently simple and effortless that it is often taken for granted. However, poor communications and the lack of a sufficiently well-developed national infrastructure are impediments that obstruct the economic progress and vitality of many second and third world nations. For example, Mexico is a country that possesses a wide range of assets that are invaluable with respect to international economic competitiveness. Located virtually at the centroid of the Americas and with major ports on both the Atlantic and Pacific coasts, its geographic position is ideal. Unfortunately, the lack of ability to move goods by road and rail within the country adds to product cost and nullifies the substantial benefits that Mexico enjoys due to significantly lower labor costs and a very favorable foreign exchange rate with many other important world currencies. The percentage of transportation costs to total product cost is substantially more than might be expected in countries in which there are adequate highways, roads, and bridges to meet the needs of commerce. For example, approximately half of the cost of Mexican forest products to the consumer comes from the cost of getting a log from the forest to the mill (Novelo-Gonzalez et. al. 1995).

Although transportation infrastructure is crucial to global economic competitiveness, other components of infrastructure are important. In Honduras, there are a wide variety of obstacles that impede economic development. The lack of available electrical energy was a most interesting and perplexing dilemma. In Honduras, most of the electrical power comes from hydroelectric generators on a major lake in the country. Recognizing that future demands for electrical energy will outpace current supply abilities, a new major hydro-electric project was proposed, approved, funded, and never built. In 1994, a severe drought in Honduras lowered the water level in the lake by about 16 meters and had a severe impact on electrical output. Electricity rationing became the norm with public utilities supplying power only between the hours of 7:00AM-7:00PM daily. Companies had to buy, install, maintain, and fuel generators to meet many of their needs and to power operations during a second or third shifts.

Other aspects of infrastructure also play pivotal roles in economic vitality, although sometimes in a less obvious way. The lack of clean water and sanitary facilities can have a major impact on community health. This factor influences worker absenteeism, student attendance in schools, and other socio-economic factors.

8. Conclusions

Successful utilization of forest ecosystem resources in a sustainable fashion is achieved when a number of important social, economic, environmental, and political factors come together. The interrelationships between these factors are complex and evolving with time. However, societies all over the world are realizing that numerous collateral social, economic, and environmental benefits are derivable from sustainable natural resource utilization. Therefore, global attention will continue to be more highly focused on this important topic.

9. Pertinent Observations from the ARW

The intellectual interactions that occur between thoughtful, scholarly people with eclectic backgrounds often surprises, if not shocks, participants in the dialogue. I learned a great many things and found myself shocked on a number of occasions during the ARW. Some of the most important issues that were either informative, thought-provoking, or perhaps, contentious are summarized below.

Perhaps the most intriguing was the relationship between culture and economics. I had always been of the opinion that these issues were perhaps tangentially connected. However, during the ARW, it was brought to my attention and I quickly came to believe that economics is at the very root of culture. With this in mind, I have been forced to rethink the mechanisms involved in shaping people's attitudes and behaviors relative to the structural use of wood in Latin America and other locations.

My historic beliefs in the value of economics as the engine to promote sustainable forest resource utilization have been strongly reinforced. Although certain people were advocating preservation to a greater or lesser degree, the flaws in their arguments became transparent. The lack of wealth in much of central and eastern Europe was shown to be the overriding factor related to the lack of ability to control pollution and other issues related to forest degradation.

I found it amazing how oblivious people are to the many subjects ancillary to their own particular area of expertise related to forest sustainable resource utilization. Contemporary professionals from throughout the world are so highly specialized as to see only a small band on the spectrum of a global problem. Also, the complexity of sustainable forest utilization became clear. After hearing lengthy discussions on many issues, it was observed that some highly important issues never even entered the forum of discussion.

Forums, such as this ARW, provide a unique opportunity for professionals from diverse professional and cultural backgrounds to discuss ideas that are relevant to topical issues.

10. References

Akerele, O., Heywood, V., and Synge, H. (eds.) (1991) The conversion of medicinal plants, *Proceedings of a conference 21-27 March 1988*, Chiang Mai, Thailand, Cambridge University Press.

Farnsworth, N.R. and Soejarto, D.D. (1985) Potential consequences of plant extinction in the United States on the current and future availability of prescription drugs. *Economic Botany* 39:231-240.

Food and Agricultural Organization of the United Nations (1993) *FAO Yearbook Forest Products*, FAO Forestry Series No. 26, FAO, Rome.

Pellicane, P.J. (1984) Forestry and forest products in Finland, *Forest Products Journal* **34**(4):9 & 69.

Pellicane, P.J., Hilson, B.O., and Smith, I. (1985) A critical appraisal of the prospects for the United Kingdom glulam industry, *Journal of the Institute of Wood Science* **10**(5):202-209.

Pellicane, P.J., Smith, G.W., and Juniper, P.N. (1990) Forest products industry in southeast Australia, *Journal of the Institute of Wood Science* **12**(1):7-13.

Pellicane, P.J. (1994) Forestry utilization in Malaysia - resource, products and research, *Journal of the Institute of Wood Science* **13**(3):413-417.

Pellicane, P.J. (1995) A perspective on forest utilisation in Honduras, *Journal of the Institute of Wood Science* **13**(5):525-532.

Novelo-Gonzales, G., D.W. Green and P.J. Pellicane. 1995. Center for the utilization of products from forest ecosystems. (abstract in Spanish) p. 202, *Proceedings of the Partnerships for Sustainable Forest Ecosystem Management: 5th Mexico/U.S. Symposium*, 165-174, Guadalajara, Jalisco, Mexico.

Pellicane, P.J., Gutkowski, R.M., and Czarnocki, J. (1996) A perspective on forestry in Poland, *Journal of Forestry* (in print) September.

Principe, P. (1989) The economic significance of plants and their constituents as drugs, *In: Economic and Medicinal Plant Research*, Vol. 3. London: Academic press.

ECONOMIC AND FINANCIAL SITUATION IN RUSSIAN FORESTRY AT THE TRANSITION PERIOD

A.P. PETROV
Prof.Dr.Ec.Sci.
Russian Institute of Continuous
Education in Forestry
Institutskaya Ul., 17
Pushkino, 141200
Moscow Region
Russia

Abstract

Economic and political reforms in Russia caused deep economic crisis in the forest sector. As a result Russian forestry suffers from shortage of financial resources to manage forests on the sustainable basis. Many silvicultural and ecological rules are not followed.

The acting financial system in forestry is based on the budget money distribution and does not stimulate effective forest management. In order to increase the role of Russian forests in the world environment protection, a new financial system must be implemented. This system must combine the principles of market pricing and state regulation. It must be used as an economical base of the ecological certification in forestry following the criteria and the indicators of the sustainable forest management.

1. Introduction

The forest potential of Russia is tremendous. The total area covered by forest is 771 mln. ha and the growing stock is 81.6 bln.cub.m. The share of the area covered by forest in the total area is 45%. 80% of the growing stock belongs to the Asian part of Russia, including the Urals.

The economical potential of Russian forests is determined by the annual allowable cut about 600 mln.cub.m. Only 1/3 of this volume is presented by the European forests. Economic and political reforms in Russia caused deep economic crisis in the forest sector. The volume of logging has reduced three times as compared with the volume in 1988 and ammounted to 125 mln.cub.m. in 1995.

R. M. Gutkowski and T. Winnicki (eds.), Restoration of Forests, 357–363.
© 1997 *Kluwer Academic Publishers.*

2. Economic Crisis

The economic crisis was mainly caused by the principal mistakes in forest industries development under the centralized planning economy when 20-30 years ago the strategy of forest sector development was chosen on the base of Asian forests exploitation. Preference was given to the Asian variant because this variant provided forest industries with high quality forest resources (coniferous species, trees of big size), pure water and cheap electric energy. Very soon the volume of cuttings in Siberia and the Far East increased considerably and caused the increase of timber harvesting in Russia (the maximum volume of logging was 350 mln.cub.m.).

In market economy the strategy of the forest industrial development based on the Asian variant became unacceptable due to economic reasons: none of the consumers is able to pay the high transport cost to bring the roundwood or wood products from Siberia to the European part. Now the forest resources of Siberia and the Far East are economically closed for the European market both in Russia and the European countries belonging to the Commonwealth of the Independent States. This factor may be considered as a main cause of the deep economic crisis in the forest sector in Russia at the transition period.

3. A New Strategy

It is critical to realize in practice the New Strategy - the "European" variant of the forest industries development based on the decision that the forest resources of the European part are able to satisfy the demands for wood in this region on account of the following reserves:

a. the full utilization of the allowable cut of deciduous forests that will increase the annual volume of harvesting by 30-40 mln.cub.m.;
b. the increase of the intermediate cuttings (thinning) by 20-25 mln.cub.m. using the new logging systems with selective felling;
c. the utilization of forest resources of group I that will increase the annual volume of cutting by 10-15 mln.cub.m.

The above mentioned goals can be achieved only on the principles of sustainable forest management in European part of Russia. Economical principle of sustainable forest management provides the forest renewal and growing with the sufficient financial assets. If the financial assets are not sufficient silvicultural practice will not correspond the ecological demands. As the results of finance shortage many silvicultural rules are not followed, in harvesting the clear cuttings predominate. For example, falling and skidding operations are performed by very heavy caterpillar tractors and machines that destroy not only the undergrowth, but the whole humus layer. Clear cuttings in the river basines considerably change the water supply and regime. There are many places where the influence of forest on ecological systems is critical and tangible.

4. Russia's Financial System for Forestry

The acting financial system in forestry could not eliminate the ineffective decisions in forests utilization and reproduction. The flow of financial assets in the forest sector is demonstrated by Figure 1. The demands of the financial system mentioned in Figure 1 are the following:

1. The amount of assets charged from the federal budget to the Federal Forestry Service in order to cover the forest reproduction and protection cost.

   Officially the amount of these assets is determined on the base of three federal programmes approved by the government:
   a. Forest fires.
   b. Forest regeneration.
   c. Soil improvement (for rural territories).

   The Federal Service has to defend the amount of financial assets annually in the Ministry of Finance. The assets recieved from the federal budget are distributed by the Subjects of the Federation (Regional Forestry Services) in accordance with the calculation.

   The problem of money distribution is estimated as the most complicated because the real criteria how to use money are absent. Money distribution is always connected with corruption and making of ineffective decisions. In their turn the Regional Forest Services distribute the money among the district forestry bodies (leskhoz) following the same methods and principles. As a result the federal financial assets are divided among 1800 forestry bodies.

   The Ministry of Finance carries out constant control on how the federal money is used (through the district departments of the Federal Treasury).
2. Local taxes charged to the district (rayon) budget.
3. Stumpage fee or leasing rent are charged to the district budget.

   The stumpage prices are fixed and approved by the district administration according to the existing legislation. The coming Forest Legislation has to change this practice transfering these rights to the Regional Forestry Service. This decision is estimated as a negative one because district administrations are not professional to set up stumpage prices taking into account all economical, ecological and social factors.

   Due to the low demand for wood in the home market the sale prices of the roundwood are not able to ensure the high level of stumpage. The share of stumpage in the sale price of the roundwood amounts to 5-10% in regions rich in forest (boreal forests in the North and Siberia) and up to 20% in regions scarce in forest.

   Many logging enterprises are still in the difficult financial situation and are not able to pay stumpage fee at all.
   The federal system of market analysis in the forest system is absent, that prevents from making proper decisions in stumpage pricing.

4. Tax on profit, which is distributed in proportion among the federal budget and the budget of the Subject of the Federation.
5. Added value tax charged to the federal budget. Now the rate of this tax amounts to 20% to the sale price of the roundwood.
6. The share of stumpage charged to the Regional Forestry Service for the forest reproduction and protection.
   The share (in percentage) is determined by the agreement between the regional administration and the Regional Forestry Service.
   In many regions this share is established by Regional Forestry Acts. The value of this share is flactuating in the big limits (from 0 to 60%).
7. The share of the stumpage charged to the district forestry body (leskhoz).
   This share as in the previous case is determined by the agreement between the district administration and the forestry body (leskhoz).
   In some regions the whole amount of the stumpage is charged to a leskhoz to cover all silvicultural cost (for example, the republic of Udmurtia). In many regions the share of charge is established by the Region Forestry Act (usually in percentage).
8. Total amount of financial assets to cover silvicultural cost (forest management, regeneration, growing and protection). The financial assets are determined by means of cost budget calculation.
9. Timber revenues from thinning and sanitary cuttings and revenues from nonwood products utilization (berries, mushrooms, animal breeding, bee keeping etc.). This revenue belongs to a district forestry body (leskhoz) and must be used only to cover the silvicultural cost. This revenue is not charged by any taxes. So the financial assets of a leskhoz are represented by:
   a) federal money;
   b) the charge of stumpage;
   c) the own revenue from thinning and nonwood products utilization.

All the above sources don't guarantee sufficient amount of money to grow and to protect forests. At present these sources cover only 20-30% of cost that are neccessary to spend following the silvicultural and ecological demands.

5. Adaptation to Market Principles

In order to manage forests on sustainable basis the acting financial system must be substituted by a new one adapted to the market principles. According to the new system the following administrative bodies have to earn money themselves selling the wood on forest sites by leasing or by auctions. The minimum stumpage price must be set up by forestry authorities. The minimum stumpage price must cover all current cost connected with the forest regeneration, growing and protection. This price will be different for different species and silvicultural conditions.

Getting these assets directly from loggers the forest administrative authorities will increase the volume of cutting. The market stumpage price must be always higher than the minimum price. Its level will be determined by the market prices for wood products and the cost of their production. The difference between the market price and the minimum stumpage price may be charged to two budgets: the federal budget and the budget of the Subject of the Federation (Figure 2) to accumulate the assets to develop the forest sector.

6. Closing Remarks

The above considered financial system must save Russian forests and increase their contribution to environment protection. This system must be used as the economical base of the ecological certification in forestry using the criteria and the indicators of sustainable forest management.

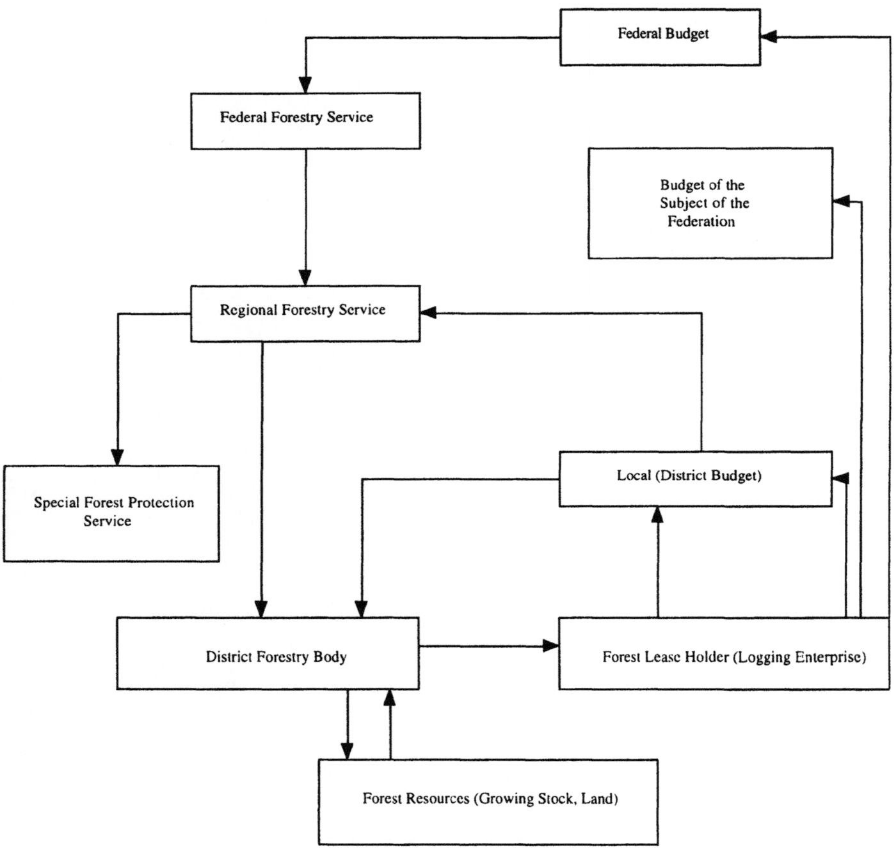

Figure 1. The Flow of Financial Assets in the Forest Sector

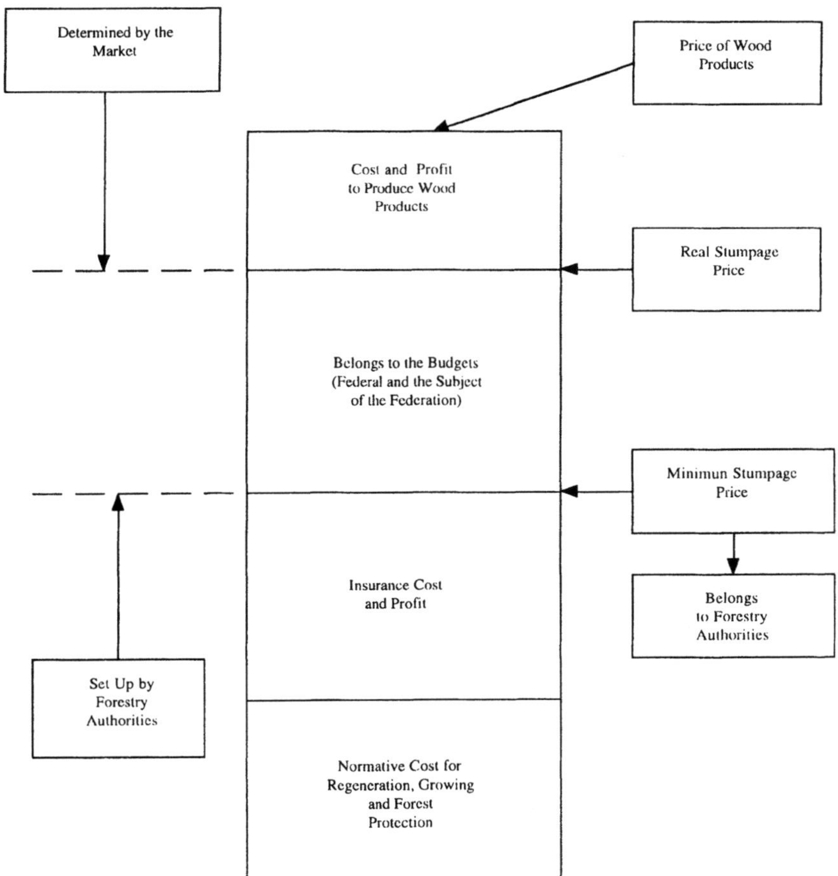

Figure 2. Financial System of Sustainable Management in Russian Forestry

PERSPECTIVES FOR THE FUTURE OF SUSTAINABLE DEVELOPMENT IN WOODPROCESSING AND MANUFACTURING

Some New Solution in Technology for Better and Cheaper Utilization of Wood Resources

M. TAMASY-BANO
Professor
University of Forestry and Wood Sciences
H-9400 Sopron, Bajcsy-Zs. u 4. HUNGARY
Phone: (36)99 311100 Fax:(36)99 311103
E-mail: mtambano@efe.hu

1. Introduction

The recurrent theme of this workshop is the sustainable development of forests focusing on Eastern and Central Europe. The ecosystem concept should by now be thoroughly familiar to all foresters. But how can one evaluate the term "sustainability" without understanding related fields? Where are the limits of those ecosystems, to what extent are these systems open? What do we know about the socio-economic operation, in the past, how we could apply some old, but forgotten ideas? What do the foresters think about joint projects with other cooperating engineers to realize the sustainability in forestry?

This paper focuses on those questions and alternates to attempt answers from the view-point of an engineer, who is teaching students in wood sciences at a Hungarian university in engineering sciences, such as applied thermodynamics, energy engineering, drying technology etc.

Some information is provided about forest resources, use of wood, forest policy and legislation, as well as wood manufacturing and educational programs at universities in Hungary, at this time.

2. Sustainable Development

Sustainable development is a global view of the world (not a view of an individual process) and it centers on ensuring that the needs of the present are met, without compromising the ability of future generations to meet their own needs (simultaneous consideration of economic, environmental and community improvement objectives). A global agenda for change was mandated within the United Nations World Commission on Environment and Development in 1983. After Brandt's Program for

R. M. Gutkowski and T. Winnicki (eds.), Restoration of Forests, 365–372.
© 1997 *Kluwer Academic Publishers.*

Survival and Common Crisis (the Brandt Commission on North-South issues), and after Palme`s Common Security (the Palme Commission on security and disarmament issues) came Brundtland`s COMMON FUTURE as a call for political action (1987). [1] In that sense industry means: "producing more with less" and energy (choice for environment and development) influence on developing as: "modern appliances should be redesigned to deliver the same amounts of energy services with only two-thirds or even one-half of the primary energy inputs needed to run traditional equipment."
Al Gore`s "Global Marshall Plan" shows the six strategic goals that simultaneously represent the most important changes to save the global environment [2]:

1. The stabilizing of world population.
2. The rapid creation and development of environmentally appropriate technologies.
3. A comprehensive and ubiquitous change in the economic "rules of the road" by which we measure the impact of our decisions on the environment.
4. The negotiation and approval of a new generation of international agreements.
5. The establishment of a cooperative plan for educating the world`s citizens about our global environment.
6. The establishment especially in the developing world of the social and political conditions most conducive to the emergence of sustainable societies.

The author has had some experiences connecting goals No 2. and No 5. involving investigation of work forestry and wood sciences, researching and developing wood and paper technology, energy engineering, system optimization, and recycling.

3. Historical Background

It is still not widely known that the concept of sustainability was probably first formulated in the field of forestry. It was the widespread scarcity of quality timber and firewood experienced in European mining districts which triggered the first regulations on the use of a particular natural resource. Obviously it was the first challenge of a crisis in resources, an environmental crisis, to be met by organizational means. There have been 250 years of different approaches to solve this problem which have produced the following conclusions [3]:

1. The substance of "sustainable use" is determined by the actual needs and views of the society (economic considerations are only a part of it).
2. Sustainability cannot be achieved through limitations and regulations: an active, generalist approach is needed on an international level.
3. Sustainability of renewable resources has to be resolved in a way which does not jeopardize the tolerance limits of the ecological (living) system.
4. In the long run, the formulation of the sustainability concept exclusively on the basis on human needs cannot bring satisfactory results. Beyond general biological (and evolutionary) aspects, sustainability has to be conceptualized based on new moral-ethical principles.

This demands not only a restructuring of economic systems but also of public thinking which may be solved exclusively in close international cooperation.

Technology has had an important role in providing sustainable aspects for the ecological system. Mining technology needed wood as a construction material as well as charcoal for metallurgy. Structures were produced from wood of high quality, which was provided by forestry activities. It was probably a very strong dependence between forestry and mining "industries," or wood technology, as well as chemistry.

The last decade brought about an unprecedented change in global attitudes towards forests, forestry and the use of forestry products in manufacturing in the past half century.

While there have always been changes in the expectations for our forests; economic development at one time, materials for a war effort the next, pressures for maintaining aesthetic opportunities or biological integrity during another period; the past half decade has coalesced these divergent interests into a more inclusive process and with more comprehensive emerging policies than ever before in the history of forestry.

This more complex, more comprehensive approach is summerized by the term sustainability.

4. Sustainability and the Second Law of the Thermodynamics

Energy is a basic system parameter of the material world (universe). Material itself is equivalent with energy as it is written in Einstein's equation (1):

$$\mathbf{E} = \mathbf{m}\,\mathbf{c}^2 \qquad (1)$$

where **c**: velocity of light propagation (in a vacuum)
E: energy
m: mass

Tentatively we can describe the processes of the changing world using thermodynamic process parameters, because most of the distinctive features of thermodynamics, such as movement, motive force, work, power, equilibrium, and potential seem to be suited for modeling the universe. Our world could be compared with a complex thermodynamic system, and therefore our universe can be investigated by the universal rules of thermodynamics.

In isolated systems there are particles with different potentials, which occur in steady movements or processes in the system up to the state of equilibrium. These processes simultaneously cause losses as internal friction, which turns into heat. We call this process irreversibility, which is measured (derived) in a unit called entropy. All real processes occur accompanied by ever higher levels of entropy. If the entropy level increases to the maximum (limit), the process breaks down, and we should open the boundaries of the isolated system to ensure new energy input and/or waste output.

Only in this way can new potential be added to keep the system running. This new energy potential comes from the surroundings which means we should enlarge the system in order to run the process again.

Any isolated system uses its own intrinsic energy to sustain itself. If all this energy has been used (for internal friction), the system achieves its maximum entropy level and dies. With an external (fresh) energy input or by exhausting the collected waste heat (friction) the system recieves new potential power to operate.

The expression for the Second Law of Thermodynamics is as follows:

$$ds = \frac{dq}{T} = \frac{du + pdv}{T} = \frac{dh - vdp}{T} \geq 0 \qquad (2)$$

Where is **T** = temperature,
p = pressure
v = volume
s = entropy
q = heat input
u = internal energy
h = enthalpy

If a comparison is made between isolated thermodynamic systems and the natural environment, the economic, social and legal systems in which the forests are managed, they will show the same effects and trend.

5. Basic Information on Forest Resources in Hungary

Hungary used to be a country with an extensive forest cover. In the 17th century, however the forested area started to decrease following a general European pattern. Besides the increase of the population and the extension of agriculture, this decrease was mainly caused by the industrialization process of the 19th century. At the time of the turn of the century, however, Hungary still had as much as 7.4 million hectares of forest.

In 1930, as a consequence of the country's territorial losses after the World War I treaties, the amount of forests dramatically fell to 1.1 million hectares, i.e., to 12 per cent of its previous level.

After World War II, the primary goal of the forestry policy in Hungary was to increase wood supply. To cope with the shortage of wood and to decrease its import, forestry had to focus on the quantity of wood production. Later, as a result of large scale, on-going national afforestations, the forested area of the country was increased by 600,000 hectare and reached the present total of 1.7 million hectare or 18.2% of the pre-WWI level. The latest figures show the total area under forest management in Hungary is 1.838 million hectares. This area includes stocked forests, temporarily unstocked areas, as well as roads, trails, clearings, fields, water surfaces and buildings, which are all integrated parts of forest holdings.

Since 1938, there have been great changes in the different ownership categories. The collectivization process which took place over the last four decades has left lasting effects. Currently, the ownership structure of forests in Hungary is a very critical issue and is undergoing dramatic changes once again. As privatization proceeds, forests are also under the gavel. Forests under nature protection are, however, not to be privatized. The majority of the forest area, 979,000 hectares is owned by 19 forest companies, now under control of the State Assets Handling Company. This company is responsible for managing state properties.

Approximately 80,000 hectares of forests belongs to other ministries, such as the Ministry of Environment and Regional Policy, the Ministry of Defense, and the Ministry of Transport, Telecommunication and Construction. There are about 22,000 hectares of forests that are owned by the state, but managed by various bodies, such as state agencies and companies, such as railways, mines, etc.

About 9,000 hectares municipal forest, formerly owned by county, city, and village councils under the socialist system, are now owned by the local municipalities, so they are no longer considered as state propertiy.

Some additional 13,000 hectares are partly owned by state companies, and partly by companies under transition. The legal status of these forests has yet to be clarified.

6. Primary Wood Processing in Hungary

Hungary's present domestic use of wood products is greater then its forests' wood production capacity. Therefore, Hungary is forced to import.

The 19 state forest companies are responsible for about 65% of the total forest land. The vast majority of primary wood processing companies have belonged to the state-owned forest companies since 1968. This structure is now changing rapidly. Wood processing plants have gradually been separated from the forestry companies and are subject to privatization. The state, private, cooperative and other wood processing plants have settled along transportation lines, as close to energy and raw material sources, as well as municipalities and markets, as possible.

The last country-wide survey of sawmills was carried out in 1979. The last reliable data on the state sector is for as late as 1990; the number of plants, their ownership and capacity are permanently changing as privatization proceeds.

The wood-based panel industry currently consists of four veneer producing plants, six plywood, two particle board, one flaxboard and one fiberboard processing plant, which processes hardboards only. MDF production has yet to be established in Hungary.

The pulp and paper industry operates 30 paper machines in several plants. Total production was 452 thousands metric tons in 1990. The state of the pulp and paper industry has brought about intensive research and analyses recently, because Hungary has great surpluses in pulpwood and a large negative trade balance in pulp and paper, which could be reduced by processing the available pulpwood surpluses. Considering that Hungarian paper consumption is far below the European average, it is assumed that the country's economic development in the future will involve much greater paper consumption than it currently does.

As far as the production technologies of the forestry industries are concerned, mechanical wood processing is most common. Apart from some small-scale distillation and conventional charcoal production, practically all industrial wood is used and processed mechanically.

7. Forest Policy and the Forestry Law

After World War II, the improvement of wood supply became the major goal of forestry policy in Hungary. Although forest policy has always aimed at improving natural environment and recreational services too, the main emphasis was put on the increase of wood production and the expansion of forest area.

The benefits of this policy included more than half million hectares of new forests which have been developed in the country since World War II. On the other hand, state companies, the disturbed balance of forestry and wildlife management and the difficulties of reforesting overly large certain areas, have created yet unresolved problems.

The previous Forestry Law was enacted in 1961. The main objective of this law was to increase forest resources, to maintain and intensify special forest functions and to develop harmonized wildlife management. The Forestry Law regulated stocked forests, clearings, roads, nurseries, open lands embraced by forests, i.e., all lands under forest management, but it ignored ownership.

The increasing public concern over the health of forests, air pollution effects, and the threat of possible climate change focused public attention on forestry issues.

The main emphasis is shifting towards the non-wood benefits of forests, while wood production is controlled by market conditions rather than central decisions. These, as well as the changing ownership structure call for adjustment in forestry policy, which finally has resulted in a new Forestry Law which was enacted not long ago.

The key issues of the new forest policy are implemented in the new Forestry Law, of which certain elements were already in effect in lower level regulations, are as follows:

- to define forests in a complex way with the priority of long term interest for human health, nature conservation and maintenance of forest resources;
- to define the role of forests in nature conservation, protection, welfare-oriented use and recreation;
- to define the role of forests in wood supply, and the necessary economic and institutional background;
- to define the prerequisites of sustainable management under the changing ownership structure, and to develop the system of guaranties for maintaining forest heritage;
- to define the desired ownership structure (the share of state forests is expected to exceed 50-55% on long run);
- to define the coordinating and controlling role of the state in ensuring sustainable management;
- to define the role of foresters, their education and training, and the conditions of their employment;
- to define the role of public relations; finally
- to define the role of international relationships.

The degree of conformity with international agreements was intended to be as high as possible. In Hungary, the rights of local people, as declared in the Hungarian Constitution, are in accordance with all international recommendations. Forests must be kept open for the public. Local people are allowed to use the forests for recreational purposes, to collect wild flowers, mushrooms, and wild fruits free of charge for their own use. Grazing livestock in forest is, however, prohibited. The new Forest Bill is expected to ensure sustainable forest management in Hungary.

8. Sustainable Forest Management

Sustainable forestry will have to be practiced under the social and economic conditions for raw materials and products which are significantly different than today.

Population growth and increased urbanization of the world's population are likely to influence the demand for forest products. Products can range from factory-built modular houses to biodegradable backing. These new materials might be less demanding on particular species, fast growing plantation-grown trees could be used more effectively, have a greater longevity in use, and increase the opportunity for recycling. The potential of biomass farms offers considerable potential for both construction material and energy. In the manufacturing sector every component of the material is viewed as a potential product. An ecological approach in industry ecology approach has potentials for energy and raw material conservation. All these aim at increasing the supply of useful products, while reducing the pressures on natural forests and the consumption of energy.

9. Conclusions

The essence of "sustainable use" is determined by the actual needs and views of the society.

Sustainability has both domestic and international dimensions; it involves both short-run needs and long-run potential; it requires for simultaneous consideration of economic, environmental and community improvement objectives; it neccesitates a framework for policy implementation which brings people together in a negotiating setting rather than a winner-take-all adversarial objectives.

This demands not only a restructuring of economic systems but also a restructing of public thinking which is possible only through close international cooperation. For sustainable forestry under increasing demand for forest products, research and technological change are needed in the area of products development, biomass farms, industrial ecology, ecosystem management, etc.

Sustainability of renewable resources has to be resolved in a way which does not jeopardize the tolerance limits of the ecological (living) system.

Sustainability cannot be achieved through limitations and regulations: an active, generalist approach is needed on international level.

Sustainability has to be based on new moral-ethical principles.

10. Pertinent Observations from the ARW

I found the workshop to be very successful and well organized. We have had memorable working conditions, an excellent group of experts representing comprehensive knowledge.

The workshop provided a great opportunity to discuss the different reasons for the environmental problems in the EU countries, which were influenced more by political than economic issues before the political changes occurred in that area.

The author emphasizes that the related or common research/developments needs recommended by the ARW should take into consideration the specific features of the region: the local socio-economic status, the cultural particularity, the accepted education system, the history of each country, as well as the level of technical development, the authoritative forest management, the use of forest products, etc. There is a high priority to transfer the know-how into the EU region directly for the actual need, step by step in according to their receptivity (more knowledge should be transfered than industrial product as a powerful help).

11. References

1. World Commission on Environment and Development (1987) Our Common Future, Oxford University Press

2. Al Gore, (1993) Earth in the Balance, Plum Penguin Books USA Inc., New York, USA.

3. Matyas, Cs. (1994) Egy megújitható eroforrás hasznosításának évszázados tanulságai,Magyar Tudomány 1994/10, 1184-1188 (in Hungarian).

4. Somogyi, Z. (1994) Basic Information on Forest Resources, Use of Wood, Forest Policy and Legislation in Hungary, ERTI (not published).

5. Kovacs, Zs (1995) Hardwood Timber Market Hungary, UN-ECE Timber Committee Market Discussion , Geneva, 14 November 1995.

6. Whaley, R.S. (1995) Research and Technological Development for Sustainable Forestry (Keynote Address) Congress Report, Volume I., IUFRO XX. World Congress, Published by The Finnish IUFRO WCOC.

THE BAVARIAN PROGRAM FOR USING RENEWABLE ENERGIES

The Biomass Network: A New Source of Energy on the Threshold to Commercial Viability

T. WEBER
Ministerialrat
Bayerisches Staatsministerium für Ernährung, Landwirtschaft und Forsten
Referat B2, Ludwigstr. 2
80539 Munich, Germany

Abstract

In Bavaria biomass is the most important regenerated resource of energy besides water. For the energetic utilization of biomass the following groups of products should be developed:

- biogenic fuels (biodiesel, nature diesel),
- biogas and
- warmth, steam and electricity based on biomass in small heating plants, heating plants, heating and electricity plants and in combustions of biomass in connection with coal.

In the long term, the future percentage of biomass used as a resource of energy could be 12%. But in a first step only 5% of the primary energy requirement should be sufficiently supplied with biomass. Biomass as a combustible has a special importance in Bavaria. Therefore residues of forest work, of the utilization of wood, of landscape conservation works, and of the agriculture, such as straw and energy crops are used. For the reason of rationalization within the utilization of biomass a self propelling biotruck, that condenses biomass, was developed.

The production and utilization of renewable resources in Bavaria give new tasks to agricultural areas and create new jobs. Moreover many positive effects for the environment and the economic welfare are given.

1. Biomass - A New Technology with a Future

The energetic utilization of biomass is an innovation with beneficial effects for

R. M. Gutkowski and T. Winnicki (eds.), Restoration of Forests, 373–386.
© 1997 *Kluwer Academic Publishers.*

- rural areas,
- the economy and
- for the environment.

Securing a source of energy which is as safe, cost-effective and as environmentally friendly as possible will remain one of the most important aims of our state's economic and social policy. In Bavaria, biomass is the most important source of renewable energy besides water. Plants for the energetic use of biomass in planning at the moment, are on the very edge of modern technology and are a necessary addition to the developments in the field of environmental technology, in which Germany is a major player world wide.

By changing from an agrarian to an industrial state during the last decades, Bavaria has already shown, that innovations with a great potential can be successfully implemented by municipalities and counties together and by the close cooperation of scientists, businessmen, the state bureaucracy and politicians.

Bavaria has been turned into a leading industrialized state without losing its rural structure and charms, which now offer us the chance to develop alternative supplies of energy. However, we can only increase the use of renewable resources for energy, if we succeed to unite the forces of free enterprise and the government. Where the development of biomass as a source of renewable energy is concerned, our motto must be "all together now."

2. Lines of Production

Today there are five lines of production in Bavaria for the energetic utilization of biomass. These are:

- biogenic fuels;
- biogas;
- gas from sewage plants and waste disposal sites;
- gas rich in hydrogen;
- biogenic solid fuels made out of biomass-residue and energy-crops.

3. The Potentials of Biomass

In the future biomass could make up about 12% of the energy generated and used in Bavaria:

- 2% of the primary energy required, can be covered by using untreated wood residue. To achieve this, we must take of all residue from the wood industry, which have not been chemically treated and use it by putting them to energetic uses.

- A further 2% of the primary energy required, can be covered by using residue from landscaping and conservation works or other biomass residue from farming.

- 2% of the primary energy required, can be covered by using wood, used biogenic lubricants and other biomass residue from the industrial sector. To achieve this aim, we have to develop an environmentally friendly multi-step system, which makes it possible to use wood, whose first use was, for example for construction purposes, for energetic purposes in a second step.

- 6 to 8% of the total Bavarian primary energy consumption can be covered by using energy crops from surplus fields.

Because of the high level of energy consumption in Bavaria, the share of energy generated from biomass cannot be pushed above 12% even over the next two decades. In countries with a lower level of industrialization and sufficient arable land, however, a much higher percentage of biomass based energy can be achieved.

4. Bavarian Policy

In his statements on the government's policy on December 8th 1994 and June 19th 1995, the Bavarian Ministerpräsident Dr. Stoiber outlined the new aim of Bavaria's energy policy. This aim is for 5 % of the required primary energy in Bavaria to be supplied through biomass by the year 2000.

This aim comprises the following operational aims:

- creating new jobs;
- safeguarding rural jobs;
- reducing carbon dioxide emissions;
- making innovative technology operable on a commercial basis;
- opening up new markets for exports;
- protecting fossil fuel resources;
- replacing energy imports by domestic energy sources;
- creating a more competitive system by creating a decentralised network of energy supply;
- developing new technologies which can be used on a help-to-help-yourself basis by less industrialized countries.

A large number of voluntary undertakings by Bavarian companies are contained in the "Umweltpakt 2000," an agreement on environmental matters. Only a few of them can be mentioned here:

- The Bayernwerk AG, the largest electricity provider in Bavaria, will conduct research whether biomass can be used in power plants in addition to brown coal.

Research results and calculations on the economic viability will then be used to determine, if biomass should be used on a day to day basis.

- A number of associations, among them several Bavarian farmers and commercial associations, as well as the municipalities of Bavaria will run parts of their fleets with biodiesel engines.

- The BayWa AG, the leading supplier of agricultural produce and fuels in Bavaria, will extend their network of biodiesel petrol stations.

- Companies have signalled their readiness to construct 240 biomass power plants including a number of heating plants and 25000 small firings operating with wood. Readiness has been signalled to build up biodiesel production capacities for about 110 000 t. p.a..

Apart from these promises, there have been a number of other undertakings and offers as well as commercial activities.

5. Today's Energetic Use of Biomass

At the moment, the following elements of Bavaria's infrastructure can be put to use for the energetic use of biomass:

Table 1. Given Infrastructure In Bavaria

Production line	Plants
biogenic fuels	– biodiesel plant, 4.000 t RME p. a. – 150 public or company biodiesel petrol stations – 21.000 t biodiesel (RME) – 11 power plants run on vegetable oils and nature diesel petrol stations with a turn over of 4.000 t. p. a.
biogas	– 130 biogas plants used in the agricultural sector
gas from sewage plants and waste disposal sites	– 250 plants using gas from sewage plant and waste disposal sites
biomass combustibles	– 130.000 stove heatings burning wood whose emissions are regularly controlled. This includes all stoves and heating systems in joineries which are used for burning leftover wood. – About 450.000 heating systems without emission control which are run on wood. This includes tiled stoves. – 1.400 heating systems using straw as fuel. – 127 heating plants and power stations.

All together, 17 000 MW are being generated out of bio-combustibles today. The exact operating hours of the above mentioned systems are not known.

6. Share of Biomass in Bavaria´s 1995 Primary Energy Consumption

Energy statistics for 1993 and 1994, which are based on information about 1992 and 1993, show a share of about 1% for biomass. As an integrated market has not yet been developed for biomass, a large percentage of the energy produced by using biomass has so far not been reflected in statistics. Furthermore, measures introduced by the Bavarian government in the early 1990s have led to a considerable growth in the use of biomass fuels. New statistics and estimates have been used to generate the following information on the percentage of biomass in Bavaria's primary energy consumption:

Biomass: Percentage of primary energy consumption in % Total 3,057%	
– biogenic fuels	0,05
– biogas	0,006
– gas from sewage plants and waste disposal sites	0,176
– firewood used in domestic stoves and heating systems	1,006
– residual wood in joineries etc.	0,686
– heating plants and power stations	0,5
– biomass from household rubbish	0,583
– sewage sludge	0,031
– charcoal	0,019

Table 2. Biomass: Percentage of Primary Energy Consumption in Bavaria

Line of production	**Quantity used p.a.**	**Share of primary energy consumed in % [1)]**
biogenic fuels		
– plant oils used in engines	25.000 t	0,049
– used plant oil in stationary incinerators	700 t	0,001
biogas	5 Mio. m³	0,006
gas from sewage plants and waste disposal sites	89 Mio m³	0,176
gas rich in hydrogen	-	-
biogenic combustibles		
– firewood used in stoves	1.250.000 t	1,006
– residual wood in joineries	850.000 t	0,686
– biomass used in heating plants and power stations (incl. non public/ private power plants)	650.000 t	0,500
– biomass from household rubbish	750.000 t	0,583
– sewage sludge [2)]	42.000 t	0,031
– charcoal	12.000 t	0,019
total percentage of biomass		**3,057**

[1)] estimate based on primary energy consumed in 1993 plus 2%: 1923 PJ

[2)] drainage with waste heat which is not used separately

The European Union and the German government have noticed that there are no reliable statistics on the use of biomass and other regenerative fuels etc. As a result there are already attempts being made by the European Union to improve the energy statistics. So far, a proposal for a German law on this matter has yet to leave the planning stage.

7. An Outline of the Situation of the Energetic Utilization of Biomass

The following reasons account for the inferiority of biomass compared to fossil fuels:

- At this moment, each and every system for the utilization of biomass is more labour-intensive or more expensive than comparable systems using fossil fuels. This is caused by the fact that possible rationalisation-measures have not been implemented so far, that systems for the utilization of biomass are only produced in small numbers and the complicated technology needed for fuel supply. Another reason for the higher costs of biomass is the filtration of flue gas.

- Prices for fossil fuels are still low.

- In parts, biomass has to be produced specially for energetic utilization. Furthermore its harvest and provision are inherently more labor-intensive than fossil fuels.

- There is wide-spread ignorance about the benefits of biomass as a fuel for the economy and the environment.

- There are no definitive industrial standards concerning the use of biogenic fuels.

- Public subsidies are not high enough to make up for all the higher costs involved in the use of biomass.

- Market demand and supply for biomass is still in its early stages: this explains the large regional differences in prices.

The situation of biomass, however, is not without hope: to take a given example, it is already cheaper to provide biomass than a comparable amount of German hard coal, and when it comes to running costs, biomass is able to match light heating oils.

Table 3. Costs of Biomass and Fossil Fuels

Source of energy	ca. DM/t (absolutely dry)[1]	ca. DM/t SKE	ca. DM/GJ
biomass pellets	110 - 160	200 - 290	7 - 10
wood chips from forestry work	70 - 120	120 - 210	4 - 8
straw	70 - 110	130 - 200	4 - 7
residue from landscape conservation	20 - 80	40 - 150	1 - 6
wood chips from residue wood	30 - 70	50 - 120	2 - 4
used wood	-50 - 20	-90 - 30	-3 - 1
German mineral coal	170 DM/t BME[2])	300	10
heating oil, 0.40 DM/kg	140 DM/t BME	270	9
foreign mineral coal	40 DM/t BME[3])	75	3
lignite	30 DM/t BME	60	2

1) In our comparison of prices, it is only possible to use the price for one ton of absolutely dry material in our calculations.
2) BME = biomass-equivalent. 1 BME = 0.55 SKE, i.e. 0,55 hard coal energy units
3) The higher calorific values of foreign coal (compared to German coal) have been allowed for.

8. The Financial Situation of the Energetic Utilization of Biomass

Currently there are the following incentive measures for the energetic utilization of biomass in Bavaria:

- Reduced taxation for biogenic fuels. Biomass fuels are taxed 0.62 DM/kg less than fossil fuels.

- A law concerning the supply of electricity generated from biomass into the public power grid guarantees a market and a fixed price for biomass electricity plants (in 1995 the guaranteed price was 0.159 DM/kW).

- The permission to grow energy crops on fields excluded from regular farming under the European program for the reduction of agricultural production.

- Public sponsorship of research projects and investment (this is part of the BAVARIAN program for the promotion of modern technologies).

9. A Strategic Concept for Raising the Level of Energy Supplied by Biomass

Using the current infrastructure as a basis, existing lines of production in Bavaria are to be maintained or extended and developed, furthermore, new lines of production are to be tested.

Research and development of new technologies and lines of production are co-ordinated by C.A.R.M.E.N., the Bavarian central coordination organization for renewable resources. One of the reasons for Bavarias success in introducing biomass as a renewable resource is the concerted action taken by the private sector, research institutions and the public sector, which have been co-ordinated by C.A.R.M.E.N..

Together with appropriate partners in other EU-member states (for example in EUBIA, the European Biomass Industry Association), C.A.R.M.E.N. is also involved in a constant process of improving the lines of production for biomass. An exhaustive report on the work of the interministerial task force on renewable resources and of C.A.R.M.E.N. was presented to the Bavarian cabinet on 05.12. 1995.

In order to meet 5% of the primary energy required by using biomass, the following measures are planned:

Table 4. Planned Measures to Meet 5% Primary Energy Requirement

Line of production	Steps to be taken
biogenic fuels	– measures to develop the market for biogenic fuels – creation of an industrial standard for biogenic fuels – the permission of using biodiesel in certain engines – extension of the network of petrol stations selling biodiesel – raising the production of biodiesel – construction of demonstration heating and electricity plants run with plant oil, especially in areas with high potentials – adaptation of engines for the use of nature diesel
biogas	– measures to promote the research into the technology involved in using biogas – construction of several plants
gas rich in hydrogen	– measures to promote research into the technology involved in using gas rich in hydrogen – construction of several demonstration plants
biomass-combustibles	– measures to develop a market for biomass-combustibles – creation of industrial standards for biomass-combustibles – R&D measures – demonstration projects for heating plants and power stations – development of a mobile system for the compression of biomass into pellets (the so called Biotruck) – test runs on using biomass as a supplementary fuel in coal power plants – incineration of wood which has been chemically treated in specially adapted waste incineration plants – incineration of sewage sludge (this might involve problems where public opinion is concerned)

9.1. TO ESTABLISH A MARKET FOR BIOGENIC FUELS, THE FOLLOWING MEASURES ARE OF PRIMARY IMPORTANCE

- The public sector, especially the Bavarian government, has to increase their pioneering role in the use of biodiesel. Wherever this is possible, diesel engines in state or publicly owned cars are therefore to be adapted to the use of biodiesel, and where it is viable, engines which can run on biodiesel should be preferred in replacements.

- The motor industry has to make possible the use of biodiesel in more engines.

  The private sector is to be reminded of the undertakings it has given in the Umweltpakt 2000, and has to be asked to increase the use of biodiesel.

- Because the first construction of a biodiesel plant in Gemünden has failed, the following plan for the production of biodiesel will be pursued: the plant at Mainburg will be enlarged, biodiesel needed in Bavaria in excess of the native production will be produced in plants outside the state while the search for a second production site within Bavaria goes on.

- Long-term strategy is to use all used plant oils (oil from deep fat fryers, lubricants etc.) for energetic purposes.

- In order to widen the use of nature diesel (unprocessed plant oil) as a fuel, it is important to convert diesel engines in gravel pits, ramblers huts in the Alps and other environmentally sensitive locations.

9.2. TO INCREASE THE USE OF BIOGAS AS A REGENERATING SOURCE OF ENERGY, THE FOLLOWING MEASURES SHOULD BE TAKEN

- construction of further agricultural biogas plants;
- construction of further demonstration plants for the fermentation of compost;
- construction of demonstration biogas systems, which show how the renewable energy resource biogas can be used.

9.3. THE USES OF HYDROGEN RICH GAS

As an energy source, hydrogen rich gas is of special interest to Bavaria. According to reports of the taskforce on hydrogen associated to the Bavarian ministry for commerce and trade, transportation and technology, the production of hydrogen from biogas compares very well indeed to other ways of producing hydrogen. It can also easily be put to use in small, decentralised plants and supply centres. Although the production of hydrogen rich gas from biomass will not account for a large chunk of the energy generated in Bavaria, it offers enormous potentials for the future, and the following steps will be taken before 1998:

- test run of a system which produces hydrogen from biomass at a test-site in Bavaria;
- planning and construction of two demonstration plants.

9.4. THE FOLLOWING MEASURES ARE FURTHERMORE BEING TAKEN BY THE BAVARIAN GOVERNMENT AND PRIVATE ENTERPRISES WITH THE AIM TO INCREASE THE USE OF BIOMASS AS A COMBUSTIBLE:

- increasing the number of small and environmentally friendly stoves;

- construction of a network of heating plants and power stations in Bavaria, with the aim of building one biomass thermal power station in every county in Bavaria;

- construction of 18 demonstration biomass heating plants and power stations using up to date technology;

- using biomass as supplement in two lignite power stations, if the test-runs in the lignite power station Schwandorf are successful;

- adapting no longer needed waste incineration plants for the incineration of old wood;

- making more use of the incineration of sewage sludge, if the public resistance to this technology should allow this.

10. Obstacles to the Thermal Use of Bio-Combustibles

At the moment, an underdeveloped market for fuels is still the main obstacle for an increased use of biomass as solid fuel in small stoves, thermal heating and power plants. The following points are crucial.

General standards for the qualities on which a system of industrial standards for fuels is to be based are still non-existent. This leads to unnecessary prolongation until administrative permits are granted, to problems when it comes to drawing up contracts and to substandard results and supplies.

- The market-situation for biogenic fuels is too complicated. This results in pricing difficulties or wrong decisions (for example: the export of old wood instead of its energetic utilization). The missing transparency of the marketplace furthermore is an obstacle to investments, as many potential investors do not have sufficient trust in the long-term availability of biogenic solid fuels or their quality.

- Regional and seasonal fluctuations in the market are too high. This leads to unwanted differences in pricing and sharply fluctuating prices, which prevent potential investments. As there is no intermediate supply system between producers and users of biogenic solid fuels, additions to biogenic fuels, which would greatly reduce the price of the latter, are not available, as they would take up too much storage facility or have to be processed to be of any use. They might be too wet or too dry, or they might have to be transported to far away consumers at steep costs.

Under the coordination of C.A.R.M.E.N., the organization for coordination of the use of regenerating resources in Bavaria, the government and the private sector have come to the following agreement concerning the creation of a market for biogenic solid fuels:

- creation of a biomass stock exchange;

- construction of a system of regional warehouses for biomass.

11. Measures of the Second Biomass Package

The realization of the first biomass package was finished in 1995. For the realization of the second biomass package the following measures are to be introduced in a combined all-out effort by the private sector and the government to raise the share of biomass in the primary energy required by 1.2% to 4.3%.

Projected increase in the share of biomass in the primary energy required[3)] (total increase by 1.202%)

– biogenic fuels	0,137
– biogas	0,003
– gas rich in hydrogen[1)]	....
– biogenic combustibles	0,082
– heating plants and power stations	0,533
– biomass in coal heating plants and cement factories[2)]	0,164
– incineration in rubbish incineration plants	0,205
– sewage sludge incineration	0,078
total percentage of projected increase	1,202
total percentage of biomass if the planned measures are successful	4,249

1) only R & D
2) projected share of 10% biomass fuels in two coal power plants
3) These estimates use the primary energy required in 1993 plus 2% as their basis.

A new catalogue of measures for the third biomass package with the aim of achieving a 5% share for biomass is therefore to be developed after a report on the measures planned and implemented between 1996 and 1998 has been compiled.

Current projections and information believe that large scale investment measures for the production of hydrogen rich gas from biomass will be among the projects included in such a future catalogue, as the first R&D projects on this technology will have been completed in the next years.

12. Principle Considerations on Biomass-Policy

The following aims and principles have been agreed upon as the foundation of the Bavarian policy towards the creation of a market for biomass:

- Measures to build up a circulation of resources are to be treated with priority.
- An equal regional development of biomass projects in Bavaria is to be aimed at.
- If possible, varying technological methods are to be tested.
- All measures have to be environmentally friendly.
- Only the end-produce of any lines of production is to be subsidised.
- The amount of grants for any project by the state is proportional to the innovation involved, that means any measure supported by the Bavarian state has to have R&D measures and the creation of a market for biomass as its primary aim.
- Any subsidies which would make biomass more cost-effective than comparable fossil fuels is to be excluded.
- Subsidies are only to be used for jump-starting biomass technology, with the exception of biomass as supplementary fuel in existing power plants no long term subsidies are to be granted.

13. Measures for the Furtherance of Biomass

In order to increase the share of biomass in the primary energy consumed to 4.3% in a second step, the following measures have to be taken:

13.1. A LEGAL FRAMEWORK FOR THE USE OF BIOMASS

- the legal and administrative framework mentioned above will also pertain for the coming years;
- binding long-term industrial standards are to be drawn up for biogenic fuels;
- further engines and types of vehicles are to be cleared for the use of biogenic fuels;
- industrial standards are to be drawn up for biogenic solid fuels.

The criteria for the annual check-up-list for small stoves and heating system is to be changed according to new insights from R&D.

13.2. AN OUTLINE FOR TAXATION ON BIOMASS

Current fiscal policy favours solar energy, heat exchange systems and heat pumps over biomass, as the tax relief of biomass in use before 1992 has been discontinued. In order to improve the fiscal situation, a law which secures the following measures should be proposed:

- Introduction of a special depreciation amount for biomass plants and systems,

- A fiscal privilege within the furtherance for the construction of housing. The aim is to entitle biomass to the same fiscal privileges as solar energy and heat exchange systems (For example: there is a current maximum subsidy of 500,- DM p.a. in Germany).

13.3. PROPOSALS FOR EUROPEAN UNION MEASURES

If put into effect through the whole of Europe, the following measures would greatly aid the cause of biomass:

- European regulations on the fiscal privilege of biogenic fuels;

- Simplification of premiums awarded for the cultivation of regenerating fuels;

- Introduction of a European tax on the emission of carbon dioxide. This tax should be levied so that it does not affect the level of income tax.

14. Protecting Jobs in Disadvantaged Areas

Using biomass as energy source offers great employment chances. Rural areas are offered the possibility of new industries and farmers are supplied with a new way of earning income, furthermore biomass creates new, safe and long term jobs which cannot be moved away from these disadvantaged areas.

Put into effect, measures planned for the second biomass package could mobilise further investments of about DM 960 Mio and generate or make safe about 3,750 jobs in Bavaria.

Roughly half of these investments are spent on construction works, most of them in rural areas.

15. Reduction of Greenhouse Effect

Renewable resources are a means of protecting our limited global resources, this is a view shared by the German chemical industry, its union and the German ministry for trade, industry and commerce in a joined paper published in the autumn of 1995.

As biomass uses all resources several times over in a circular process, it is a way of using solar energy in a way that reduces the greenhouse effect. The energy balances of biomass are positive.

By using biomass as fuel, an important step towards fighting the greenhouse effect can be made. The emission of carbon dioxide is reduced by 3t if one replaces 1t of hard coal with biomass, if biogenic fuels are used as a replacement for normal diesel, the creation of about 2.5t of carbondioxide emissions can be avoided. There are other positive effects for the environment, such as reduction of dangerous emissions, a reduced danger for soil and water, beneficial side effects for wild animals and birds and the country-side in general.

Even today, biomass used as a source of energy has reduced the emission of carbondioxide by about 4.5 Mio t compared to the emissions which would have been caused by fossil fuels. If the share of biomass generated energy in the primary energy required can be raised from today's 3% to 4.3% this should lead to a further reduction of carbon dioxide emission of roughly 1.8 Mio t.

Technical Excursion
to
The Bavarian Forest

EXCURSION WITH FURST THURN UND TAXIS FORESTRY FORSTAMT THIERGARTEN

W. DUFFNER
Furst von Thurn und Taxis
Emmeramsplatz 5
D-93047 Regensburg
Germany

Abstract

The ARW delegation participated in a day long technical excursion to the nearby forest of Thurn und Taxis in the area of Regensburg. A sequence of stops provided informative observations and discussion of modern forest growth and management practices. These stops are described below.

Figure 1. Dr. Winfried Duffner in Discussion with NATO Delegation

R. M. Gutkowski and T. Winnicki (eds.), Restoration of Forests, 389–392.
© 1997 *Kluwer Academic Publishers.*

1. Wernereck Section

Observed: Complete natural regeneration consisting of Norway Spruce, White fir and Beech (Picea abies, Abies pectinata, Fagus silvatica) produced by classical shelter wood system. The result will be uneven aged mixed stand that meets high silvicultural standards. Natural regeneration requires consequent control of deer population. Currently, we have to kill 16 roe deer per 100 ha per year.

2. Schusterrinne Section

Observed: A 35 year old stand of Norway spruce (Picea abies) in optimal growing condition after a fully mechanized thinning with a harvester, 300 crop trees per ha have been selected and artificially pruned up to height of meters. The first log should produce clear lumber. The philosophy is: planting less trees per ha id est. 1,600 trees or reduce natural regeneration by early spacing in order to encourage diameter growth and reduce the rotation age for spruce from 120 years to 80 years. Wider spacing means longer and thicker branches which produce black knots. Therefore, artificial pruning is a necessity and an integral part of the system.

Spruce stands are highly susceptible to snow and windbreak. Trees with a height to diameter ratio of less than 80 are considered to be sturdy enough to withstand snow and wind damage. Early spacing to encourage diameter growth produces sturdy stands.

3. Roßstich Section

Observed: FMG Super EVA one-grip harvester working in first thinning. Scandinavian type one-grip harvesters are excellent tools for a proper silviculture: They enable us to do early thinnings producing pulpwood and small chip-n-saw logs with a contribution to profit. this was not possible before with motor-manual logging. This was the reason why overstocked unstable young and intermediate spruce stands were common all over Germany. The situation changes quickly. Harvesters put our young stands in optimal growing condition.

4. Roßstich Section

Observed: Mechanical pruning - Demonstration of a simple electrically driven milling system to prune crop trees up to the height of 5 meters at cost of DM 3 per tree.

5. Ratzelbuche Section

Obseved: A 40 year old and a 32 year old well structured spruce stand with a caring mixture of beech in the understory. Early spacing carried out by firewood collectors and two subsequent thinnings have formed a stand which shows optimal value growth. This concept of stand education is designed to produce 75% of the total volume growth as heavy high quality timber while only 25% come from the early thinnings as pulpwood and small chip-n-saw logs. All thinnings are carried out in the first half of the rotation, in the second half no measure will be taken unless natural regeneration will be started by single tree selection and opening up the canopy. The neighboring stand is 32 years old, it has been spaced and thinned even earlier and has reached the same diameter as the 40 year old stand beside in shorter time.

6. Ill Johannisthiergarten Section

Observed: Complete high quality natural regeneration under the shelter of the old grown stand which is ready to be harvested. Our intention is to have a stock of natural regeneration of beech which gives us many options to create different stand types: keep the pure Beech stand or enrich the natural beech with 100 to 200 conifers like douglas fir so that the natural beech serves as "caring" mixture.

7. Pfestling Section

Observed: Demonstration of precommercial spacing using motor mowers. This job needs special skills and training which should be carried out by specialists. Precommercial spacing is expensive. It costs DM 800 to 1,000 per ha, but it is crucial to produce high quality stands.

8. Schiederbauerwiese

Observed: A 20 year old very dense pure spruce stand. 7,000 trees per ha have been planted. The stand counts still 5,000 trees per ha. These dense stands are very susceptible to any kind of damage like windfall, snow break and subsequent beetle infestation. This example proves that early quality management in spruce stands is indispensable.

9. Kohlstatt Section

Observed: A 27 year old Douglas fir stand with beech in the understory. The natural regeneration of beech was 15 years old when 125 Douglas fir trees per ha have been planted. Douglas fir represents the economic value while the beech in the understory, the main species of natural forests in the region has more ecological functions like soil tree interactions, habitat for wildlife and last but not least its contribution to aesthetics of our forests. In our densely populated country recreation is an important function of our forests. Even private forests are open to the public.

10. Hochholz Section

Observed: Demonstration of fully mechanized single tree logging with a Valmet 911 one-grip harvester. The single tree selection has the purpose to open up the canopy for natural regeneration and to release crowns of selected seed trees. This operation proves that single tree selection with harvesters even in heavier timber can be carried out without causing too much soil compaction and skinning remaining trees. Skidding is done with a clambunk skidder.

In order to avoid stand and soil damage it is crucial to establish a network of logging paths in a distance of 20 meters. The reach of the harvester boom is 10 meters. The movement of all vehicles like harvesters and forwarders is limited to the logging path. They have to move straight forward, no curving through the stands or even bending towards the stand are admitted.

Our 10 year experience with Scandinavian harvesters as well as our first attempts to use one-grip harvesters on excavator basis in heavy timber have convinced us that traditional high standard silviculture and high mechanized logging methods fit together excellently. the harvester turned out to be a tool of silviculture rather than a mere logging equipment.

Working Group
Reports

Working Group No. 1

Working Group Topic: Environmental Factors in Forest Depletion

Participants: Dr. David Ellsworth (USA)
Ing. Vladimir Henzlik (Czech Republic)
Professor Jozef M. Pacyna (Norway)
Dr. Svetlana Sergeichik (Belarus)
Professor Tomasz Winnicki (Poland)

Overview of Discussion:

We discussed major environmental factors leading to forest depletion in terms of reduction in forest quality on the basis of damage to vegetation and to the ecosystem and potential loss in productive forests. Our consensus is that air pollution at the local and regional-scale air pollution is the dominant environmental factor affecting forests within the natural limits imposed by the climate and geography of the region, although we identified a number of other major factors that are also important in forest depletion and may interact with air pollution.

The major source of pollutants is the utilization of fuel combustion technologies that results in local and regional-scale emission of toxic compounds and pollutants with a high capacity to alter the function of the forest ecosystem. Current technology for energy production via fuel combustion is resulting in exceedence of air pollutant levels such as ozone doses and critical loads of acidic compounds like sulfuric and nitric acids. Air pollution can damage forests and limit forest productivity and hence it is imperative to find options to control air pollution emissions.

We discussed options of emission abatement, energy conservation and alternative energy sources to reduce the air pollutant load in Europe. We especially advocate the return to small, local sources of energy production such as wind and hydro-power. We also discussed primary and secondary technologies that can be implemented in combust theion of current fuel sources for reducing pollutant emissions. Short-term solutions to the current environmental problem in Eastern Europe are dependent on technology transfer from other nations in Western Europe and North America, changes in the management, economic and bureaucratic structure at the local and national level, and the development of a strong financial support structure for alternative energy utilization.

Issues:

The following environmental factors are currently leading to forest depletion in Central and Eastern Europe: 1) Air pollution and mass transport of pollutants by weather patterns, 2) Weather extremes and anomalies, 3) Insect and disease outbreaks, 4) Depletion of soil quality from mining and erosion, 5) Urbanization and human pressures on the forest, and 6) Inappropriate management practices.

R. M. Gutkowski and T. Winnicki (eds.), Restoration of Forests, 395–397.
© 1997 *Kluwer Academic Publishers.*

We consider air pollution as the major factor influencing forest depletion in the region. Combustion of fuels for power generation, transportation, heating and waste elimination results in local and regional emission of pollutants, which can be concentrated in central and eastern Europe. Air pollution has both acute and chronic impacts on forests and interacts with other major environmental factors at a local and regional scale, hence an improvement of the forest situation will require reductions in emissions of air pollutants.

There are several technological solutions to reducing pollutant emissions from current energy sources such as the introduction of modern combustion processes, and broader application of secondary solution such as the simultaneous treatment of components of flue gases. Substitution of currently used fuels containing high amounts of sulfur materials by low-or no-sulfur fuels should proceed these other technological solutions. Energy conservation via recycling and energy recapture are major non-technological solutions that cam be employed. Alternative fuel sources such as geothermal heat, passive solar technology, hydro and wind power generation, biomass and waste fermentation and methane and natural gas burning need to be encouraged.

Most of these alternatives can be implemented at the local scale in a relatively short time providing appropriate financial support and technology transfer are provided. There are still significant gaps in our knowledge of environmental factors and the nature of their impacts on forest ecosystems: including 1) the role of nitrogen compounds in acidification and forest nutrient balance, 2) interactions among different pollutants in forest ecosystems including CO_2 and ozone, and acidic and toxic components, and 3) knowledge of damage detection and forecasting of pollutant effects on forests. Significant gaps in our technology to conserve energy and reduce emissions include energy storage technologies, desulfurization of fuels, and application of low NO_x burners still need to be investigated more aggressively.

Scientific Conclusions Reached (Preliminary):

1) Air pollution provides an additional stress on forests and is detrimental to the ecological and economic vitality of forests. Improvement of the situation of forest depletion in eastern and central Europe requires consolidated action by multi-disciplinary measures involving emission reductions, education, proper management, and economic incentives and financial investment.

2) Computer tools such as GIS and remote sensing technologies need to be applied more widely in forest damage assessment and forecasting and in local and regional planning on the basis of forest damage inventories.

3) Reductions in emissions should improve the forest situation although the dose-response relationship of such improvement requires further scientific investigation.

4) Cost-effective coal liquidification and gasification should be considered as an intermediate step to desulfurization for reducing toxic gaseous emissions in central and eastern Europe.

5) While abatement of emissions should be encouraged as a short-term solution to reducing air pollutant impacts on forests, as much activity as possible should be moved from abatement activities to initial desulfurization of fuel sources and reductions in N emissions from furnaces.

6) Energy conservation via recycling and energy recapture need to be encouraged by public education. Alternative fuel sources such as geothermal heat, passive solar technology, hydro and wind power generation, biomass and waste fermentation, and methane and natural gas burning need to be developed and/or more widely employed.

Potential Collaborative Actions:

- Revisions of international agreements with respect to monitoring and evaluating consequences of air pollution on forests.
- Creation of regional study centers for research on special problems of regional concern and for education at a number of different levels from the general public to a higher rank.
- Continued encouragement of interdisciplinary and international effort on forest damage monitoring.

Recommended Research and Development Needs (NOT by Priority):

- There is a strong need for the creation of an international financial system of loans, credits or waivers for supporting industrial restructuring to reduced emissions technologies. International joint-ventures for investment into clean energy-oriented industries require encouragement.
- There is a need for improvement of tools and concepts for assessing forest damage from ar pollution, including refinement of the concept of critical loads and interactive pollutant stresses. Continued, interdisciplinary research of relevance to forest management is needed involving interactive effects of air pollutants and acidic compounds on forests, including mobilization of toxic metals, behavior of nitrogen compounds in the ecosystem, and air pollution-stress interactions.
- Oxygen enrichment in combustion air needs much farther research so that it can be more widely applied.
- Alternative, cleaner energy production technologies such as a solar energy, and energy storage and delivery systems require a high priority for research.

Working Group No. 2

Working Group Topic: Forest Growth, Expansion and Management

Participants: Dipl. Eng. Florin Atanasiu (Romania)
Mr. Austin Brady (United Kingdom)
Professor Dr. Ladislav Paule (Slovakia)
Professor Kazimierz Rykowski (Poland)
Ing. Otakar Schwarz (Czech Republic)
Mr. Gerard Zak (USA)

Overview of Discussion:

A wide ranging discussion covering the questions of forestry objectives, and the purpose of forest management. Forestry is influenced by a range of environmental and economic and social factors many of which change must faster than the normal forestry timescale. Growth, expansion and management can relate to a range of inputs and outputs. Some dialogue centered on the objectives for forest management, and the nature and role of facts, figures and outcome measures applied to forestry.

Issues Identified:

- Impact and justification of roads.
- The future benefits of tree breeding.
- Reversing damage of pollution how and why?
- Can forestry be truly multifunctional?

Objectives for forestry should be share and accepted by society and stakeholders. Current policy decisions are based only on the limited information available. The many products of forestry - not just wood-need to be considered and quantified.

Scientific Conclusions Reached (Preliminary):

Roads:
Adding roads changes the forest. Road requirements are dictated by current technology - but we could consider new technologies. Lower density of roads with more long range harvesters, even airships! Road design should relate to objectives. Road density can influence forest damage and improve forest treatment in unexpected ways. What can we learn from the relationship between forest authorities and logging companies?

Future trees:
Tree breeding should shift its emphasis towards producing more quantity of wood and not be narrowly focused on characteristics which we may no longer want when those trees reach maturity. Ecological and market conditions change faster then breeding programs. Improving plantations which are already highly productive will give the best net benefit.

R. M. Gutkowski and T. Winnicki (eds.), Restoration of Forests, 399–400.
© 1997 *Kluwer Academic Publishers.*

Improved plantation forestry can provide productive capacity to "shelter" sensitive forests from timber demand. Management systems for forestry are likely to be unstable if based only on wood. Such a system cannot be reliably manipulated using only one or two elements. Continued under-cutting in Europe is likely to lead to an increased risk of catastrophic forest stand collapse, on a large scale.

Non-timber Management:
We need to try and balance the positive management against the negative ecological view. Steps to repair the damaged or destroyed forests of Central Europe need clear objectives and justification about what is indigenous or natural. It may not always be best to rebuild a previous version of the forest but perhaps jump start natural succession processes for the future.

Multi-functional forestry:
We need to understand more about the driving forces which are pushing forestry in different directions. The process of adopting increasingly complex system to pursue higher values must be balanced by low input low output systems to fulfill other objects. We cannot maximize all outputs on all sites at one same time. Pollution and population growth are key driving forces pushing all forestry into a narrowing range of opportunities and outputs.

Forestry feedback:
We must feedback the consequences of unchecked damage and excessive pollution such that decisions are taken with full knowledge of the consequences. We should not adapt our forests to fit in with a pattern of use and abuse which we know will ultimately destroy them.

Potential Collaborative Actions (Preliminary):

To consider the range of criteria and outcome measures which would give a true and comprehensive picture of forest performance. To consider the basis on which regional scale forest areas could be identified i.e. social/economic areas, impact zones, ecological zones, water catchments at both European scale and national scale.

Recommended Research/Development Needs by Priority (Preliminary):

That the Helsinski forestry principles should form the basis for the development of a small number of high order forest performance indicators, for use in conjunction with other new indicators.

That the development of a transnational regional study should be instigated immediately to identify the current extent and nature of forestry and the potential impact it has or could have within each zone, under a range of management scenarios.

Working Group No. 3

Working Group Topic: Forest Utilization for Raw Materials and Value Added Products

Participants: Professor Milos Drdácký (Czech Republic)
Dr. Marian A. Gizejowski (Poland)
Professor Richard M. Gutkowski (USA)
Professor Joze Kusar (Slovenia)
Professor Audronis Kvedaras (Lithuania)
Professor Julius Natterer (Switzerland)
Professor Helmuth Resch (Austria)
Professor Wolfgang Winter (Austria)

Overview of Discussion:

Direct and indirect interaction links between the utilization of forest raw material resources and value-added products were addressed. Different aspects related to engineering applications of wood and wood-based products were given the priority in the discussion.

The situation in homebuilding industry in Poland and other participating countries was overviewed in details. General needs for housing as the dream of having one's own place in the private ownership exist. A great demand for a new housing concept in PC countries is observed. The post World War II growth of construction industry was hasty and the consequent product of living space was a basic utility, neither quality nor value related. Heavier construction technologies were allowed to control the building market. A move is advocated to go away from heavy construction technologies and the architecture of residence estates typical for the previous decades. The need for affordable (low income) housing can be materialized by means of value-added lightweight technologies.

To create the housing lobby an adequate financial supporting system is needed; loans however are only possible if a value exists in the end product. People want to live in a "warm environment"; 30-50% of the total population in the Central Europe accept timber solutions provided that there is a quality. Thus, the quality assurance based on common standards and followed by certification urgently needed to add value to the native and mixed building technologies; by definition, they are environmentally friendly.

Timber is one of the readily available and renewable raw materials. The value-added timber housing has to account for the most suitable mode of construction. This one however can not be standardized. The question of traditional construction modes arose in the discussion; whether 2x4 frames with panels and hollow walls to be filled by insulation or post-and-beam systems would be better to adopt. This in turn will influence the manufacturing processes, their technological advancement and the capital invested in the industries supplying structural and non-structural timber products.

Presently, timber products, as building materials, have only a 5-10% market share in Europe's countries, and even less in Central and Eastern Europe's countries, with

R. M. Gutkowski and T. Winnicki (eds.), Restoration of Forests, 401–407.
© 1997 *Kluwer Academic Publishers.*

comparison to a 70% share of cement-based products and a 20% share of steel products. Issues related to the advantages vs the reasons of depreciatory treatment of timber in building industry were discussed. For economic reasons timber is not an interesting material for monopolized industry of building materials; it costs more than minerals or gravel, and is not available in unlimited quantities at one site. Timber has to compete on the basis of
sustainability and efficiency to accomplish the socio-demands of civilized world.

As an example, the need for research investigations into the technological aspects of "zero energy" and ecosystem oriented housing concept options were emphasized. Not only economic and technical questions but also psychological questions need to be addressed about the "warm environment" in wooden homes and the fears about safety in the case of fire or earthquake. Although, for the housing sector, the North America's and North Europe's examples prove at least the technical equivalence of timber construction with masonry, still reservations exist towards the application of timber as a structural component of heavier loaded structures, e.g. multi-story buildings for living and offices. Not only multi-story buildings, but also other examples anticipated in the discussion (roof structures for industry, sport and culture; transportation and bridge infrastructure) showed that even the most complicated tasks could be achieved in timber. Many successful projects of these type exist in Western Europe.

The lack of understanding still existing among the lay public, lack of knowledge among professionals and decision makers, and finally building by-law restrictions, energy price policies and lack of ecoindexation (e.g. CO_2 indexation) are the main factors preventing timber from gaining a wider acceptance. Relevant collaborative actions possible for undertaking in timber related areas discussed were: change in psychology, followed by understanding and engineering knowledge; cross-border cooperation based on informative techniques, removal of trade barriers and introduction of macro-oriented timber product supply; cooperation enhanced by quality assurance based on uniform standards and certification; all this could help the growing interest in forest resources and to explore new forest value-added products.

The importance of the following activities are under scored - informative and promotional activities for the whole cross-section of society, university timber engineering education, continuing education of practicing professionals, interdisciplinary research, international cooperation in transfer of proven educational methods and technology.

Issues Identified:

1) Criteria for the eco-value oriented utilization of building materials.
   Why timber is an eco-advantageous and customer friendly material?
   How to develop cost-effective methodologies to evaluate the ecological, energy and engineering consequences of wood usage in structural applications?

2) Prospects for intensification of the development of timber structures in Central and Eastern Europe's countries.
   Are there any factors (economic, technical, psychological) to be responsible for the depreciatory treatment of timber structures in PC countries?
   Can human needs for home units and desire of property be materialized in timber?

3) Value-added home units - the customized type of construction, best suited for individual countries and for different regional areas, based on the harmonized use of native and mixed technologies, and combined with use of different construction materials locally available.

4) Affordability and easy available methods of financing are stimulating factors for timber housing development.

5) Quality assurance and certification of timber products and timber structures.

6) Psychology of building concepts and of living in homes.

7) Trading methods and pricing policy.

8) Conditions to ensure a greater share and sustainability for wood-based technologies. Can timber be a suitable material for sustainable housing technology development? What is the range of technically justified structural applications for timber? Is the timber technology sufficient enough to compete and win in major structural applications?

9) Equal treatment of timber as a structural material in building by-law regulations.

10) Creation of a collaborative mood among builders to enhance the use of timber in addition to other building materials; creative and innovative investigations into the application of new engineered wood composite structures.

11) Development of new market products and technologies to account for more efficient laminated beams, and also reconstructed beams from used timber, wood wastes, undervalued and low value tree species.

12) Optimization and modernization of existing manufacturing technologies to account for easy and non-toxic disposal of their end products. What should be optimized in the development of more efficient systems? Is it energy conservation? lowest cost?; Are they ecological priorities? socio-objectives of the full employment or a strategy for the realignment of non-employed labor?

13) Intensification of development of engineered timber structures for heavy loaded structural forms, longer spans and space behavior applications.

14) Harmonization of all the procedures applied in National Standards with those in Eurocodes which facilitates the macro-oriented trade and co-operation; future adaptation of European Standards as guiding codes for quality of wood products and structures, and for the unified methodology of testing, design, erection and maintenance.

15) Importance of education, international collaboration and technology transfer in the

area of timber engineering in order to develop the skill for conceptual work and creativity in engineering applications of timber.

Scientific conclusions reached (Preliminary):

1) From the environmental priority point of view, timber should be one of the Europe's main building materials. Public relations should be created outlining the ecological, economic and technical advantages of wood-based technologies
   a) ability to fulfil the basic socio-demands: ecosystem sustainability, low energy input in production, CO_2 conservation, source of labor, employment opportunities and security, psychological close-to-nature balance inherited from an old craftsman tradition;
   b) reduction in transport and erection costs (material and labor locally available; elimination of heavy on-site equipment);
   c) importance of human skill and art;
   d) the best load capacity/self-weight ratio (potential for challenge in structural applications).

2) Timber housing, to become a value-added, needs to be subjected to the following developments:
   a) environmentally clear and low-energy consumption wood manufacturing processing industries for structural and non-bearing components such as lumber, timbers, laminated and reconstructed composite beams, and different panel products;
   b) greater attempt at recycling of used timber and the development of new engineered wood-based products, with particular emphasis put on such components as wood related wastes, and under-utilized and low value species;
   c) improvement and innovation in window construction technologies to protect wood from the deterioration processes due to moisture penetration and solar radiation; come back from plastics to the native material most desirable from the point of view of future waste disposal problems;
   d) introduction of new wood preservation techniques with organic components which can be burnt for disposal and/or energy without pollution.

3) The reinstatement of the position of timber structures, in general, can be achieved by:
   a) quality housing and low rise buildings for schools, hospitals and trade;
   b) persuading the diversity of engineered structural applications; the demanded range of applications should include mustoryorey buildings for living and offices, long span roof structures for exhibition and sports buildings, bridge infrastructure; peaceful coexistence with other non-timber structural forms;
   c) educational activities in the area of architecture and engineering to develop the skills of creativity and conceptuality in timber engineering profession; this is the challenge beyond 2000;
   d) demonstration projects and general promotional activities in order to address the whole population is seen to be working for the future success;

e) creation of an collaborative mood among professionals (architects, structural engineers and contractors) in order to settle peacefully all the ecological and technical problems; in this positive climate, wood will be used to the greater extend if compared with other materials; perhaps an "eco-rating" system for building materials is needed (to reflect renewability, energy of roduction, CO_2 absorption etc.);
f) development of different composite structures through the utilization of wood in addition to other building materials traditionally used in heavy constructions;
g) application of non-destructive testing procedure for timber-based products to assure the uniformity in quality and endurance of wood products as to the same extend as being accepted for reinforced concrete and steel; to account for both the quality in manufacturing process and the quality in examining in-place performance is desirable;
h) educational and technology timber know how transfers to ensure the cooperation within the Central and Eastern Europe's building markets; the latter are being recaptured by lobbies other than related to timber;
i) development of the trading and employment policies in the region and on the macro-scale for the United Europe; concepts of self-sufficiency and open border trade require a constant assessment, and economic analysis and answers;
j) Harmonization of National Standards with Eurocodes would facilitate the macro-regional trade, cooperation and development in the area of timber structures.

4) The government support for forest utilization oriented education, training and research programs, and activities aiming in providing loans for new and better industrial developments are essential for the multi-functional forestry, sustainable timber industry and society as a whole. In education, the university training of students and continuing education of engineers for timber structures need to be a priority.

5) An additional exercise aiming in persuading the establishment of forestry trade associations would facilitate the monitoring of costs, creating the price policy, helping the educational and research activities aimed in setting an ecologically oriented (green peace approved) certification of forest products.

Potential Collaborative Actions (Preliminary):

1) There is a strong need to develop a system of artificial intelligence based indicators, accounted for interdisciplinary knowledge and measures, to value all the structural materials being currently in use; a definite answer, from the evaluation supported by a system of scientifically elaborated indicators of all the assets and outputs, would advocate the superiority of wood over the other structural materials.

2) We need to investigate and introduce into practice new wood based products and structures, for which also proper standards and testing procedures should be developed.

3) The cooperation in the development of product standards and a framework of building recommendations such as the EN-norms and the Eurocodes; adaptation of European Standards for describing the characteristics and testing methods of wood products, also unifying the design, assembling and maintenance procedures would strengthen the position of timber structures and intensify the timber usage for engineering applications.

4) Knowledge on timber in modern houses needs to be transferred from NATO countries to Central and Eastern European countries where an extreme demand for housing exists.

5) Exchange of experience among cooperated countries in historic wood construction practice involving crafts, built heritage and wood products can be a base for new development as well as public awareness programs and governmental policies.

6) In education and training, a priority is designated to encourage professionals, decision makers and the lay public, both youth and adults, for activities carried out into recapturing of interest and knowledge of different groups of society with respect to all the aspects of ecosystem, forestry resources, timber product utilization and timber engineering.

Recommended Research/Development Needs (NOT by Priority):

1) To investigate the methodologies and system of indicators for auditing evaluation of major constructional materials currently being in use.

2) To develop an educational program for a broad spectrum of society, through the European Union funding, to address all the energy, ecological and engineering aspects of timber utilization, including the forestry, processing and engineering application.

3) To gain a wider acceptance of timber structures by professionals and customers, develop a system of finance, credits and waivers to support the reorientation of university engineering courses and continuing education training scheme for practicing engineers towards all the aspects related to wood as an organic raw material, wood processing industry and engineering applications of wood and wood-based products.

4) To gain a wider acceptance of timber structures by professionals and customers, develop a system of finance, credits and waivers to support the reorientation of university engineering courses and continuing education training scheme for researchers and practicing engineers towards all the aspects related to wood as an organic raw material, wood processing industry and engineering applications of wood and wood-based products.

5) To launch an explanatory program in PC countries to account also for humanity oriented sides of NATO activities, such as Ecology, Science, New Technology,

Economics and Cultural Links, besides the ongoing informative activities which are at present concentrated only on NATO's military and defense role.

Working Group No. 4

Working Group Topic: Socio-economic Considerations for Forest Resources

Participants: Professor Mestwin St. Kostka (Poland)
Professor Patrick Pellicane (USA)
Professor Margit Tamasy-Bano (Hungary)
Dr. Theodore Weber (Germany)

Overview of Discussion:

- discussion of the definition of a social scientist - that being one who analyses interactions between social institutions
- discussion surrounding the basic concept of supply/demand
- how does society place economic value on forest resources
- the difficulties surrounding the lack of integration of language (e.g. economic, biological, technical, etc.) in forest resources
- the social impact of economic development: jobs, rural development, and environmental benefits
- how does one create an open-door to professional interchange

Issues Identified:

- how does culture impact socio-economic issues
- culture is based on economics
- importance of land tenure on forest resource policy
- multiple use - viewing policy through utilities and money
- how to focus on fundamental policy targets
- how can NATO be an instrument to influence forest resource policy?
- the need to identify a strategy for the resolution of problems related to environmental degradation
- the importance of forest-based economic vitality with respect to ecosystem health and vitality
- the role of sustainable forest ecosystem utilization in saving forests
- the affect of regional circumstances on attitudes about forest resources

Scientific Conclusions Reached (Preliminary):

- sustainable forest ecosystem utilization can provide a vehicle for substantial economic and social benefit
- eastern and central European countries lack the financial capital necessary to affect sustainable forest utilization
- taking a long-term perspective is vital to achieve sustainable forest ecosystem management and utilization

R. M. Gutkowski and T. Winnicki (eds.), Restoration of Forests, 409–410.
© 1997 *Kluwer Academic Publishers.*

Common or Related Research/Development Needs by Priority (Preliminary):

- Legal reform land tenure
  tax code modification
  modification of property rights
- regional self-sufficiency with respect to energy production
- action to establish mechanisms to determine the value (integrating social and economic considerations) of natural resources
- make a cost/benefit analysis of resource preservation
- NATO should facilitate the development of a civil organization to advocate political change necessary to realize sustainable forest management and utilization

Joint
Working Group
Reports

JOINT MEETINGS OF WORKING GROUPS

Working Groups: 1 and 2

Working Group Topics:

Environmental Factors in Forest Depletion and Forest Growth, Expansion and Management

Participants:

Dr. David Ellsworth (USA)
Ing. Vladimir Henzlik (Czech Republic)
Professor Jozef M. Pacyna (Norway)
Dr. Svetlana Sergeichik (Belarus)
Professor Tomasz Winnicki (Poland)
Dipl. Eng. Florin Atanasiu (Romania)
Mr. Austin Brady (United Kingdom)
Professor Dr. Ladislav Paule (Slovakia)
Professor Kazimierz Rykowski (Poland)
Ing. Otakar Schwarz (Czech Republic)
Mr. Gerard Zak (USA)

Common or Related Issues Identified:

Both groups recognize that there are a large number of issues that are of great mutual concern and that areas of overlap and complementarity exist in the individual group reports.

Inappropriate management is recognized as an important environmental constraint on forests.

There was much discussion involving utilizing biomass sources as a clean fuel alternative (with gasification), although it is recognized that there will be important ecological limitations to short-rotation, biomass plantations on many sites.

Common or Related Scientific Conclusions Reached:

- We advocate the introduction of clean-burn technologies in many forms, including coal, biomass, etc. and in some cases further development of these technologies.

- There is currently not an adequate transnational understanding of the scale and impacts of air pollutants on forest growth and performance that would motivate the implementation of emission control and improved management. Longer-term, internationally-coordinated data on forest growth and performance measures from forests across Europe are required to provide a more objective assessment of the current forest growth situation than currently exists.

R. M. Gutkowski and T. Winnicki (eds.), Restoration of Forests, 413–414.
© 1997 *Kluwer Academic Publishers.*

Common or Related Research/Development Needs by Priority (Preliminary):

- There must be a specific international body for coordinating management of forests based on new, integrative performance indicators of forests that incorporate a wide variety of values of the forest including wood production, economic factors, social and recreational value, etc.

- A model must be developed for integrated assessment of forest performance factors starting at the transnational scale and ultimately to be implemented within local management zones. The model will require several years of development, data collection, and validation before it can be implemented.

JOINT MEETINGS OF WORKING GROUPS

Working Groups: 1 and 3

Working Group Topics:

Environmental Factors in Forest Depletion and Forest Utilization for Raw Materials and Value Added Products

Participants:

Dr. David Ellsworth (USA)
Ing. Vladimir Henzlik (Czech Republic)
Professor Jozef M. Pacyna (Norway)
Dr. Svetlana Sergeichik (Belarus)
Professor Tomasz Winnicki (Poland)
Professor Milos Drdácký (Czech Republic)
Dr. Marian A. Gizejowski (Poland)
Professor Richard M. Gutkowski (USA)
Professor Joze Kusar (Slovenia)
Professor Audronis Kvedaras (Lithuania)
Professor Julius Natterer (Switzerland)
Professor Helmuth Resch (Austria)
Professor Wolfgang Winter (Austria)

Common or Related Issues Identified:

Education activity links
Displacement of people - forestry create the job potential
Energy resources give CO_2 - carbon storage capacity of forests
Timber as a forest value maintained
Proper documentation of interdisciplinary issues on forestry

Common or Related Scientific Conclusions Reached:

Develop forestry managers - user related quality criteria required
Documentation and forest connected to the product
Marketing products as eco-friendly materials
Document environmental effects on wood properties

Common or Related Research/Development Needs by Priority (Preliminary):

Forest lobby image creation
Value of a well managed forest
Forest advocacy as a function of forest related trade association
Dissemination of forestry information
Documentation of the forest state for GIS visualization
Create the documentation for energy consumption indicators of different products

R. M. Gutkowski and T. Winnicki (eds.), Restoration of Forests, 415.
© 1997 *Kluwer Academic Publishers.*

JOINT MEETINGS OF WORKING GROUPS

Working Groups: 1 and 4

Working Group Topics:

Environmental Factors in Forest Depletion and Socio-economic Considerations for the Forest Resource

Participants:

Dr. David Ellsworth (USA)
Ing. Vladimir Henzlik (Czech Republic)
Professor Jozef M. Pacyna (Norway)
Dr. Svetlana Sergeichik (Belarus)
Professor Tomasz Winnicki (Poland)
Professor Mestwin St. Kostka (Poland)
Professor Patrick Pellicane (USA)
Professor Margit Tamasy-Bano (Hungary)
Dr. Theodore Weber (Germany)

Common or Related Issues Identified:

- The need for rational assessment tool and procedures for both forest assets and depletion
- Economic vehicles to promote behavior that will be environmentally-benign
- Need for a common language for discussing forest economics issues

Common or Related Scientific Conclusions Reached:

- A need exists for tools and methods to rationally and fairly determine the value of forests, assets and depletion
- That cost/benefit ratios are a valuable tool in decision making regarding forest management and utilization
- At the current level of understanding and over short-term, cash (e.g. dollars, deutschmarks, etc.) are a reasonable means of quantifying the value of forest depletion
- Methodologies for the measurement of such things as forest depletion and related issues are not adequate to provide useful information
- Create a methodology to link the environmental damage with its cause (i. e. damage with instrument of damage)
- A need exists to better inform professionals of the spectrum of fields that are directly involved to sustainable forest activities and to become conversant with the appropriate terminology

Common Research/Development Needs Identified (Not by Priority):

- The introduction of international capital ventures to undertake major environmentally-related activities

R. M. Gutkowski and T. Winnicki (eds.), Restoration of Forests, 417–418.
© 1997 *Kluwer Academic Publishers.*

- Communication links needed among specialists
- Facilitate the availability of low-interest loans to individuals and small business for energy-saving technologies
- Tax credits for energy investments
- Development of a rational, long-term stategy to discourage energy inefficiencies

JOINT MEETINGS OF WORKING GROUPS

Working Groups: 2 and 3

Working Group Topics:

Forest Growth, Expansion, and Management and Forest Utilization for Raw Materials and Value Added Products

Participants:

Dipl. Eng. Florin Atanasiu (Romania)
Mr. Austin Brady (United Kingdom)
Professor Dr. Ladislav Paule (Slovakia)
Professor Kazimierz Rykowski (Poland)
Ing. Otakar Schwarz (Czech Republic)
Mr. Gerard Zak (USA)
Professor Milos Drdácký (Czech Republic)
Dr. Marian A. Gizejowski (Poland)
Professor Richard M. Gutkowski (USA)
Professor Joze Kusar (Slovenia)
Professor Audronis Kvedaras (Lithuania)
Professor Julius Natterer (Switzerland)
Professor Helmuth Resch (Austria)
Professor Wolfgang Winter (Austria)

Common or Related Issues Identified:

- How to achieve public relations which show forestry as a positive force.
- We need to interest the younger generation in the benefits of forestry.
- There is little joint working despite the many common interests of different forestry groups in Europe.
- How to promote timber as an eco-friendly material.

Common or Related Scientific Conclusions Reached:

- It is difficult to predict future needs in relation to timber use and quality.
- Timber is not currently compared on a full cost basis with other materials.
- We need markets to drive forestry activity.
- Not all forests will benefit from utilization.

Common or Related Research/Development Needs by Priority (Preliminary):

- We need to develop methodologies to properly account for the energy and ecological consequences of timber use compared with other materials.

- We need to develop models of forest processes based on the definition of forest types at the regional scale, regardless of national boundaries. We need to apply such models

R. M. Gutkowski and T. Winnicki (eds.), Restoration of Forests, 419–420.
© 1997 *Kluwer Academic Publishers.*

to inform forest planning by identifying and predicting the ecological, social, economic, etc.. consequences of the available options.

- We need to consider how to apply such models so that they accommodate the views of other interest groups and of the public.

- We need to develop an accurate picture of the current forestry situation to facilitate delivery of future objectives.

JOINT MEETINGS OF WORKING GROUPS

Working Groups: 2 and 4

Working Group Topics:

Forest Growth, Expansion, and Management and Socio-economic Considerations for the Forest Resource

Participants:

Dipl. Eng. Florin Atanasiu (Romania)
Mr. Austin Brady (United Kingdom)
Professor Dr. Ladislav Paule (Slovakia)
Professor Kazimierz Rykowski (Poland)
Ing. Otakar Schwarz (Czech Republic)
Mr. Gerard Zak (USA)
Professor Mestwin St. Kostka (Poland)
Professor Patrick Pellicane (USA)
Professor Margit Tamasy-Bano (Hungary)
Dr. Theodore Weber (Germany)

Common or Related Issues Identified:

- Defining values of intangible, or non-commodity factors (aesthetics, ecology, heritage etc..)
- We cannot use just money for this kind of comparison, we need other factors to "balance" this...
- Objectives should be established first → what people want or 'need'
- Has the public lost touch with forestry and natural resources?
- Economic and ecological concepts are unclear to many decision makers

Common or Related Scientific Conclusions Reached:

- Many societies rely too heavily only on financial indicators to aid decision making.
- Many societies are increasingly urbanized + have lost understanding of natural systems.
- Multifunctional forestry must include both economic and natural valuation systems + allow decisions based on these.

Common or Related Research/Development Needs by Priority (Preliminary):

1. Need more focused approach to methodology to value forestry elements, activities and resources on a range of scales not just money.

2. Need to understand the relationships between these elements.

R. M. Gutkowski and T. Winnicki (eds.), Restoration of Forests, 421–422.
© 1997 *Kluwer Academic Publishers.*

3. We need an understanding of the public's perceptions and arguments to help us to calibrate or validate the model we develop.

4. We could start by introducing one simple element such as water into existing forestry models.

JOINT MEETINGS OF WORKING GROUPS

Working Groups: 3 and 4

Working Group Topics:

Forest Utilization for Raw Materials and Value Added Products and Socio-economic Considerations for the Forest Resource

Participants:

Professor Milos Drdácký (Czech Republic)
Dr. Marian A. Gizejowski (Poland)
Professor Richard M. Gutkowski (USA)
Professor Joze Kusar (Slovenia)
Professor Audronis Kvedaras (Lithuania)
Professor Julius Natterer (Switzerland)
Professor Helmuth Resch (Austria)
Professor Wolfgang Winter (Austria)
Professor Mestwin St. Kostka (Poland)
Professor Patrick Pellicane (USA)
Professor Margit Tamasy-Bano (Hungary)
Dr. Theodore Weber (Germany)

Common or Related Issues Identified:

- Forestry related industries attractive to redisplacements of people (job creation)
- Encourage of the exploration and utilization of undervalued species
- Current price policies and factors affecting

Common or Related Scientific Conclusions Reached:

- Exploration and utilization of undervalued species
- Growing forestry areas increase the sense of secureness of local communities
- Development of low-energy wood technology processing industry

Common or Related Research/Development Needs by Priority (Preliminary):

- Development of trade forestry organization in PC countries to facilitate the monitoring of costs and creation of stable price policy for timber based products (basis for long term pricing development)

R. M. Gutkowski and T. Winnicki (eds.), Restoration of Forests, 423–424.
© 1997 *Kluwer Academic Publishers.*

- Harmonization of macro-codes to standardize the procedures and facilitate off-border trade and development

- Sustainable forest product certification (a "green" oriented concept) and macro-marketing oriented education

Joint Meeting of Working Group Leaders-Report

JOINT MEETING OF WORKING GROUP LEADERS

Working Group Topic Titles:

Environmental Factors in Forest Depletion
Forest Growth, Expansion and Management
Forest Utilization for Raw Materials and Value Added Products
Socio-economic Considerations for the Forest Resource

Participants:

Dr. David Ellsworth (USA)
Mr. Austin Brady (United Kingdom)
Dr. Marian A. Gizejowski (Poland)
Professor Patrick Pellicane (USA)

PRELIMINARY CONCLUSIONS:

Primary Issues Identified:

1. Lack of coordination between forestry interests and between other interest groups over a range of key forestry + environmental matters.

2. Lack of understanding of forestry processes amongst professionals, public and decision makers.

3. Lack of understanding over the ecological and environmental consequences of timber use vs other materials

4. Difficulty of assigning values to a wide range of forestry outputs.

Primary Scientific Conclusions:

1. We need to encourage cleaner technologies or the forests damage continues.

2. Urbanized societies have lost understanding of natural systems

3. Insufficient understanding of scale and impacts of air pollutants on forest growth and performance.

4. Insufficient documentation to asses the ecological costs and benefits of timber compared with other materials.

5. A stable environment is important for social well being.

R. M. Gutkowski and T. Winnicki (eds.), Restoration of Forests, 427–429.
© 1997 *Kluwer Academic Publishers.*

Primary Potential Collaborative Actions:

1. We need to develop models of forest processes based on the definition of forest types at the regional scale, regardless of national boundaries. We need to apply such models to inform forest planning by identifying and predicting the ecological, social, economic, etc. consequences of the available options.

2. To encourage the development of tools to rationally determine the value of all forest assets and their depletion.

3. We need to develop methodologies to properly account for the energy and ecological consequences of timber use compared with other materials.

Short Term Benefits of the ARW:

- Promoting dialogue and achieving mutual understanding between the CP countries and the rest of NATO.

- Identification of many of the salient issues relating to the saving and better utilization of the forests of C + E Europe.

- Highlighting and publicizing the evolving role of NATO.

Long Term Benefits of the ARW:

- Setting the foundation for future collaboration between the participating nations over the key issues identified.

- Establishing a professional dialogue between key interdisciplinary individuals with many shared objectives. This is the crucial first step towards realizing an important NATO goal of **"Title"**.

PRELIMINARY RECOMMENDATIONS:

Primary Recommended Research/Development Needs (by Priority):

1. To develop a scheme to promote the continued activities of ARW participants in order to ensure the identified objectives are pursued.

2. To research the methodologies to build the models of forest processes identified above.

3. To develop the alternative processes and mechanisms required to deliver real reductions in forest depletion.

4. To conduct an energy and ecological audit of all major constructional materials currently in use.

5. To explore new methods of using under-utilized and low value species.

6. Develop regional, interdisciplinary centers for research and education on all issues of forestry.

Final Report
of
The Whole Delegation

FINAL REPORT OF THE WHOLE DELEGATION

Primary Issues Identified:

The contribution which science and technology can make to solving the problems facing Central and Eastern Europe's forests will remain limited unless the following issues are addressed;

1. There is a lack of coordination between forestry interests and between other interest groups over a range of key forestry and environmental matters. This limits the flow information and obstructs the development of a consensus for future action.

2. There is a lack of understanding of forestry processes amongst professionals, the wider public and amongst decision makers. Wood as a consumer product is separate from the forest in the mind of the public, this undervalues the importance of forest management and of forestry as a profession.

3. There is a lack of understanding of the comparative ecological and environmental consequences of using timber as compared to other materials.

4. There is some difficulty in assigning values to the wide range of benefits and outputs resulting from forest management, beyond those immediately connected with timber production.

5. Forestry and environmental management has ethical, moral and human dimensions which may not be adequately accommodated in some current decision making processes which rely too heavily on more easily quantified elements such as wood volumes or monetary values.

Primary Scientific Conclusions:

1. A stable environment is essential for long-term social and economic well-being.

2. Cleaner technologies for energy production and other industrial processes must be developed and adopted or continued forest damage is inevitable. There has been insufficient recognition of and response to the scale and impact of forest damage at the regional, national and international levels.

3. Many urbanized societies are losing understanding of natural systems, such as forests, and concerted efforts are needed via education and information to recapture the public interest.

4. There is insufficient documentation to accurately assess the ecological, energy and

R. M. Gutkowski and T. Winnicki (eds.), Restoration of Forests, 433–436.
© 1997 *Kluwer Academic Publishers.*

engineering costs and benefits of wood compared with other building materials.

Primary Potential Collaborative Actions:

1. There is a strong need to develop models of forest processes based on the definition of forest types at the regional scale, regardless of national boundaries. We need to apply such models to inform forest planning by identifying and predicting the ecological, social, and economic consequences for a wide range of forest management and utilization options.

2. Interdisciplinary tools need to be developed to help rationally determine the value of all forest assets and outputs. This will help to ensure that the potential gains and losses from management interventions and from other human-induced environmental change are comparable over a number of different measurement criteria. Consequently, decisions can be made which extend beyond the consideration of wood volumes or monetary values alone.

3. We need to develop cost-effective methodologies to properly account for the energy, ecological and engineering consequences of wood use compared with other materials.

4. We need to develop mechanisms to engage professionals, decision makers and the wider public in a process of education and information. This should relate both to the extensive body of scientific knowledge and to more accessible forest and environmental information.

5. To investigate the development of standards and testing procedures to support the wider adoption of new wood-based products and structures. To assist in knowledge transfer on the use of timber in modern houses to Eastern European countries where demand for new housing is high.

6. Exchanging experience and knowledge of historic and traditional wood construction practices involving crafts, built heritage and wood products. This may be a base for new development as well as raising public awareness, and linking with other government policy areas.

Short Term Benefits of the ARW:

Promoting a continuing dialogue and achieving mutual understanding between the Cooperation Partner countries and the NATO member countries.

Identification of may of the salient issues relating to saving and better utilizing the forests of Central and Eastern Europe, and starting the process of addressing them.

Highlighting and publicizing the evolving role of NATO in helping Member and

Cooperation Partner nations to work together to address social, environmental, economic, technical and cultural problems.

Long Term Benefits of the ARW:

Establishing a professional dialogue between key interdisciplinary individuals with many shared objectives. This is the crucial first step towards realizing the important NATO goal of saving and better utilizing Central and Eastern Europe's forests.

Setting the foundation for future collaboration between the participating nations in continued efforts to address the key issues identified.

Demonstrating the NATO and Partner nations can work together to focus efforts to solve issues of environmental security.

PRELIMINARY RECOMMENDATIONS:

Primary Recommended Research/Development Needs:

1. To develop a scheme to promote the continued activities of ARW participants in order to ensure that the issues and objectives identifies at the ARW are pursued.

2. To research the methodologies necessary to build the models of forest process identified above.

3. To secure funds or develop a system of finance, credits and waivers to support the development of regional projects designed to implement changes in the management and structure of industrial and energy facilities in Eastern Europe, and so effect a major reduction in air pollution.

4. To conduct an energy, ecological and engineering audit of major constructional materials currently in use, for the purpose of confirming or disputing the superiority of wood as a more environmentally preferable building material over a wide range of circumstances.

5. To develop new methods for obtaining the highest social and economic value from forests while maintaining their health and ecological integrity, with particular emphasis on underutilized and low value species. The impact of human population and consumption should be considered as part of such evaluations.

6. Develop regional, interdisciplinary centers and/or projects for research and education on all issues of forestry and wood engineering. To include participation from multiple Universities and other institutions, and support the reorientation of

university courses and continuing education courses for practicing engineers, architects, etc. towards the better understanding of wood based industries and the better application of wood-based products.

7. To develop a program of education and information to raise the awareness and understanding of forestry and the environment across a broad spectrum of society, with particular emphasis on younger people.

Potential Follow-up to the ARW:

1. To start a program of work which addresses all seven of the primary research and development needs identified above.

2. Set up an Internet Website as a point of contact for further communication and ideas. The site should provide a means of tapping into the extensive library of existing research findings. The site could also be accessed by the growing number of public and student Internet users.

3. NATO to support a study tour of forestry and environmental problems within the Cooperation Partner nations, undertaken by small expert group as a direct follow-up to the ARW.

4. NATO to support the introduction of new technology to large scale forest and environmental planning, such as satellite imagery and Geographic Information Systems.

5. NATO to coordinate and support a bid for European Union funding to enable the outcome of the ARW to be implemented.

6. NATO to support the development, with forest agency and industry collaboration, of an environmental youth program to facilitate visits, camps and voluntary work in forests of participating nations.

Summary Observations and Conclusions

BACKGROUND TO THE CONCLUSIONS AND RECOMMENDATIONS OF THE INVITED EXPERTS

The conclusions and recommendations of the Invited Experts are presented in the written reports provided in the preceding pages. As described in the Preface, the conduct of the ARW was carefully ordered and this publication is prepared as "Written Record" as opposed to a "Proceedings". The text material is presented in the same order as the events which took place. The intent is to allow the reader to connect them in the same progression as occurred for the participants themselves. In this way the path to the findings can be traced, hopefully providing insight to the logic and thought processes that led to the results. The reader is advised to review the material within that perspective. The reader can gain insight to the pre-ARW study tour, read the presented Situation Papers which gave a setting, study the related State-of-the-Art Papers considered in discussions, examine the "advance" Position Papers, be informed of what was seen in the excursion to the Bavarian Forest, and follow the evolution to the observations which ultimately followed as a Final Report of the Whole Delegation.

Extreme care and effort was taken to enable an evolution in thoughts and ideas to occur, gradually moving from expression of individual input to consensus findings of a diverse delegation. Hopefully, the reader found it interesting to examine closely the progression of outcomes resulting from the meetings of individual Working Groups to those of arising from the joint meetings of paired Working Groups. The latter reflect the consequence of the interaction of the various groups of specialists. This provided opportunity for persons from the ends of the spectrums (WG1 and WG4) to meet as well as from more closely related areas (WG1 and WG2, WG2 and WG3, WG3 and WG4). This approach was taken to allow of potentially polarized groupings to have dialogue, as well as the potentially more aligned groupings. It is believed that this allowed all passions, biases, and alliances to surface and have a mechanism to be voiced.

Finally, the findings from the Meetings of the Whole Delegation are presented. First, a draft was prepared in the Joint Meeting of the Working Group Leaders and a summary presentation made to the Whole Delegation. All Invited Experts provided reaction in the "one by one" processes used. Specifically, in round 1 of input, each person gave individual reaction and input to the document, one person at a time. Then, in round 2 of input, each person gave individual reaction, one person at a time, to what had been said in the first round. The Working Group Leaders revised the report to reflect this input. It was sent to all delegates by mail for final input and written expression of dissent on any points made. The Working Group Leaders had been delegated authority to accept or reject any final input, but could not reject any dissent. However, no dissent was received. The outcome of that last phase is what is presented herein as the Final Report of the Whole Delegation, as prepared by Mr. Austin Brady on behalf of all the WG leaders. Intentionally, the Co-Directors provided on site input only as peers in the "one by one" input process. To conclude the publication, the Co-Directors provide the following post ARW summary observations, conclusions and recommendations regarding the workshop itself and the Final Report of the Whole Delegation.

R. M. Gutkowski and T. Winnicki (eds.), Restoration of Forests, 439–443.
© 1997 *Kluwer Academic Publishers.*

SUMMARY OBSERVATIONS OF THE CO-DIRECTORS

Conduct of the ARW

The ARW was highly successful due in great part to the enthusiastic participation of the Invited Experts. The assembled delegation did, in fact, constitute a group with a wide range of expertise by subject matter and by involvement/experience in many forest environmental issues. The interest, commitment and dynamism of the individuals in the delegation was also universally strong. From the outset, an outstanding "chemistry" of the entire delegation unfolded whereby the focus was on the commonality of the issues and challenges, at least considerably much more so than the differences. The overriding thrust appeared to be a drive for factual information, completeness and identification of common ground. Of course, discussion on some points was sharp or vocal at times, but never divisive. As members of Working Groups themselves, the Co-Directors observed that identification of common issues came about naturally, almost readily in all phases of the Working Group individual dialogue and joint interaction.

The venue was in the Lower Bavaria region of Germany located at the edge of the Bavarian Forest and in a unique forest house, within a retreat like setting. Local community and government officials were strongly involved and face to face contacts during prior site visits were important. Day 2 was an intensive scientific excursion directly into the nearby forest, directed by the Chief Forest Engineer (also an Invited Expert) of the Forest of Thurn und Taxis. It was superbly done, exposing the CP participants to the most current scientific, genetic and harvesting practices consistent with environmental issues affecting growth and production, as well as demonstration of the most modern equipment available. The host indicated the ongoing dialogue with the experts was invaluable to his perspective of forest management directions, too. The forest region of Bavaria provided an ideal case study of how, in reality, science and technology have been used to address conflicting forest environmental vs. economics issues. The headquarters of Germany's largest environmental protection organization was only 12 km away. That community held an evening reception and guided tours of its environmental "eco center" and the group was warmly received by the organization.

The ARW was highly scientific and the ASI publication will demonstrate that reality. The main topic was centered in the setting of regional issues of environment and forest management, growth and utilization. The basic science and engineering needs must reflect the regional scope of the applied problems. Regionalism brings in the socio-political, economic, governmental, historical and other non-scientific aspects because the technological problems cannot be solved without cooperation. Indeed, the ability to cooperate is prerequisite to the technical tasks. Despite the extreme range of expertise and subject matter (from genetics to ecology to civil engineering to industrial pollution control to economics etc.), the camaraderie created in the meeting was extremely high. The interface of scientists, technologists and industrialists led to interesting common ground and many complementary conclusions.

NATO should continue to conduct more ARW's on broad environmental issues. They encourage regional networking and interfaces. The reaction from the attendees was very favorable to this point. Attendees hope NATO can expand its scope so as to use developments in science and technology as tools to address the greater national-regional cooperation needs. As a continuation of addressing this challenging topic, a second extended ARW should be held at the same venue in Bavaria. In a full ARW, typically, only one ore two persons from any one nation can be involved because of NATO limits on delegation size and its dispersion rules. Preceding the next ARW with a multi-country mini scientific tour would be invaluable. Many delegates recommended that a one day workshop conducted by a few representative experts should be held on a sequential basis in a series of CP countries. This would allow the many key professionals within each CP country involved in this regional problem to participate in addressing the paths to solutions and help plan the ARW.

Scientific Content of the Final Report of the Whole Delegation

The results in the Final Report of the Whole Delegation appear earlier in this Written Record. Some important summary points are made follow.

Working Group 1 - Environmental Factors in Forest Depletion

It is clear that the basic cause of the forest environmental problem is energy modes used in the region. Fossil fuel sources are critical to meeting current energy demand but extensive emission of air pollutants is occurring. A need for modernization of fuel burning technologies and plant facilities is recognized (and actually started in some countries). But despite such abatement, depletion of fossil fuel sources appears inevitable (and sooner rather than later). That reality alone is incentive to reduce their use. This Working Group strongly focused on strategies for replacing them and its recommendations reflect that philosophy.

Working Group 2 - Forest Growth, Expansion and Management

It was universally agreed that, by contemporary standards, the "resource" in "forest resource" infers its multipurpose socio-economic value, i.e the human sensation factors as well the forest products market. Despite the long backlog of environmental conditions and events highly destructive to forests, there is merit and opportunity to meet the multiple forest resource demands of the public. The prevailing mood was that a look past the present forest generation and approaches to use is critical. In other words, if limited fiscal resources are involved, pilot efforts to produce a new generation of forests may be more productive than saving some presently threatened stands. However, the caveat is that one cannot predict the future demand for the forest resource (human or material) for a future that far away. Regardless, a path toward eliminating destructive patterns is essential.

Working Group 3 - Forest Utilization for Raw Materials and Value Added Products

In spite of worldwide volatility between forest protectionists and forest utilizers, this delegation appeared to welcome the evolving potential for the introduction of environment based rating of materials of construction. If universally applied, it was felt the wood based materials have numerous relative advantages in regard to environmental considerations in the production of material, renewability of the resource, reduction of day to day energy consumption and long term carbon dioxide conservation. Compelling housing and commercial building construction needs exist in Central and European CP countries and there seemed little or no exception to a belief that greater and fairer pricing access to forest productions by the public would undoubtedly create the introduction of timber solutions, provided high quality resulted. Quality could be best fostered by meeting western code standards and freer import/export trade policies to foster competitive interest in quality and joint transnational (within Central/Eastern Europe) ventures.

Working Group 4 - Socio-economic Considerations for the Forest Resources

The trend toward market based economies is evident in many emerging CP countries. New principles are being learned and tried and outcomes are early and tenuous to date. The goal is to stabilize the economies (some being almost impoverished) and move to future growth via commercial industry. In the area of forestry and the value of forests to the products making industries reflect that, but are magnified by the criticality and complexity of land ownership (access to the forests or other land) in the post communist period). Beyond that, the biological material wood instills even greater challenges vis-a-vis the marketing of end products. The planning and preparation for the future use of the forest involves the needs within of the above three WG subject matters. This introduces communication challenges between scientists, technologists, economists and so forth. The ability of these parties to effectively speak or understand each other clearly is a challenge in itself, and within the setting of emerging CP countries learning or being directly exposed to new ideas and concepts, the difficulty is greatly magnified.

Final Report of the Whole Delegation

Within the contexts of the Position Papers, presented Situation Papers and State-of-the Art papers, and the individual and joint Working Group reports, the focus was on detailed scientific, technological and technical content. Extensive information about the issue of "Science and Technology to Save and Better Utilize Central and Eastern Europe's Forests" is included as well as approaches to help accomplish it. The networks of experts that are either enhanced, created or encouraged by this ARW (either via direct participation or the reading of this Written Record) is the most important path to implementing the recommendations of the WG reports. Beyond consolidating the findings into broader issues, objectives and goals, the Whole Delegation elected to focus its conclusions and recommendations on more philosophical content and larger "trends" that must be effected on a regional basis. The Final Report focuses on those items which cannot be effectively implemented without collaboration and on matters that are on a large enough scale to warrant follow-up by NATO. In summary, the Co-Directors make the following important points about the Final Report of the Whole Delegation.

Forest environmental issues are major issues and widespread within the CP countries of Central and Eastern Europe. One critical thrust needed is "communication". Communication is needed in various forms, including : 1) networks a) of specialists, b) via international and international agency and government officials, c) of interdisciplinary education, research and development (industry), d) via alliances of interested youths, 2) contemporary computer and electronic based technology sharing and interaction as well as GIS and GPS database sharing, A second critical thrust is a needed "head-on" resolve and action to accept the energy based challenges ahead, including a) dramatic reduction of fossil fuel use in favor of environmentally safer alternative fuels, b) evolution of energy and ecological rating criteria for materials of construction, c) strengthened monitoring of air pollution levels and their consequent effect on forests as tied to planned reduction in fossil fuel usage. Financial and fiscal aspects comprise a third tenet of needed progress. These include : a) facilitation of cross border markets for forest products, b) development of a regional trade association for forest based industry, c) fiscal mechanisms to enable regional management of industrial energy use and pollutant output.

SUMMARY CONCLUSIONS OF THE CO-DIRECTORS

Some key broad conclusions should be mentioned. The underlying topic of this workshop (forest growth and utilization, amidst environmental impacts) is technically challenging and provocative in many (if not all) forested areas of the world. This delegation addressed it fully and in an intensely cooperative nature, One factor was an apparent commonality among the represented CP nations that forests are well worth saving and invaluable for societal as well as material utilization, the two being strongly related in their post communist era circumstances. The extension of networking within the central and eastern Europe regions should be fostered. The extensive expertise represented by the combined scientific and technological community of these countries is enormous. Efforts to consider consortia across borders should be strongly fostered. This ARW is evidence that the representatives from the CP countries have much common ground and a will to face a seemingly insurmountable environmental security problem, namely the life of their valued forests.

CLOSING REMARKS

The ARW was a rewarding outcome to a concerted effort to bring the interested professional experts together on a vital, compelling regional environmental challenge. The dual goals of scientific progress and multi-national collaboration were achieved. A strong base for continued work is established by the extensive and well documented reports of the needs and the priorities. A noteworthy fact is that the delegation discussed fiscal aspects primarily in the setting of the desire for flexibility to bring international financing and collective effort more to the fore. Although lack of funds is pervasive, the dialogue was more on the willingness for each nation to "do its part" with whatever resources available. The key need was identified as cross-border flexibility and a fostering of political will to work together. The scientists and technologists themselves had much in common and an "esprit de corps" that was remarkable in its visibility. We believe it is in NATO's interest to act on the suggested potential follow-up to the ARW, as the ideas and recommendations were the result of the common ground and the strong sense of a commitment to working together that were established at this ARW.

Co-Directors
Professor Richard M. Gutkowski
Professor Tomasz Winnicki

R. M. Gutkowski and T. Winnicki (eds.), Restoration of Forests, 445.
© 1997 *Kluwer Academic Publishers.*

LIST OF CONTRIBUTORS

Dipl. Eng. Florin Atanasiu
National Institute of Wood
R-79688 Bucharest
Sos. Fabrica de Glucoza 7
ROMANIA
Phone: 40 1 240 4660
Fax: 40 1 240 7985 (688 4985)
Email: inlrom.@ul.ici.ro

Mr. Austin Brady
Forestry Commission
Cuckney Road
Carburton, Worksop
Nottinghamshire, S80 3BP
UNITED KINGDOM
Phone: 44 1909 472965
Fax: 44 1909 482875
Email: sherwood@dial.pipex.com

Dr. Mykola Chernyavsky
Ukrainian State University of Forestry & Wood Technology
Pushkin str. 103
290057 Lviv
UKRAINE
Phone: 380 322 767 696
Fax: 380 322 423 494
Email: OK@ratusha.Lviv.uk

Professor Milos Drdácký
Head of the ARCHISS Department
Institute of Theoretical and Applied Mechanics
Vysehradska 49
128 49 Prague
CZECH REPUBLIC
Phone: 42 2 29 33 47
Fax: 42 66 96 28 21

Dr. Winfried Duffner
Furst von Thurn und Taxis
Emmeramsplatz 5
D-93047 Regensburg
GERMANY
Phone: 49 941 504 8121
Fax: 49 941 504 8129

Dr. David Ellsworth
Brookhaven National Laboratory
BNL Forest FACE Project Office
located at: School of the Environment
Duke University
Durham, NC 27708-0328
USA
Phone: 919 613-8089
Fax: 919 684-8741

Dr. Marian Gizejowski
Inst. Konstrukcji Budowlanych
Ploitechnika Warszawska
al. Armiii Ludowej 15
00-637 Warszawa
POLAND
Phone & Fax: 48 22 256 532
Email: IKB1@omk.il.pw.edu.pl

Professor Richard M. Gutkowski
Department of Civil Engineering
Colorado State University
Fort Collins, CO 80523
USA
Phone: 970 491-8291
Fax: 970 491-2788
Email: rgutkowski@vines.colostate.edu

Ing. Vladimir Henzlik
LESPROJEKT
Forest Management Institut
Nabrezni 1326
250 44 Brandys n. L.
CZECH REPUBLIC
Phone: 42 202 804 481
Fax: 42 202 802 434

Mr. Nikolay Kaletnik, Head
Scientific-technical Administration
Ministry of Forestry of Ukraine
Kreshchatik 5
252005 Kiev
UKRAINE
Phone: 380 44 228 78 58
Fax: 380 44 228 77 94
Email: yyy@milg.kiev.ua

Professor Mestwin St. Kostka
Technical University of Bialystok
Institute of Informatics
Wiejska 45 A
15-351 Bialystok
POLAND
Phone: 48 85 428 206
Fax: 48 85 422 393
Email: KOSTKA@ii.pb.bialystok.pl

Professor Joze Kusar
Vicedean of Faculty of Architecture
University of Ljubljana
Zoisova 12, SLO-1000 Ljubljana
SLOVENIA
Phone: 386 (0) 611 2574 15
Fax: 386 (0) 611 2574 14
Email: joze.kusar@uni-lj.si

Professor Audronis Kvedaras
Vilnius Technical University
Saultekio Aleja 11
LT-2054 Vilnius
LITHUANIA
Phone: 370 2 766 650
Fax: 370 2 700 112
Email: as@rasa.vtu.lt

Professor Julius Natterer
Construction en Bois
Ecole Polytechnique Federale de Lausanne
GCB (Ecublens), CH1015 Lausanne
SWITZERLAND
Phone: 41 21 693 23 95
Fax: 41 21 693 23 91

Professor Joze Pacyna
Norwegian Institute for Air Research
P.O. Box 100
Instituttveien 18
N-2007 Kjellar
NORWAY
Phone: 47 63 89 8000
Fax: 47 63 89 8050
Email: Jozef.Pacyna@nilu.no

Professor Dr. Ladislav Paule
Faculty of Forestry
Technical University
SK-96053 Zvolen
SLOVAKIA
Phone: 42 855 635 ext. 538
Fax: 42 855 22654, 51327
Email: paule@vsld.tuzvo.sk

Professor Patrick Pellicane
Department of Forest Sciences
Colorado State University
Fort Collins, CO 80523
USA
Phone: 970 491-7752
Fax: 970 491-6754
Email: patp@cnr.colostate.edu

Professor Anatoly Petrov
Education and Training Centre for Forest Specialists
Institutskaya Str. 17, 14200 Pushkino
Moscow District
RUSSIAN FEDERATION
Phone: 7 095 584 3644
Fax: 7 095 511 4520

Professor Helmuth Resch
Inst. f. Holzforschung
University BOKU
Gregor-Mendel Str. 33
A 1180-Wien
AUSTRIA
Phone: 43 1 47654 or 4251
Fax: 43 1 3105 175

Professor Kazimierz Rykowski
Instytut Badawczy Lesnictwa
(The Forest Research Institute)
Bitwy Warszawskiej 1920 r. 3
00-973 Warsaw
POLAND
Phone: 48 22 224932
Fax: 48 22 224935
Email: KARYK@ikp.atm.com.pl

Ing. Otakar Schwarz
Krkonose National Park Administration
543 01 Vrchlabi
CZECH REPUBLIC
Phone: 42 438 285 411
Fax: 42 438 285 422

Dr. Svetlana Sergeichik
Surganova Street 2A
Central Botanic Garden
Belarusian Academy of Sciences
220012 Minsk
BELARUS
Phone: 375 172 68 53 37
Fax: 375 172 68 44 83

Professor Margit Tamasy-Bano
University of Forestry and Wood Sciences-Sopron
H-9400 Sopron
Baicsy Zs. u. 4
HUNGARY
Phone: 36 99 311 100

Dr. Theodore Weber
Bayerisches Staatssministerium fur
Ernahrung, Landwirtschaft und Forsten
Ludwigstrasse 2, D-80539 Munchen
GERMANY
Phone: 49 89 218 219
Fax: 49 89 218 22677

Professor Dr. habil Gerd Wegener
Universitat München
Istitut fur Holzforschung
Winzerstrasse 45
D-80797 Munchen
GERMANY
Phone: 089 306 3090
Fax: 089 306 309 11

Professor Tomasz Winnicki
Environmental Engineering Dept.
Technical University of Wroclaw
Wybrzeze Wyspianskiego 27
50-370 Wroclaw, POLAND
Phone & Fax: 48 71 22 98 92
Email: winnicki@pwr.wroc.pl

Professor Wolfgang Winter
Technical University Wien
Trawerkslehre, Ingenieurholzbau
Karlsplatz 13/254, A-1040 Wien
AUSTRIA
Phone: 43 1 58 801 3467
Fax: 43 1 50 442 11

Mr. Gerard Zak
Atlantic Environmental Services, Inc.
P.O. Box 297
188 Norwich Avenue
Colchester, CT 06415
USA
Phone: 860 537-0751
Fax: 860 537-6347

SUBJECT INDEX

MIX
Papier aus verantwortungsvollen Quellen
Paper from responsible sources
FSC® C105338
www.fsc.org

If you have any concerns about our products,
you can contact us on
ProductSafety@springernature.com

In case Publisher is established outside the EU,
the EU authorized representative is:
Springer Nature Customer Service Center GmbH
Europaplatz 3, 69115 Heidelberg, Germany

Printed by Libri Plureos GmbH
in Hamburg, Germany